THE POWER GRID

THE POWER GRID

Smart, Secure, Green and Reliable

Edited by

BRIAN W. D'ANDRADE

Exponent Inc., Washington, DC, United States

ACADEMIC PRESS

An imprint of Elsevier

elsevier.com

Academic Press is an imprint of Elsevier
125 London Wall, London EC2Y 5AS, United Kingdom
525 B Street, Suite 1800, San Diego, CA 92101-4495, United States
50 Hampshire Street, 5th Floor, Cambridge, MA 02139, United States
The Boulevard, Langford Lane, Kidlington, Oxford OX5 1GB, United Kingdom

British Library Cataloguing-in-Publication Data
A catalogue record for this book is available from the British Library

Library of Congress Cataloging-in-Publication Data
A catalog record for this book is available from the Library of Congress

ISBN: 978-0-12-805321-8

For Information on all Academic Press publications
visit our website at https://www.elsevier.com/books-and-journals

Publisher: Joe Hayton
Acquisition Editor: Lisa Reading
Editorial Project Manager: Ashlie Jackman
Production Project Manager: Kiruthika Govindaraju
Designer: Greg Harris

CONTENTS

LIST OF FIGURES

LIST OF TABLES

LIST OF CONTRIBUTORS

William H. Bailey Exponent Inc., Bowie, MD, United States

Shoham Bhadra Exponent Inc., Natick, MA, United States

Cathy Chen Exponent Inc., New York, NY, United States

Benjamin R.T. Cotts Exponent Inc., Bowie, MD, United States

Brian W. D'Andrade Exponent Inc., Washington, DC, United States

Abdel-A. Edris IEEE Fellow, San Jose, CA, United States

Kevin L. Graf Exponent Inc., Menlo Park, CA, United States

Michelle Kuykendal Exponent Inc., Phoenix, AZ, United States

Noshirwan K. Medora Exponent Inc., Phoenix, AZ, United States

Sonal K. Phan Exponent Inc., Atlanta, GA, United States

Balasubramanian Pinnangudi Exponent Inc., Phoenix, AZ, United States

Matthew Pooley Exponent Inc., New York, NY, United States

Jay R. Prigmore II Exponent Inc., Warrenville, IL, United States

Ahmad Shahsiah Exponent Inc., Menlo Park, CA, United States

Daniel Shai Exponent Inc., New York, NY, United States

Nareg Sinenian Exponent Inc., New York, NY, United States

Adam Sorini Exponent Inc., Menlo Park, CA, United States

Ernesto Staroswiecki Exponent Inc., Menlo Park, CA, United States

PREFACE

The evolution of the infrastructure that supports and enriches the lives of billions of people on a daily basis is an extraordinarily remarkable and complex process. Smart cities, smart transportation, smart homes, and of course, the smart grid, are key components of that evolutionary process. It is beyond the capacity of a single book to completely cover any smart system in its entirety, let alone all of them, so the authors of this book have reviewed a number of topics, from the fundamentals to the possibilities. The authors, mainly forensic engineers from Exponent Inc., offer their unique perspectives formed by having led the investigations of failure events. They relish the opportunities to develop technology that is safe, green, secure, and more efficient.

This book addresses several important issues to a smart, green grid in chapters on energy storage, power conversion, and energy transmission. Energy storage is important because clean energy generation systems, such as solar power, may produce more or less energy than the instantaneous demand; hence, one of the major engineering developments in the last decade has been the proliferation of energy storage systems as a means of realizing a more agile and efficient electric grid. Power conversion is important because it enables the integration of renewable sources of energy such as wind and solar. Energy transmission is important because the optimal locations for renewable energy sources are found in rural or offshore locations that are long distances away from urban areas.

Additionally, this book provides insight into some of the side effects of making a smart system. The voluminous data require innovative techniques to analyze and the complexity of the controls to operate a smart grid opens up an exponentially increasing attack surface for bad actors to infiltrate and manipulate the grid, a key element of modern civilization.

It is my hope that you glean a deeper knowledge of smart grid technology from this dense collection of information.

Brian W. D'Andrade

1

EVOLUTION OF THE TRADITIONAL POWER SYSTEM

Ahmad Shahsiah
Exponent Inc., Menlo Park, CA, United States

CHAPTER OUTLINE

1.1 Introduction

1.1.1 Background

The power system in traditional form consists of three major components: generation, transmission, and distribution. Subtransmission levels can also be considered as a separate component in traditional power systems. In this system, the electricity is generated in masses, brought to the main distribution hubs, and then distributed to the end-users, much like any other commodity. However, unlike other commodities, the electricity should be consumed instantaneously at the same time that it is being generated. In other words, the power generation and the demand must match at any moment. Automatic closed-loop control

The Power Grid. DOI: http://dx.doi.org/10.1016/B978-0-12-805321-8.00001-X

systems at the generating units ensure matching load and generation. That is, of course, in case of availability of enough spinning reserve to quickly match the peak demand with the generation. Load forecasting, using historical data, provides information for the system operators to ensure availability of enough base generation and spinning reserves in order to supply loads while maintaining the system's frequency and reliability. Combined with the variable cost of electricity from different generating plants, where the system operators decide how to match the next increment of demand with the lowest cost of generation, electricity becomes a sophisticated market. Additionally, local and federal government regulations add to the complexity of this market and its operation. In such a system, complex analytical tools to predict real-time behavior of the system become essential.

Over decades the traditional power system has been studied, understood, and modeled, and sophisticated tools and methods have been developed to analyze and operate this system reliably. In a large traditional power system, the flow of electricity and contribution of each generator to the loads can be described by a system of equations known as the "load flow" equations. The equations are nonlinear and numerous (as many as the number of nodes in the system). The resulting nonlinear matrices that describe the system of equations can therefore be large. Although these matrices are usually sparse and can be presented with good approximations in the form of decoupled systems of matrices, solutions require advance numerical methods. For a large power system, such as the North American grid, the sheer number and nature of nodes combined with the speed at which the demand and generation changes at any moment challenges even the most advanced computers and methods of numerical solution. For this reason, and perhaps because of the associated cost, real-time simulations often include generation and transmission systems, which comparatively include fewer components than the distributions systems. Load-flow studies in the distribution systems are often done offline and in discrete studies for scenarios such as maximum loads as part of the reliability and protection studies of the systems.

For the reasons mentioned above, some aspects of the traditional power systems (particularly on the distribution side) are managed conservatively and therefore not optimally due to the sheer size and complexity of this system. This results in either oversizing or overloading equipment to meet the peak demand. This, combined with other factors such as reliability and cost of operation, has prompted application of newer systems with better and more efficient usage of the existing resources. A more efficient system utilizes new technologies to optimize generation and delivery of the electricity to end-users. This is widely known as the smart grid. In general, the traditional large power systems have centralized generation with limited control over the flow of electricity. Failures and blackouts in such large interconnected systems are not uncommon. A smart grid will have distributed generation, better monitoring, and control capabilities and will provide more efficient delivery of the electricity from the generating units to the end-users.

1.1.2 History and Outlook

Power systems in traditional forms have been in operation for more than a century. The first three-phase line in the United States became operational in 1893. Steam-driven generators became operational in the same year. Steam-driven generators still account for more than 80% of the US electric energy generation. Gas turbines are also used in some cases to meet the peak demand. Steam plants are fueled primarily by coal, gas, oil, and nuclear fuel. Coal has been the most widely used fuel for this type of generator in the United States due to its abundance. In 2014 the total net electricity generation from all fuels was 4,093,606 gigawatt-hours of which 1,581,710 gigawatt-hours generated from the coal which is approximately 39% of the net generation. Other main sources of generation include natural gas, nuclear, and hydrogenation with approximately 28%, 19%, and 6% share of generation, respectively, in 2014. The share of the renewable sources in total generation of electrical energy in 2014 was approximately 7%, which includes wind, utility-scale solar, geothermal, and biomass according to the US energy information administration (U.S.Energy_Informaion_Administration, 2016). This data shows that a large portion of the electricity in the United States still depends on the concentrated generation (mainly at large steam-driven power plants) with transmission, and distribution of electricity via the existing and aging traditional power system. The distributed and renewable sources are central to a fully implemented smart grid. Although the outlook is toward increasing the renewable sources and moving away from large centralized generation, as it is evident from the presented data, transition from the traditional power system to a fully implemented smart grid is still a distant future. This is further discussed later in this book in Chapter 10, Prognostics for the Power Industry.

1.1.3 Future Trends and Drives

The smart grid has received a lot of attention in recent years although studies for such systems go back more than a decade.

The main drive for the industry to move toward a smart grid is the newer and cheaper technologies and government incentives, particularly in the area of renewable energy. The existing traditional grid has been retrofitted with new data acquisition and control systems. At the distribution level, Supervisory Control and Data Acquisition (SCADA) systems have been implemented in many large metropolitan areas that allow monitoring or control of the distribution grid. More information from the status of the system is available through a wide application of sensors, and more components of the distribution systems can be monitored or automatically controlled now than ever before. Smart meters have been installed in large scales by many utilities that allow collection of the load data easily and efficiently. Vast amounts of data collected by smart meters can be used for many applications, including energy efficiency planning. Despite all the advancements in this area, many challenges still lie ahead for a full smart grid. Among these challenges are the legal aspects of a fully implemented smart grid. In a distributed generation system, for example, failure of local generation units can affect the loads in the same island or section of the system. Unless the legal ownership of the distributed generation equipment and the associated responsibilities and liabilities are fully defined, claims of damages as a result of equipment failure in such systems cannot be easily resolved.

1.1.4 Outline of the Chapters

Transition to a smart grid starts with integration of new technologies into the existing power systems. It is therefore important to understand the fundamentals of operation and function of the traditional power systems to overcome technical challenges during the transition. In other words, it is necessary to understand the fundamentals of the traditional power systems in order to understand various technical aspects of the transition to a smart grid. This chapter is an introduction to some of the fundamental concepts of the traditional power systems. Detailed aspects and applications can be studied using other references. Topics introduced in this chapter have been taught by the author for several years in academia for different engineering courses. The selected topics to be discussed are chosen to best represent fundamental knowledge required to understand the operation of traditional power systems as well as providing foundation for more advanced topics that are presented in other chapters of this book. The reader can return to this chapter for reference while reading through other chapters.

The next section of this chapter presents power system fundamentals, including the steady-state operation, transmission lines, and generating units of traditional power systems. Power flow, fault calculations, and reactive power compensation are introduced next in this section. Some of these topics are discussed in other chapters of this book in connection with the smart grids. The traditional distribution systems are also discussed briefly, which includes distribution transformers. Other aspects of the distribution system such as reliability and aging equipment are discussed later in Chapter 10, Prognostics for the Power Industry. An introduction to transients in power systems is presented in this section with an emphasis on fault calculation methods related to the unsymmetrical faults. More information are presented in Chapter 2, Transmission Grid Smart Technologies. The power electronics and energy storage systems has a lot of application in the newer systems such as alternating current/direct current (AC/DC) converters and plug-in hybrid electric vehicles. Advanced topics relevant to this section are presented in later chapters including Chapter 2, Transmission Grid Smart Technologies, Chapter 3, Advances in Power Converters, Chapter 4, Smart Grid Energy Storage, Chapter 6, HVDC Transmission for Renewable Energy Integration, and Chapter 7, Electric and Plug-in Hybrid Electric Vehicles and Smart Grids. These include advances in power converters, battery, and storage systems. Other relevant topics presented in this book include Chapter 5, Electromagnetic Interference Considerations for Electrical Power Systems, Chapter 8, Cybersecurity for the Smart Grid, and Chapter 9, Big Data and Monitoring the Grid.

1.2 Electric Power System Fundamentals

1.2.1 Background

Traditional distribution of electric power includes a network of power lines and cables carrying electricity to the end-users at specific voltage levels. This network is managed and maintained by the utility companies. In other words, the utility companies bring the electric power close to the end-user via the distribution network. In a traditional power system, the voltage level is reduced from transmission to distribution level using service transformers and delivered to the consumer at the main service panel. In the North American grid this power is distributed to the end-users via three or four wire systems of distribution network consisting of three phases or three phases and a neutral. In this system the amplitude of the voltage in each phase of the three-phase circuit changes 60 times per second from a negative peak to a positive peak in the form of a sinusoidal wave (60 Hz frequency). The voltage level and waveform at each phase are identical in shape and amplitude; however, they are different in phase angle. The power systems characteristics can be expressed by different measures such as the system voltage level, configuration, capacity, etc. In this chapter we discuss the fundamental concepts governing operation of a traditional power system with passive loads. Steady-state operation of such system is discussed. Some aspects of transient operations such as fault calculations are discussed as well.

1.2.2 Characteristics and Fundamentals

What follows is the fundamentals of mathematical representation and characterization of the main variables in power systems, namely, voltage, current, and impedance. It is assumed that these variables have negligible harmonic contents. This is not always a realistic assumption especially in case of the electronic load currents. However, it is still an acceptable approximation for many applications even with growing usage of power electronics and DC systems in modern power systems.

1.2.2.1 Phasor Representation and Passive Components

Variables such as voltage and current can be mathematically represented as vectors. In power systems these vectors are often referred to as phasors. Phasors in polar form are represented by a magnitude and an angle. In phasor notation, the magnitude of the current or voltage are shown only with their root mean square (RMS) value. The RMS value of any periodic waveform can be computed from the integral shown here where $v(t)$ is the waveform (voltage for example) and T is the time period of the waveform.

$$V_{\mathrm{RMS}} = \sqrt{\frac{1}{T}\int_0^T v^2(t)dt}$$

For a sinusoidal waveform $v(t) = V_{max}\sin(\omega t + \varphi)$, the RMS value is obtained by dividing the maximum value of the variable by the square root of 2. Once we have computed the RMS value, we can represent the phasor form of an alternating variable as the RMS value and the phase angle as shown here:

$$P[v(t)] = V_{\mathrm{RMS}} \angle \varphi$$

When adding two sinusoidal waveforms at the same frequency, the result is another waveform with the same frequency but with a different magnitude and phase angle. The new magnitude and phase angle can be calculated by adding two phasors, similar to adding two vectors with certain magnitudes and angles.

Like any other vector, phasors can be represented in rectangular form as well, and therefore can be represented by a real part plus an imaginary part multiplied by the imaginary number j or $\sqrt{-1}$.

Another important concept here is the Ohm's law, which relates current, voltage, and impedance, and is valid for the voltage and current phasors. Writing Ohm's law as the voltage phasor divided by the current phasor, we obtain another phasor called the impedance. The angle of the impedance phasor is an important parameter, as the cosine of this angle is called the power factor. We will talk about the power factor in detail later in this chapter.

Impedance can be written in rectangular form as $R + jX$ where R is the resistance and X is the reactance.

$$Z = R + jX = \frac{V}{I}$$

Also, current phasor divided by the voltage phasor is another phasor called admittance and has the unit of 1/Ohms, also known as Siemens, and is denoted by the letter Y. Admittance is a measure of how easily current can flow in an AC circuit. Admittance can be written in rectangular coordinates as $G + jB$, or it can be written in polar form as the reciprocal of impedance. Admittance in rectangular form has two components: G, or conductance, is the real part of the admittance vector and B, known as the susceptance, is the imaginary part. Ohm's law can be rewritten for admittance as the current equals admittance times the voltage. Again, admittance is the current phasor (I) divided by the voltage phasor (V) and is the reciprocal of impedance.

$$Y = G + jB = \frac{I}{V}$$

For a resistor, the admittance equals the reciprocal of resistance, which is equal to a conductance G. For an inductor, the admittance equals $-j$ times one over the angular frequency ($\omega = 2\pi f$), times the inductance L. For a capacitor, the admittance is j times ω times the capacitance, C.

For parallel components, the admittances can be added so that the equivalent admittance of a parallel combination equals admittance 1 plus admittance 2, etc. It does not matter whether the components are the same or different. For example, you could have two capacitors or a capacitor and an inductor in parallel. For series impedances, you would simply add impedances, which is equivalent to adding the reciprocal of admittances.

Impedance can be represented by the impedance triangle, where the x axis represents the real part of impedance, also known as a resistance, and the y axis represents the imaginary part called the reactance. Fig. 1.1 shows vector representation of the impedance. The angle between the real and imaginary part of impedance is called θ in this case.

Let us take another look at our three basic electrical components—resistors, capacitors, and inductors—to determine how they can be represented in the form of impedance. Ideal resistors have only a real part to their impedance and thus, have no reactance. This means that in polar notation, resistance can be written as R with phase angle of 0 degrees. In the rectangular form, ideal resistors have no imaginary component in their impedance vector, and therefore their impedance can be written as $R + j0$ ohms, or simply R ohms.

The ideal capacitor has no resistance, and therefore the impedance can be written as a reactance at an angle of -90 degree in the polar form. The reactance of an ideal capacitor is calculated as $1/jC\omega$, where ω is simply 2π times the frequency, and C is the capacitance. Note that the capacitor will have a negative reactance when written as the imaginary component in rectangular form with the j operator in the numerator. Thus, the loads with negative reactance are called capacitive loads. An example of the relation between the voltage and current waveforms for a capacitive load is shown in Fig. 1.2.

You can see in Fig. 1.2 that in a capacitive load the current waveform is ahead of the voltage waveform. As shown in the

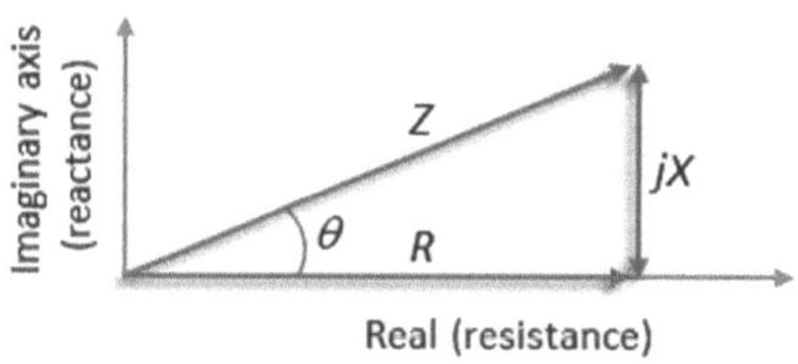

Figure 1.1 Vector representation of the impedance phasor.

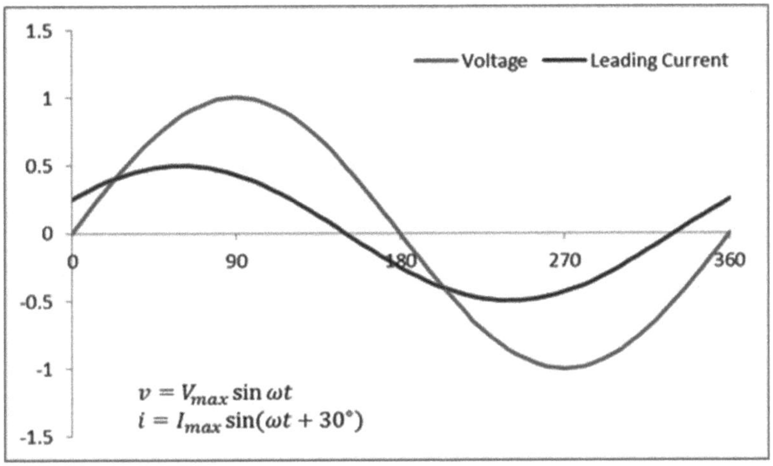

Figure 1.2 Example of the relation between the voltage and current waveforms in a capacitive load.

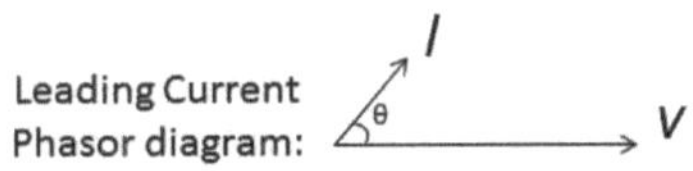

Figure 1.3 Phasor representation of the voltage and current for a capacitive load.

diagram, if the peak of voltage starts at time 0, you can see that peak of the current waveform has already started before time 0. This represents a capacitive load and therefore in capacitive loads the current leads the voltage. The impedance phasor in this case will have a negative phase angle since it is the result of the voltage phasor with angle 0 (the reference angle) divided by the current phasor with a positive phase angle. Thus, impedance with negative phase angle is capacitive. This is also consistent with the definition of reactance of a pure capacitor that we saw before which is $1/jC\omega$ or $-j/C\omega$; therefore, the phase angle of a pure capacitive impedance is -90 degree in the polar form. Phasor representation of the voltage and current for an example capacitive load is shown in Fig. 1.3.

The ideal inductor has no resistance but does have a reactance of $L\omega$, where ω is 2π times the frequency and L is the inductance. In the polar form, the reactance of an inductor can be written as $L\omega$ at an angle of 90 degree, or in rectangular coordinates, as $jL\omega$. Note that the inductor will have a positive reactance when written as the imaginary component in rectangular form using the j operator. An example of the relation between the voltage and current waveforms for an inductive load is shown in Fig. 1.4.

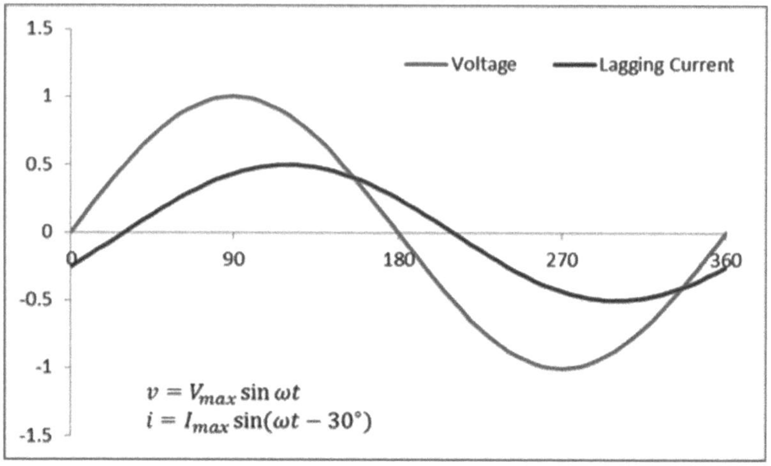

Figure 1.4 Example of the relation between the voltage and current waveforms in an inductive load.

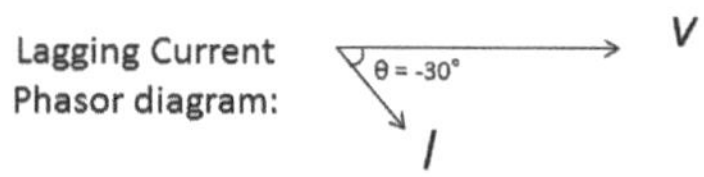

Figure 1.5 Phasor representation of the voltage and current for an inductive load.

You can see in Fig. 1.4 that the peak of the current waveform starts some time after time 0. Therefore, in an inductive load, the current lags the voltage. The impedance phasor in this case will have a positive phase angle since it is the result of the voltage phasor with angle 0 (the reference angle) divided by the current phasor with a negative phase angle. This is also consistent with the definition of reactance of a pure reactor that we saw before which is the operator j multiplied by the inductance multiplied by $2\pi f$; therefore, the phase angle of a pure inductive impedance is 90 degree in the polar form. Phasor representation of the voltage and current for an example inductive load is shown in Fig. 1.5.

Overall, since impedance in AC circuits replaces the general concept of resistance in DC circuits, when dealing with AC circuits, Ohm's law needs to be rewritten as $V = I \cdot Z$ where Z represents impedance instead of resistance; V and I are the voltage and current phasors, respectively.

1.2.2.2 Power in AC Circuits

Let us assume an electrical system consisting of sinusoidal waveforms of voltage and current. For such electrical system, we define the power factor as the cosine of the difference in phase angle between the current and the voltage waveforms.

Later we will show what this means physically. Power factor is an important concept in power systems and is a function of the impedance of the load since the voltage and the current phasors are related to the load impedance based on Ohm's law; the impedance phasor is the voltage phasor divided by the current phasor. Therefore, the difference in phase angle between the current and the voltage are reflected in the phase angle of the impedance phasor. Now we can define the active power as the product of the RMS voltage, RMS current, and the power factor. Here we show that this active power is, in fact, the same as the average AC power.

Consider the general voltage and current waveforms shown below for an inductive load (θ is a positive value) in a single-phase AC system:

$$v_a(t) = V_{max}\cos\omega t$$

$$i_a(t) = I_{max}\cos(\omega t - \theta)$$

The instantaneous power is simply the product of the two waveforms, which is also a waveform that is a function of time:

$$p_a(t) = v_a(t) \times i_a(t) = V_{max}I_{max}\cos\omega t \bullet \cos(\omega t - \theta)$$

Using trigonometric identities, it can be shown that the instantaneous power can be written as the sum of two terms:

$$p_a(t) = \frac{V_{max}I_{max}}{2}\cos\theta \bullet (1 + \cos 2\omega t) + \frac{V_{max}I_{max}}{2}\sin\theta \bullet \sin 2\omega t$$

The second term involves products of sine functions, the average of which over a complete period of the sine wave is 0. The first term in the instantaneous power equation does not average to 0 over its complete time period because of the term $(1 + \cos 2\omega t)$. In fact, the average of the first term over a full cosine period is $\frac{V_{max}I_{max}}{2}\cos\theta$, where $\cos\theta$ is the power factor. The product of the peak voltage and the peak current divided by 2 is the same as the product of the RMS voltage and the RMS current for the sinusoidal waveforms. Therefore, the average of the instantaneous power equation is nothing but the algebraic products of the RMS voltage, the RMS current, and the power factor. This is known as the active power.

Similarly, the reactive power is defined as the multiples of the RMS voltage, the RMS current, and $\sin\theta$. Note that the reactive power is not the mathematical average of the instantaneous power. Therefore, it is not a useable power by the load. However, more current will be drawn from the line as a result of the reactive power of the load which results in Ohmic losses.

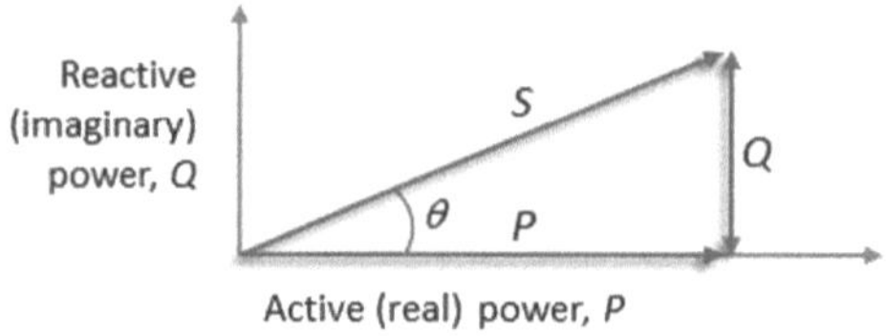

Figure 1.6 Complex power triangle.

We can define a complex power triangle by plotting the active and reactive power as shown in Fig. 1.6. The active power is a scalar value and is plotted on the horizontal or real axis as it is the real component of power. The reactive power is plotted on the vertical or imaginary axis.

The active and reactive powers can be presented as components of a vector quantity known as the apparent power, S. Thus, the apparent power S, or the complex power, being a phasor, can be presented in the following rectangular and polar forms:

$$S = P + jQ$$

$$S = |S| \angle \theta, |S| = \sqrt{P^2 + Q^2}, \quad \theta = \tan^{-1} \tfrac{Q}{P}$$

The apparent power, also known as the complex power, is measured in volt-amperes. The unit of the active power, P, is Watts. The reactive power, Q, is expressed in Vars. We previously defined the active power as the product of the RMS voltage, RMS current, and the power factor or $\cos\theta$. We also defined the reactive power as the product of the RMS voltage, RMS current, and $\sin\theta$. Therefore, it can be shown that the magnitude of the apparent power is simply the product of the RMS voltage and RMS current. The angle of this vector as shown in Fig. 1.6 is the cosine inverse of the power factor ($\cos\theta$). Note that the angle θ is the same as the phase angle of the impedance phasor.

We now define the positive power as the "consuming" power and negative power as the "generating" power. A resistive or inductive load, therefore, has a positive active power. The reactive power is negative for a capacitive load and therefore a capacitor can be considered as a generator of the reactive power. The complex power vector can be shown in terms of the voltage and current phasors in the following form where V is the voltage phasor and I^* is the conjugate vector of the current phasor.

$$S = V \bullet I^*$$

The I^* is a vector with the same magnitude of the current phasor and a phase angle that is multiplied by -1. The application of the conjugate current vector in the above formula makes the definition of the lagging (inductive) and leading (capacitive) loads consistent with the definition of the positive and negative powers as shown in Fig. 1.6.

1.2.3 Steady State Operations

1.2.3.1 Three-Phase Circuits

A balanced three-phase source consists of three single phase sources with equal magnitude and 120° phase shift between each source. A fourth conductor in such system is called the neutral, which is also connected to the ground reference. Fig. 1.7 shows a conceptual representation of such system.

As a result, two types of voltages can be defined in the three-phase system: line-to-line voltage, also known as the line voltage, and line-to-neutral voltage, also known as the phase-to-neutral or phase voltage. Waveforms and phasor representations of these voltages are shown in Fig. 1.8. Note that the magnitude of the line voltage is $\sqrt{3}$ times the magnitude of the phase voltage. Also note that the phase-to-neutral voltage at phase "a" is considered as the reference (zero phase angle) for the phasors shown in Fig. 1.8. Choosing a different reference will result in different phasor angles shown in the figure.

Three-phase loads can be connected together in two different forms: wye and delta. Fig. 1.9 illustrates a wye-connected three-phase load. In this system current through each phase of the load, or the phase current (I_φ), is the same as the current through each of the three-phase lines from the source that are also known as line currents (I_L). Also, voltage magnitude across each phase of a Wye-connected load to neutral, or the phase-voltage (V_φ), is equal to the line-to-neutral voltage (V_{LN}) of the line supplying the load. The

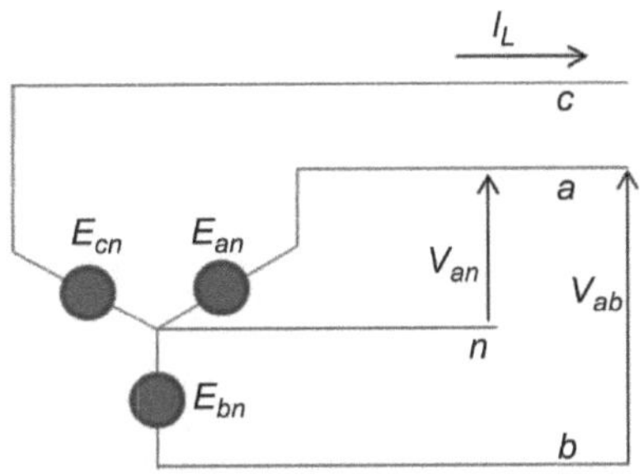

Figure 1.7 Conceptual representation of three-phase 4-wire system.

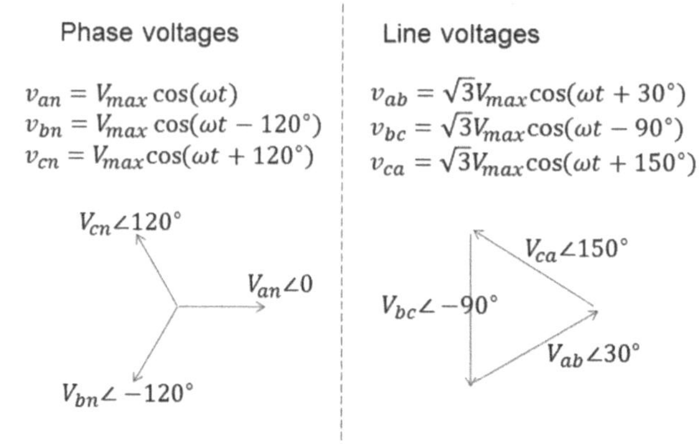

Figure 1.8 Waveform and phasor representation of the phase voltages and line voltages.

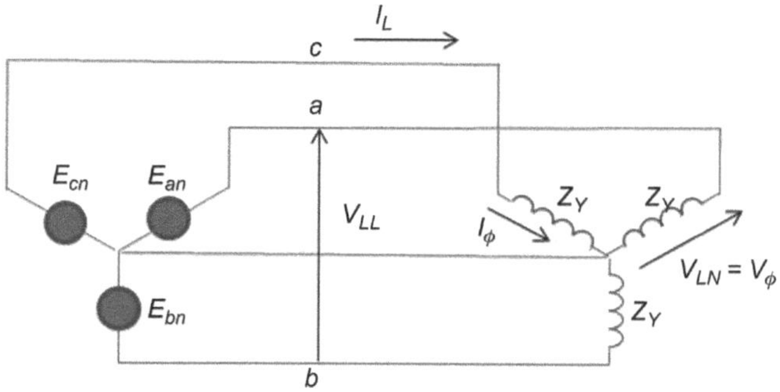

Figure 1.9 Wye-connected load.

line-to-neutral voltage is equal to the line-voltage (V_L) divided by $\sqrt{3}$. The line voltage is sometimes referred to as the line-to-line voltage (V_{LL}).

In wye-connected loads: $V_\varphi = \frac{V_L}{\sqrt{3}}$; $I_\varphi = I_L$

Similarly, in a delta-connected load, voltage across each phase of the load, or the phase voltage (V_φ), is the same as the voltage across each two phases of the three-phase line known as line voltages (V_L). However, current magnitude through each phase of a delta-connected load, or the phase-current (I_φ), is equal to the line-current (I_L) divided by $\sqrt{3}$. This is illustrated in Fig. 1.10.

In delta-connected loads: $V_\varphi = V_L$; $I_\varphi = \frac{I_L}{\sqrt{3}}$

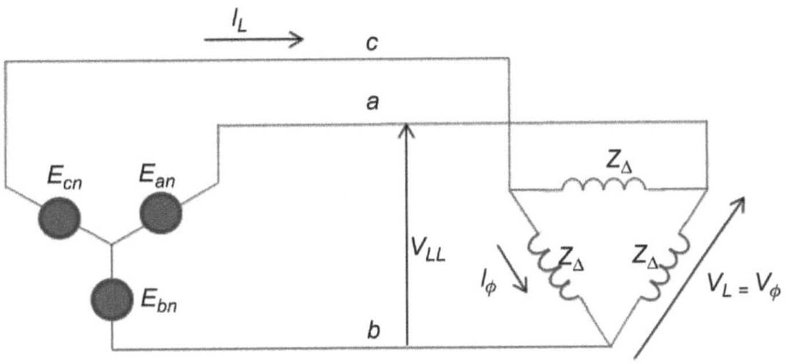

Figure 1.10 Delta-connected load.

The impedance per phase of a balanced wye-connected load (Z_Y) can be replaced by its equivalent delta-connected load (Z_Δ) using the following formula:

$$Z_Y = \frac{Z_\Delta}{3}$$

The active power, P, of the three-phase system is three-times the active power of each single phase. Remember that for the power through each phase, the phase voltage and phase current should be considered for each phase of the three-phase system. Therefore:

$$P_{3\phi} = 3 \times V_\phi I_\varphi \cos\theta$$

By replacing the corresponding values of the phase voltage and phase current for a balanced wye-connected or delta-connected load, it can be shown that the three-phase power can be calculated by multiplying $\sqrt{3}$ by the line voltage, the line current and the power factor. This is a more common formula since it is independent of how the load is connected. Similarly, reactive power, Q, can be calculated, except $\sin\theta$ replaces $\cos\theta$ in the equation.

$$P_{3\phi} = \sqrt{3} V_L I_L \cos\theta$$

$$Q_{3\phi} = \sqrt{3} V_L I_L \sin\theta$$

1.2.3.2 Power Transformers

Distribution power transformers are efficient devices that change the distribution line voltage to lower voltage levels to be delivered to the customer's main service panel. However, like any other device, there are losses associated with these transformers. The transformer losses include copper losses, and core losses. Copper losses are resistive heating losses in primary and secondary windings of the transformer. Core losses include hysteresis and eddy current losses at the transformer core. Additionally, a nonideal transformer will have a leakage reactance. The leakage reactance is associated with the magnetic flux lines that escape the core and pass through only one of the transformer windings. These flux lines produce self-inductances in primary and secondary windings. In an ideal transformer the magnetic flux created by the primary winding travels through an ideal core with infinite magnetic permeability and all the flux lines remain within the core and cross the secondary winding where the conversion of energy occurs. Since we do not live in an ideal world, this is not the case in a real distribution transformer. Therefore, there is an inductive reactance associated with a real transformer due to the leakage flux within the transformer core. Fig. 1.11 shows per-phase model of a nonideal transformer.

With reference to Fig. 1.11, V_p is the voltage-to-neutral at the transformer primary winding and V_s is that at the secondary winding. The parameter "a" is the primary to secondary turn ratio and $R_{eq} + jX_{eq}$ is the total equivalent leakage reactance as seen from the primary side of the transformer. Parameter R_c represents the core losses that include eddy-current and hysteresis losses inside the transformer core. Parameter X_m represents magnetization of the transformer core. This is due to the current from the primary windings that is spent on establishing magnetic flux inside the transformer core. Although small compared with load current, magnetization current exists due to limited magnetic permeability of the transformer iron core.

In an ideal transformer the leakage reactance will be 0 and thus $R_{eq} + jX_{eq}$ will be equal to 0. Also core and magnetization losses will be 0 which translates into infinite values for R_c and X_m. Therefore, in an ideal transformer the only relation between the primary and secondary voltage under any load condition will be defined by the transformer turn ratio.

In most of the distribution transformers core losses and magnetization current are small enough that can be ignored.

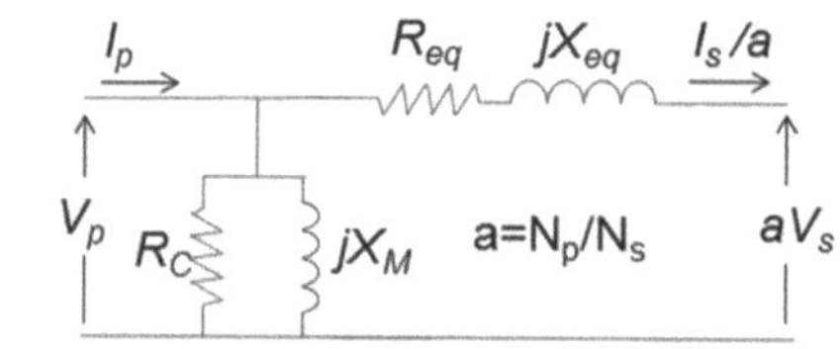

Figure 1.11 Per-phase model of a nonideal transformer.

Figure 1.12 Pole-top, single-phase transformers wired for a three-phase service.

Leakage reactance, however, cannot be ignored. Also, the three phases of a distribution transformer on both primary and the secondary can be either wye- or delta-connected. A distribution transformer can be a single three-phase transformer or three single-phase transformers wired for a three-phase service. Fig. 1.12 shows such installation using the pole-top, three single-phase bank.

Let us now consider different combinations of the primary and secondary winding connections in three-phase distribution transformers. Later we will show how to use the per-phase model to solve the power system equations that include a three-phase transformer. For that, we will introduce the concept of "per-unit calculations" so we do not have to worry about the transformer winding configuration when dealing with models of a large system that includes transformers with different types of connections. But for now, let us consider the wye–wye (Y–Y) connected transformer illustrated in Fig. 1.13. The turn ratio of a three-phase transformer is considered the ratio of the voltage across its primary terminals to that at the secondary terminals of the transformer. Therefore, the turn ratio of the three-phase transformer is the ratio of the line-to-line voltages at the primary and the secondary side of the transformer. If the turn ratio of each single-phase is "a," then the turn ratio of the three phase transformer connected in Y–Y will also be "a" as illustrated in Fig. 1.13.

Similarly, in a three-phase transformer that is connected in the form of delta-delta ($\Delta-\Delta$), the turn ratio of the three-phase

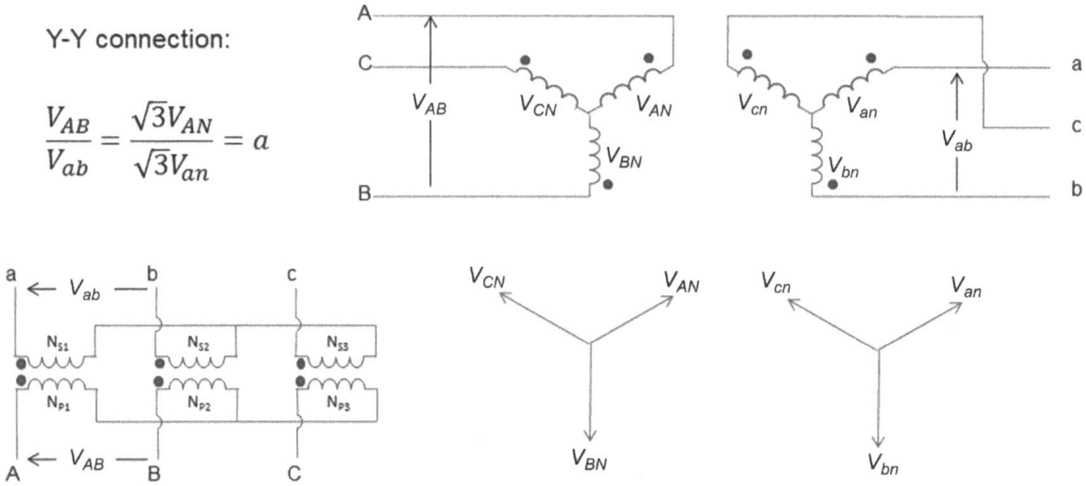

Figure 1.13 A three-phase transformer connected in Y—Y.

transformer is the same as the turn ratio of each single-phase transformer. This is illustrated in Fig. 1.14.

Now, consider a scenario in which the three-phase transformer is connected in either Y–Δ or Δ–Y. In this case, the ratio of the line-to-line voltage on the primary to that on the secondary side of the transformer will be different from the turn-ratio of each single-phase transformer. Additionally there will be a 30 degree phase shift in the line-to-line voltages at the two sides of the transformer. In accordance with the American Nation Standards Institution (ANSI) the windings in a Y–Δ or Δ–Y three-phase transformers are built and labeled in a way that the high voltage side is always 30 degree ahead of the corresponding low voltage side. This is illustrated in Fig. 1.15 that shows a Y–Δ connected transformer. Terminals designated in capital letters are on the high-voltage side of the transformer. It can be seen from the phasor representation that the V_{CA} phasor, for example, is 30 degree ahead ($+30^{\circ}$ counter-clockwise) compared to the V_{ca} phasor.

In balanced three-phase systems we can solve the circuit per phase since the only difference in phases of the circuit can be considered a 120 degree shift in phase-angle. However, with a three-phase transformer in the circuit, finding the per-phase equivalent may not be as straight forward.

As an example in a single-phase transformer we know that the load impedance, as measured from the secondary side of the transformer, is multiplied by the square of the

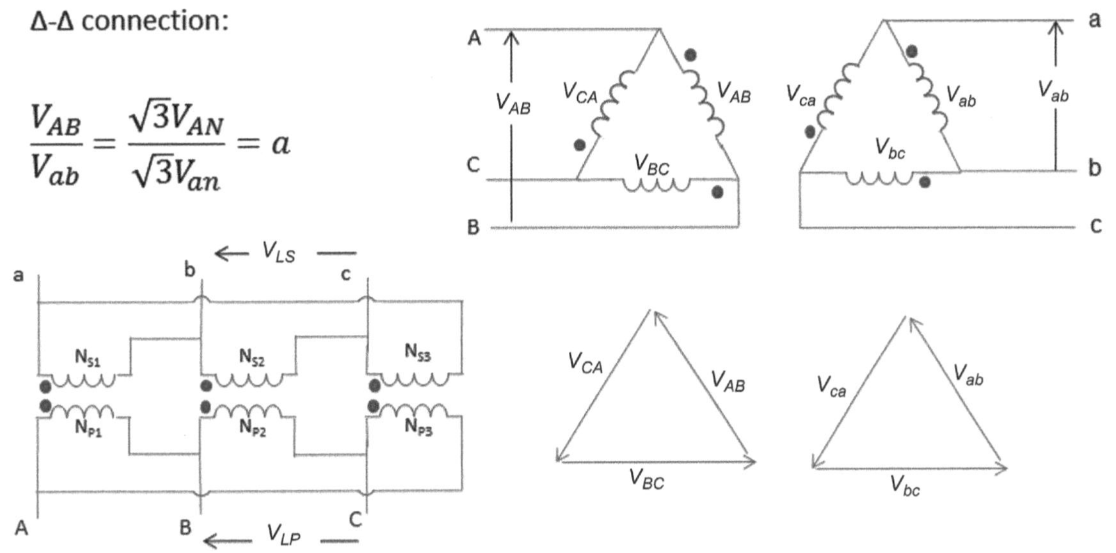

Figure 1.14 A three-phase transformer connected in Δ−Δ.

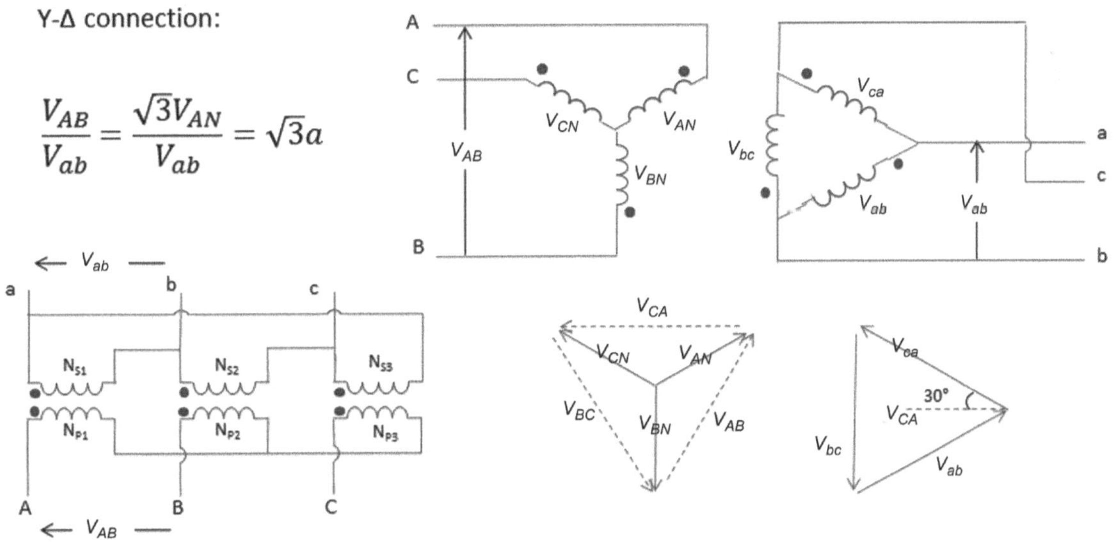

Figure 1.15 A three-phase transformer connected in Y−Δ. High voltage terminals are labeled in capital letters. Line-to-line voltages on the high voltage side are 30 degree ahead of the corresponding low voltage side.

transformer turn ratio or a^2 where a is the primary to secondary turn ratio of the transformer. Now consider the system illustrated in Fig. 1.16. In this example we will see how to find the load impedance as measured from the high-voltage side

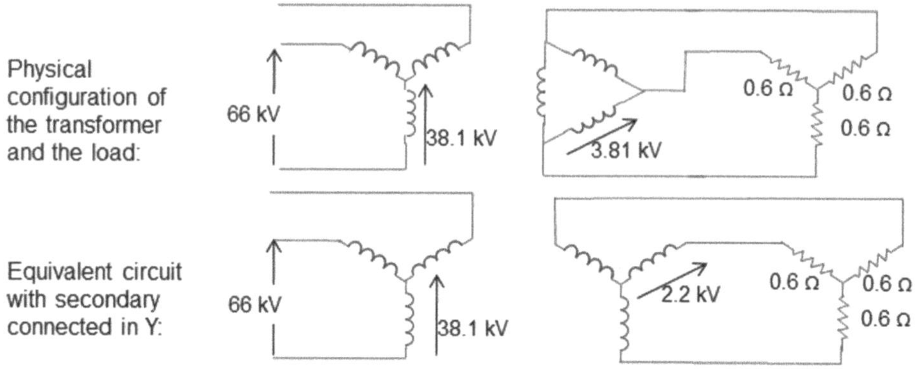

Figure 1.16 Example of a Y–Δ transformer with a Y-connected resistive load and the equivalent Y–Y transformer.

of a Y–Δ three-phase transformer with a Y-connected resistive load at the low voltage side. We will show later that the same example can be solved much easier using the per-unit calculations.

We first have to find the per-phase circuit. This would have been easier if the transformer was connected in the form of Y–Y in which case, phase-to-neutral at each phase could be considered the per-phase model. The equivalent Y–Y connected transformer can be replaced with the existing transformer with physical Y–Δ connection. In the equivalent circuit, the line-to-line voltage across the transformer terminals must remain unchanged. Therefore, voltage across the transformer windings of each single-phase bank in the equivalent circuit must be divided by $\sqrt{3}$ if the banks are Y-connected. In other words, the equivalent Y–Y transformer can be assumed to be made of three single phase banks with the turn ratio of $38.1/\left(\frac{3.81}{\sqrt{3}}\right)$ or $a = 38.1/2.2$. The per-phase turn ratio can then be used to find the load resistance from the high voltage side. We know that the load seen on the high voltage side of a transformer is equal to the square of the transformer turn ratio. Therefore:

$$Z = 0.6 \times a^2 = 0.6 \times \left(\frac{38.1}{2.2}\right)^2 \approx 180\Omega$$

1.2.3.3 *Per-Unit Calculations*

As the name implies, per-unit value of a quantity is defined by dividing the quantity by a base value. For example, per-unit voltage of a line is the value of the voltage, whether line-to-line or line-to-neutral, divided by the same base value of the voltage (base line-to-line or line-to-neutral voltage). The per-unit quantity can be defined for any parameter in the power system. Once turned into per-unit, circuit calculations can be done in the same manner as the quantitative values. After solving the circuit in per-unit, the actual values of interest can be found by multiplying the per-unit quantities by their corresponding base value. This is of course in case that all per-unit quantities are derived from a common set of base quantities. These base quantities normally include the apparent three-phase power in kVA_{base} or MVA_{base}, the line-to-line voltage in kV or $kV_{LL-base}$, and the per-phase impedance or Z_{base}. Two of the base quantities, normally the MVA (or kVA) and the kV_{LL}, have to be specified and the rest can be calculated using the following formulas in a three-phase system:

$$\text{Base current in amperes} = I_{base} = \frac{kVA_{base}}{\sqrt{3} \times kV_{LL-base}}$$

$$\text{Base impedance in ohms} = Z_{base} = \frac{kV_{LL-base}^{\;2}}{MVA_{base}}$$

Note that base voltages on the two sides of a transformer are different since the nominal voltage levels are different. The ratio of the base voltages on the two sides of the transformer is equal to the transformer's nominal line-to-line voltage ratio. Also note that per-unit impedances of different equipment may be expressed in different bases. In order to solve an interconnected power system, all per-unit impedances have to be in the same base. The following formulas are used to convert a per-unit impedance in an old base (z_{old}) to that in a new base (z_{new}) where kV_{LL-old} and kVA_{old} are the base line-to-line voltage and apparent power in the old base and kV_{LL-new} and kVA_{new} are the base line-to-line voltage and apparent power in the new base system.

$$z_{new} = z_{old} \times \left(\frac{kV_{LL-old}}{kV_{LL-new}}\right)^2 \times \left(\frac{kVA_{new}}{kVA_{old}}\right)$$

Often the equipment impedance in per-unit is presented by the manufacturer of the equipment. However, each manufacturer reports the per-unit impedance of the equipment in the VA base of the equipment itself. For example, the per-unit impedance of a transformer is presented in the VA rating of the transformer and per-unit impedance of an induction motor or a generator are presented based on their own rated VA. When analyzing the equipment in the same circuit, per-unit impedances of the equipment must be converted to a new value using a common base. This can be done using the above-mentioned base-change formulas.

It should also be noted that a transformer percentage reactance is the same as its per-unit reactance presented in percentage (%x). The reason is that this reactance is measured by the manufacturer of the transformer using a factory test known as the short-circuit test. In this test the transformer secondary terminals are short-circuited and reduced voltage is applied to the primary side until nominal current flows through the shorted terminals of the transformer secondary. The applied voltage at the primary will be a small percentage of the nominal primary voltage of the transformer otherwise the shorted transformer will be destroyed due to overheating. It can be shown that this small percentage of the applied voltage to the primary side is numerically equal to the transformer's leakage reactance in per-unit. This value is generally reported on the transformer nameplate as the percentage reactance of the transformer.

In order to show application of the per-unit calculations let us go back to the example shown in Fig. 1.16. Let us assume that the transformer is rated for 75 MVA. This is considered as the base impedance although any other arbitrary base MVA can be considered in the following calculations. We can find the load impedance value as measured from the high-voltage side by first calculating the per-unit impedance of the load and then multiplying it by the base impedance of the high-voltage side:

Low-voltage side base impedance: $Z_{base-low} = \frac{3.81^2}{75} \approx 0.194\ \Omega$

Load impedance in per-units: $z_{pu} = \frac{0.6}{0.194} \approx 3.1$ pu

High-voltage side base impedance: $Z_{base-high} = \frac{66^2}{75} \approx 58.1\ \Omega$

Load impedance seen from the high-voltage side: $Z = z_{pu} \times Z_{base-high} = 3.1 \times 58.1 \approx 180\ \Omega$

This is the same value that was previously calculated using the equivalent Y–Y transformer method and the per-phase turn ratio of the equivalent circuit.

Fig. 1.17 shows another example of a radial line with two parallel generators and two transformers. The single-line diagram of this system is also shown in this figure with all the impedances presented in per-units with common base kVA of 100 and 9600 V voltage base at the transmission line. It can be seen that after obtaining the single-line diagram of the system

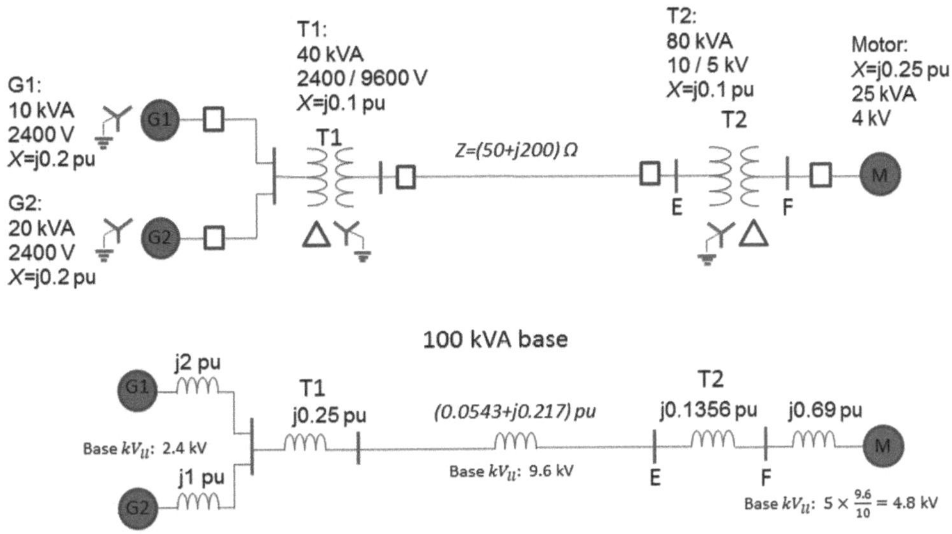

Figure 1.17 Example of a radial power system with different components and the equivalent single-line diagram in per-unit.

with all impedances expressed in the same base unit, the circuit can be calculated for voltages at different points within the system for different load currents. The calculated voltages in per-unit can be converted then into numerical values using the appropriate base voltage.

1.2.3.4 Distribution Line Model

Lines fewer than 50 miles (80 km) are considered short lines and can be modeled by only inductance and resistance of the line. For longer lines, the capacitance of the line cannot be ignored. For the lines longer than 150 miles, distributed models should be used to provide acceptable accuracy. Also, reflection of the voltage and current waveforms become important for switching transients and lightning surges in the long lines. In this section we will only discuss short line models since the distribution lines fall within this general category. Fig. 1.18 shows the per-phase model of a distribution line in which V_S and I_S are the sending-end voltage and current; and, V_R and I_R are that of the receiving-end. Resistance and inductive reactance of the line are shown in this model by R_L and X_L, respectively. It should be noted that per-phase parameters of short transmission lines are considered balanced even if the phase conductors are not spaced equilaterally and the line is not transposed.

Voltage regulation of a line is defined as the percentages of voltage rise at the receiving end of the line when the full load is removed (no-load condition) while the sending-end voltage remains constant. This is shown by the following formula in which $|V_{R,NL}|$ in the receiving-end voltage magnitude at no-load conditions and $|V_{R,FL}|$ is the receiving-end voltage magnitude at full-load conditions.

$$\text{Percent } voltage \text{ regulation} = \frac{|V_{R,NL}| - |V_{R,FL}|}{|V_{R,FL}|} \times 100$$

In a distribution line the effect of power factor on voltage regulation can be shown by phasor representations. With reference to Fig. 1.18, it can be seen that the sending-end voltage can be calculated as follows if the current phasor at the receiving-end (load current), voltage phasor at the receiving-end (load voltage) and the line impedance ($Z = R + jX$) are known.

$$V_S = V_R + RI_R + jXI_R$$

Phasor diagrams of the above equation are shown for three different load power factors in Fig. 1.19. In these diagrams the receiving-end voltage (V_R) is considered as the reference phasor

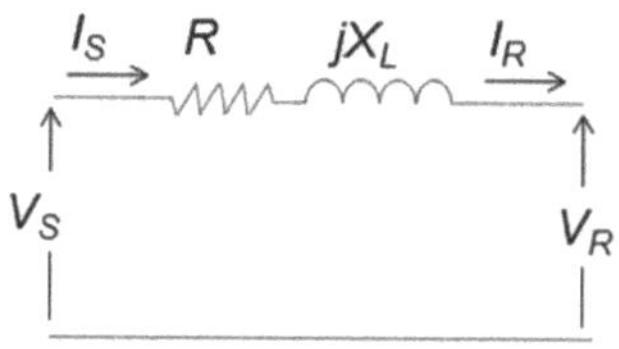

Figure 1.18 Circuit model of a distribution line.

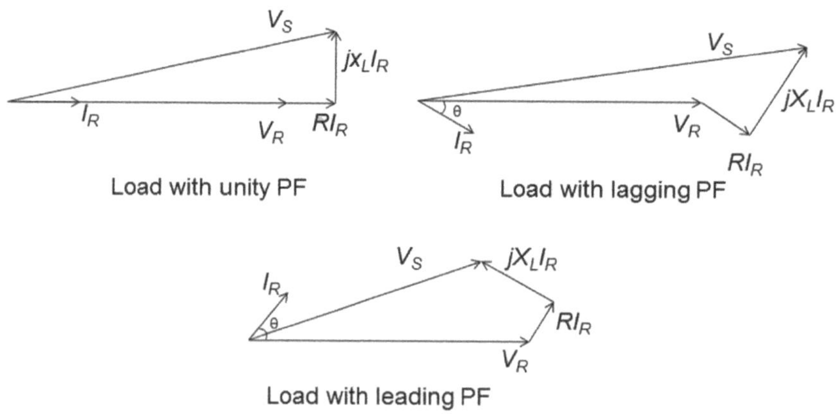

Figure 1.19 Phasor diagrams of a line for three different power factors at the receiving end of the line.

with the phase angle of 0 and the rest of the voltage and current phasors are plotted with reference to that. The angle θ is the current angle with respect to voltage, the cosine of which is the power factor.

It can be seen from the phasor diagrams that the magnitude of the sending-end voltage (V_S) could be potentially larger or smaller than the magnitude of the receiving-end voltage (V_R) depending on the power factor of the load at the receiving end.

Let us now consider the power transferred through the line. With reference to the model shown in Fig. 1.18 it can be seen that the active power delivered per phase of the receiving end can be calculated as below:

$$P = V_R I_R \cos\theta$$

where $\cos\theta$ is the load power factor. Using trigonometric identities illustrated in Fig. 1.20 for a loss-less line (negligible line resistance) and an inductive load (lagging power factor), it can be shown that $X_L I_R \cos\theta = V_s \sin\delta$; therefore:

$$I_R \cos\theta = \frac{V_s}{X_L}\sin\delta$$

Replacing the $I_R\cos\theta$ term in the previous equation:

$$P = \frac{V_s \bullet V_R}{X_L}\sin\delta$$

where δ is the difference in phase angles of sending-end voltage and the receiving-end voltage. This also shows that the dotted vertical line shown in Fig. 1.20 of the phasor diagram is proportional to the active power that is delivered to the load. It is apparent from the above equation and from the phasor diagram shown in Fig. 1.20 that the active power delivered to the receiving end of the line can be changed with the load power factor, or $\cos\theta$, even if the magnitude of the receiving-end voltage remains the same.

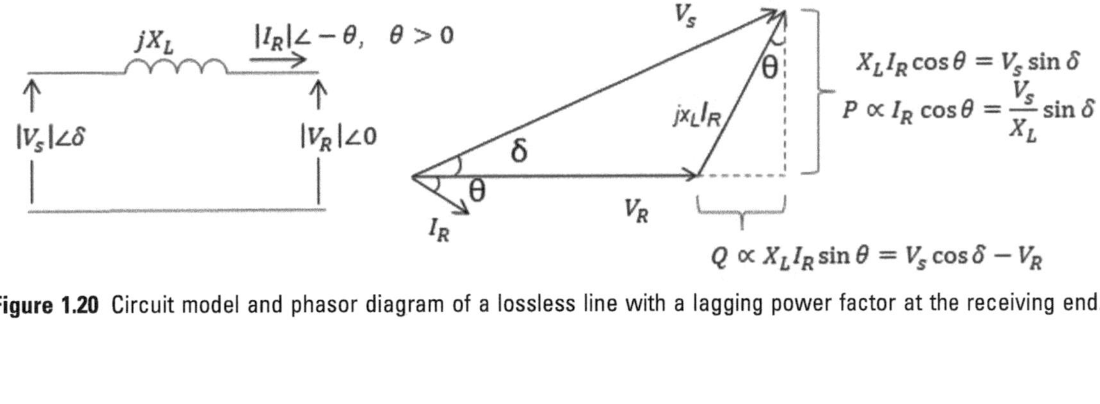

Figure 1.20 Circuit model and phasor diagram of a lossless line with a lagging power factor at the receiving end.

Similarly, it can be shown that the reactive power delivered at the receiving end of the line is proportional to the dotted horizontal line in Fig. 1.20 and can be found from the following equation:

$$Q = \frac{V_s \bullet V_R}{X_L} \cos \delta - \frac{{V_R}^2}{X_L}$$

This shows that the power delivered to a load via the line can also be controlled by changing the phase angle of the voltage at the sending end of the line without changing the voltage magnitude. This is a useful concept especially when controlling power through radial line from a generator unit.

1.2.4 Transients and Abnormal Conditions

1.2.4.1 Three-Phase Unbalanced Circuits

So far we have assumed that the three phases of the AC system are balanced. This means that voltage and current waveforms in all three phases are sinusoidal with the same magnitude and exactly 120 degree phase shift. Therefore, it was sufficient to solve the system for one phase only. The rest of the phases would have had the same voltage and current magnitude only 120 degree shifted in phase angle. In a traditional distribution system this is a correct assumption for the voltage waveforms. The utility companies ensure that the voltage delivered to the end-users in the form of balanced 60 Hz sinusoid with a nominal magnitude within acceptable tolerances. The current waveform, however, is a function of the load, and the utility companies have limited control over the loads. Although the single-phase loads are distributed evenly among the three phases of the distribution lines, these loads are rarely balanced. Therefore, the three-phase system of a distribution line cannot be considered balanced. Also, in abnormal conditions such as unsymmetrical faults, fault analysis cannot be performed assuming a balanced three-phase system. In these conditions, the AC system is assumed to comprise the sum of three fictitious balanced circuits known as symmetrical components. As the name implies, each component can be considered a balanced three-phase system and solved per-phase. Actual value of the parameters is then obtained by vector addition of the calculated individual components.

1.2.4.2 Symmetrical Components

Unlike balanced three-phase systems that we have discussed so far, in some applications the three phases may have different phase currents. In other words, the three-phase currents are not symmetrical. This condition could occur in power systems in the case of an unsymmetrical fault in which each phase will draw different current depending on which phase or phases have been involved in the fault. In this case, the circuit can be analyzed using the concept of symmetrical components. It can be mathematically proven that any imbalanced three-phase phasor can be written as the sum of three balanced components of positive sequence, V_{a1}, negative sequence, V_{a2}, and zero sequence, V_{a0}. As an example, for three imbalanced voltage phasors, voltage in each phase can be written as:

$$V_a = V_{a0} + V_{a1} + V_{a2}$$

Fig. 1.21 shows vector representation of the sequence systems. Note that each sequence component is a balanced three-phase system. The negative sequence system will have three balanced vectors, V_{a2}, V_{b2}, and V_{c2}, for each phase that are 120 degrees apart in the clockwise direction. The positive sequence system will have three balanced vectors, V_{a1},, V_{b1}, and V_{c1}, for each phase that are 120 degrees apart in the counter-clockwise direction. Three phases of the zero sequence system are in the same direction.

Breaking an imbalanced system into its symmetrical component networks means that the system can be solved separately for each sequence network. The advantage of this method is that each sequence network can be treated as a symmetrical three-phase system, the same way that we solve balanced three-phase systems. Once the sequence networks are solved and the desired phasor in each sequence network is calculated, the sequence phasors can be added vectorially to obtain the final phasor in the original unbalanced system.

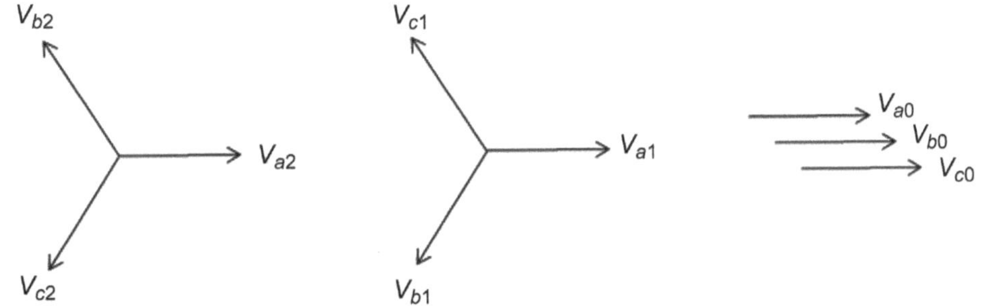

Figure 1.21 Symmetrical components of a systems of three imbalanced phasors.

Let us look at how we can determine the sequence components of an unsymmetrical system of voltages, namely, V_a, V_b, and V_c. We first define the vector operator "a" which is a vector with a magnitude of one and phase angle of 120 degrees.

$$a = e^{j120^\circ} = -0.5 + j0.866$$

It can be observed that when the operator "a" is multiplied by a phasor, the magnitude of that phasor remains the same, and its phase angle shifts by 120 degrees. Similarly it can be shown that a vector multiplied by the "a^2" will result in another phasor with the same magnitude and its phase angle shifts of 240 degrees or minus 120 degrees.

We also know that the sum of all three-sequence phasors in each phase will result in the original phasor in that phase as shown below. Now, using the operator "a" we can define the sequence phasors for each phase. For example, the three zero-sequence phasors will always have the same magnitude and the same angle.

$$V_a = V_{a0} + V_{a1} + V_{a2}$$

$$V_b = V_{b0} + V_{b1} + V_{b2}$$

$$V_c = V_{c0} + V_{c1} + V_{c2}$$

The positive sequence voltage, V_{b1}, has the same magnitude as V_{a1}, and its angle is 120 degrees behind the angle of the phasor V_{a1}. Therefore V_{b1} can be written as a^2 multiplied by V_{a1}. Similarly, other phases in each sequence network can be written in the form of the operator "a" or "a^2" multiplied by the sequence vectors in the phase a.

$$V_{a0} = V_{b0} = V_{c0}$$

$$V_{b1} = a^2 V_{a1}$$

$$V_{b2} = aV_{a2}$$

$$V_{c1} = aV_{a1}$$

$$V_{c2} = a^2 V_{a2}$$

Adding the sequence phasors obtained in this way for each phase and equating them with the actual phasor in that phase results in three equations that can be written in the following matrix form.

$$\begin{bmatrix} V_a \\ V_b \\ V_c \end{bmatrix} = \begin{bmatrix} 1 & 1 & 1 \\ 1 & a^2 & a \\ 1 & a & a^2 \end{bmatrix} \begin{bmatrix} V_{a0} \\ V_{a1} \\ V_{a2} \end{bmatrix}$$

By solving the system of equations in this matrix, three sequence phasors can be found for the voltage phasor V_a:

$$\begin{bmatrix} V_{a0} \\ V_{a1} \\ V_{a2} \end{bmatrix} = 1/3 \begin{bmatrix} 1 & 1 & 1 \\ 1 & a & a^2 \\ 1 & a^2 & a \end{bmatrix} \begin{bmatrix} V_a \\ V_b \\ V_c \end{bmatrix}$$

As an example, consider the following three phase current phasors presented in rectangular form in a Y-connected unbalanced load:

$$I_a = 44 - j33$$

$$I_b = -(32 + j24)$$

$$I_c = -40 + j25$$

We can simply substitute the three unbalanced vectors into the matrix equation for symmetrical components and solve for the first row in the matrix for I_{a0}, the zero sequence current vector. Note that components of this matrix equation are all phasors. Similarly, the positive, I_{a1}, and the negative, I_{a2}, sequence currents are calculated:

$$\begin{bmatrix} I_{a0} \\ I_{a1} \\ I_{a2} \end{bmatrix} = 1/3 \begin{bmatrix} 1 & 1 & 1 \\ 1 & a & a^2 \\ 1 & a^2 & a \end{bmatrix} \begin{bmatrix} I_a \\ I_b \\ I_c \end{bmatrix}$$

$$\begin{aligned} I_{a0} &= \frac{1}{3}(I_a + I_b + I_c) = \frac{1}{3}\left((44 - j33) - (32 + j24) + (-40 + j25)\right) \\ &= -9.33 - j10.67 = 14.17\angle -131.2^\circ \end{aligned}$$

$$\begin{aligned} I_{a1} &= 1/3\left(I_a + aI_b + a^2 I_c\right) = \frac{1}{3}\left((44 - j33)\right. \\ &\quad \left. - (-0.5 + j0.866)(32 + j24) + (-0.5 - j0.866)(-40 + j25)\right) \\ &= 45.141 - j27.62 = 52.92\angle -3146^\circ \end{aligned}$$

$$\begin{aligned} I_{a2} &= 1/3\left(I_a + a^2 I_b + aI_c\right) = \frac{1}{3}\left((44 - j33)\right. \\ &\quad \left. - (-0.5 - j0.866)(32 + j24) + (-0.5 + j8.66)(-40 + j25)\right) \\ &= 8.192 + j5.287 = 9.75\angle 32.84^\circ \end{aligned}$$

To understand how the symmetrical components method can be used to analyze unsymmetrical fault currents consider a single phase to ground fault. It is assumed that phase "a" is shorted to the neutral while the condition of the other phases remains the same. Phase "a" will carry the fault current to ground, I_{af}. Let us assume that there is also a fault resistance between the faulted phase and the ground, R_f.

We can write fault currents in each phase as the sum of its sequence components, based on the definition of symmetrical components. We know that the total fault current in phases "*b*" and "*c*", I_{fb} and I_{fc}, are equal to zero since the fault has only occurred in phase "*a*," which provide us with the equations shown here for I_{fb} and I_{fc}:

$$I_{fb} = I_{fa0} + a^2 I_{fa1} + a I_{fa2} = 0$$

$$I_{fc} = I_{fa0} + a I_{fa1} + a^2 I_{fa2} = 0$$

$$(a^2 - a) I_{fa1} = (a^2 - a) I_{fa2}$$

Equating these two equations, and solving, results in $I_{fa1} = I_{fa2}$. This means that the positive and negative sequence components of the fault current in phase "a" are equal. Substituting into one of the previous equations results in equal zero sequence fault current as well:

$$I_{a0} = I_{fa1} = I_{fa2}$$

Therefore, in the case of a single phase to ground fault, all three sequence currents of the faulted phase are equal to each other. Also, writing the sequence component equation for the voltage vector in phase "*a*", V_a, and substituting the fault current, I_{fa} with its components results in the following equation:

$$V_a = V_{a0} + V_{a1} + V_{a2} = (I_{fa0} + I_{fa1} + I_{fa2}) R_f = 3 R_f I_{fa0}$$

Let us now assume three Thevenin equivalent circuits for each sequence network. The positive sequence will have a Thevenin equivalent voltage, E_1 and Thevinin equivalent impedance, Z_1. Similarly, the negative and zero sequence networks will have equivalent impedances of Z_2 and Z_0, respectively. Note that there is no Thevenin equivalent voltage source for the negative and zero sequence networks. That is because no component of the voltage other than the positive sequence will have a nonzero value due to rotational sequence of the generator units in a power system. Therefore, the generator units will always create symmetrical positive sequence voltage phasors.

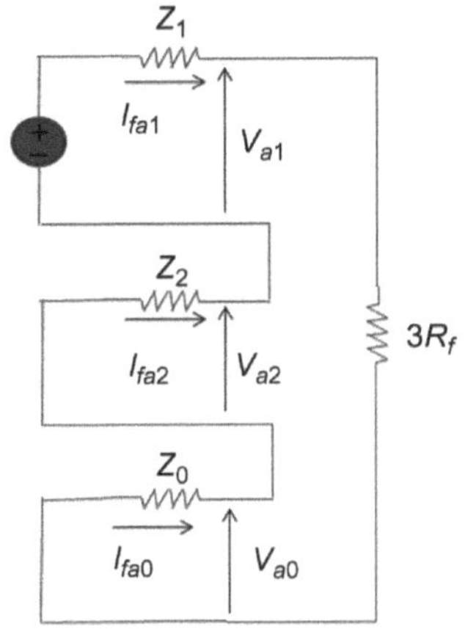

Figure 1.22 Thevenin equivalent circuit of the sequence networks for a single-phase-to-ground fault.

Let us now return to the problem of the single-phase fault. We found that in this fault condition $I_{a0} = I_{fa1} = I_{fa2}$. We also found that $V_a = V_{a0} + V_{a1} + V_{a2} = 3R_f I_{fa0}$. Thevenin equivalent circuits of the three sequence networks will satisfy this condition if all three sequence networks are connected in series as shown in Fig. 1.22. Therefore if the sequence reactances of a system, or their equivalent Thevenin network reactances, are known, the circuit shown in Fig. 1.22 can be solved to find the sequence fault current in the case of a single-phase-to-ground fault. Once the sequence currents are calculated in this way, the total single-phase fault current will be three times of the sequence fault current since all three sequence fault currents are equal in the case of a single-phase-to-ground fault.

1.2.4.3 Unsymmetrical Faults and System Protection

Let us now see how we can apply the symmetrical components method to calculate a single-phase-to-ground-fault currents. Assume that the positive, negative, and zero sequence reactances of a 20 MVA, 13.2 kV synchronous generator are 0.3 per-unit, 0.2 per-unit, and 0.1 per-unit, respectively. The generator is solidly grounded. We would like to calculate the magnitude of the fault current if a line to ground fault occurs on phase "*a*." Assuming a voltage reference, the generator voltage will be 1 per-unit with angle 0. Therefore, the zero sequence fault current, I_{a0}, can be calculated as below.

$$I_{a0} = \frac{1\angle 0}{j0.1 + j0.2 + j0.3} = 1.67\angle -90^\circ = -j1.67 \text{ per-unit}$$

The total fault current is three times the zero sequence current. Therefore, the fault current is calculated as five per-units with the angle of -90 degrees.

$$I_a = 3I_{a0} = 5\angle -90^\circ \text{ per-unit}$$

The actual fault current can be calculated by multiplying the per-unit current with the base current. The base current, I_{base}, is calculated from the nominal system voltage and the system's apparent power. Magnitude of the total fault current is 5 times I_{base}.

$$I_{base} = \frac{20,000}{\sqrt{3} \times 13.2} = 874.8 \text{ A}$$

$$I_{af} = 5 \times 874.8 = 4374 \text{ A}$$

This information is used in the system-protection design. A protection system should not operate under normal load conditions but it has to operate under fault conditions. Different fault condition scenarios are normally studied during the design stage and the results are used for the system-protection design.

1.2.5 Reactive Power Compensation

We previously defined active and reactive powers in the AC circuits. Recall that the power factor is the cosine of the difference in the phase angle between the voltage and current waveforms. Also remember, when a load is inductive, the current lags the voltage. When the load is capacitive, current leads the voltage. This can be understood in terms of the angle between the current and voltage waveforms. In doing so, it is useful to recall the unit circle from trigonometry. Recall that positive angles result in counterclockwise motion on the circle. Therefore, in a capacitive load that the current vector leads the voltage vector, the current vector would be oriented in a counterclockwise direction with respect to the voltage. Conversely, since the current vector lags the voltage vector in an inductive load, the current vector would be oriented in a clockwise direction with respect to the voltage. Note that as an accepted convention, we measure the angles with respect to the voltage. If the load is purely inductive, the current vector is a 90 degrees clockwise (that is negative) with respect to the voltage vector. If the load is purely capacitive, the current vector is 90 degrees counterclockwise (or positive) with respect to the voltage vector.

Power factors closer to 1 are desirable because the active power, *P*, will be higher for higher power factors. The active power is the power that creates useful work. It is apparent from the power triangle that higher power factors will result in smaller power angle, theta, which will result in less reactive power and more active power for the same apparent power. Utilities typically penalize customers who operate with lower power factors because utilities charge only for the active energy usage (kwh). Therefore, more reactive power by a load will not cost the consumer any more, however it will result in higher current through the distribution feeders as a result of higher apparent power. This extra current will result in Ohmic losses for the utilities. Therefore, the utility creates incentives for customers to increase the power factor by penalizing lower power factor loads.

Consider a problem in which it is desired to bring the power factor of a system shown in Fig. 1.23 to 0.9 lagging with a parallel capacitor. We are asked for the approximate reactive power that is required from a parallel capacitor to bring the power factor of the combination to 0.9 lagging.

The existing power factor is given and we are asked how we can increase it to a higher value. The power triangle shown in Fig. 1.24 demonstrates two different conditions with two different apparent powers, S1 and S2. S1 represents the system prior to the power factor correction with a given power factor of 0.7. The reactive power for this condition is calculated as Q1. S2

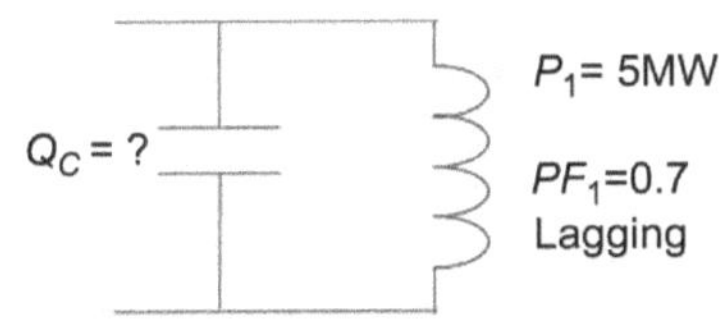

Figure 1.23 A 5 MW system presented by an inductive load with a power factor of 0.7 and lagging to be compensated for reactive power with an external capacitor.

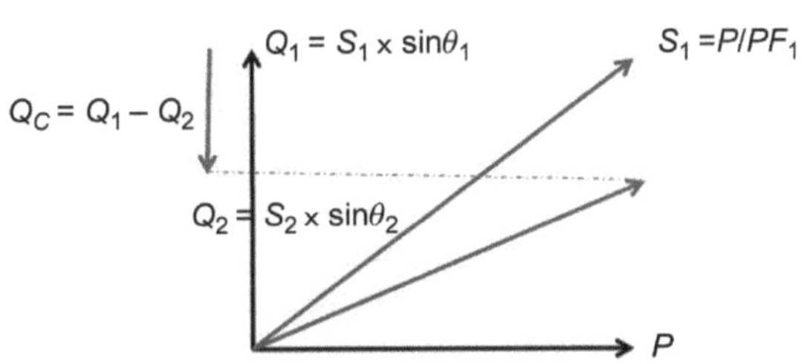

Figure 1.24 Vector representation of the power factor compensation shown in the previous problem.

represents the system after the power factor correction to the new power factor of 0.9. The corresponding reactive power in this case is Q2 and as shown in Fig. 1.24. In order to correct the system power factor, a capacitor has to be added, with the reactive power equal to the difference of the calculated reactive powers in these two scenarios. Note the active power does not change since the active power is a function of the load and an added capacitor will not create additional active power in this system.

Presented concepts in this section are by no means a comprehensive review of the concepts that power engineers encounter in traditional power systems. However, in this authors' view, the presented concepts are essential for fundamental understanding of the power systems. For further information, readers can refer to the *Power Systems Analysis* by John J. Grainger and William D. Stevenson, Jr. (John J. Grainger, 1994).

1.3 Evolution and Outlook

1.3.1 Existing Grid Versus Smart Grid

There are many publications available about the characteristics of a smart grid, each with different visions on how to evolve from the existing traditional power system into an efficient and automated smart grid. The next chapter of this book provides more details on this topic. However, almost all experts agree that such evolution is a gradual process and starts with application of new technologies in the existing traditional power system. New technologies such as remote monitoring and control systems have long been implemented in power systems. However, such technologies have been used less in power distribution, especially as they are related to the end-user applications, compared to other sections of the traditional power systems.

Since 2009, there has been widespread adoption and installation of advanced smart meters at the consumer service locations in distribution systems thanks to federal funds directed at modernizing the nation's electric power systems. In the area of power generation, new technologies have been implemented in the area of renewable energies, most notably wind and solar. Although the technology has been available in this area for a long time, renewable technologies have been recently supported by the new regulatory and tax environments. Such renewable energy technologies are more visible to consumers than transmission and distribution technologies, while transmission and distribution technologies have been driven the improvements in reliability and life expectancy of these systems. The application of new technologies includes integration of digital capabilities with enhancements in digital data storage and data analytics, advanced measurement capabilities, smart inverter technologies, etc. Application of new technologies and more will be discussed in later chapters of this book.

1.3.2 Evolution and Challenges

It is clear that a fully implemented smart grid that is more reliable, adaptive, self-monitoring, and self-healing, with fully integrated distribution generation and that is predominantly supplied by renewable energies, is still a distant future. Significant achievements, however, have been made toward this goal in recent years thanks to government and regulatory supports. Challenges in this area include technical and organizational challenges. However, in this author's view, technical challenges are not as significant as organizational challenges. For the past several decades, technology has been ahead of the power systems. Government incentives have largely justified the movement to attain a smarter power grid which has led to an accelerated rate of change in the existing traditional system. Challenges remain in managing implementations of the new technologies in the existing grid as well as understanding and managing the organizational changes that come with it. Transition of the utility organizations, from the traditional system in which all variables are known and well understood, to a new system with variables that may not even be identified yet, poses a significant challenge. Some of the new variables include engaging customers more in the operation of energy distribution, as well as effectively participating in the regulatory framework of the future power system. For further information about these new challenges, readers can refer to a published article in the September/October 2016 issue of *Institute of Electrical and Electronics Engineers (IEEE) Power & Energy*, titled, "Becoming the Utility of the Future, Risks, and Opportunities" (Richard E. Brown, 2016).

1.3.3 Emerging Standards and Regulations

Regulatory acts have traditionally played a significant role in the development and reshaping of the power grid. In 1978, the Public Utility Regulatory Policies Act required utility companies to buy from qualified energy generators, ending the utility monopoly on power generation. Deregulation laws in the early 2000s reshaped power generation and distribution in an effort to move toward an open market electricity system. In 2006, the Federal Energy Regulatory Commission (FERC) mandated the utility companies to allow generation companies to transmit and sell energy using their transmission systems. Over the next several years, further standards were published by FERC on sharing information in the wholesale energy market and addressing regional and interregional costs. Meanwhile, the federal government provided funds to modernize the nations electrical grid. Federal funds and regulatory incentives have largely contributed to the high rate of evolution in the existing infrastructures of the traditional power system, as well as new advanced installations; mainly in the area of renewable energies. Technical committees of engineering societies have been active in standardizing technology applications, as well as sharing new methods and best practices, which includes publications by the IEEE Smart Grid Committee. More information is available through the IEEE website.

References

Grainger, J.J., Stevenson Jr, W.D., 1994. Power System Analysis. McGraw-Hill, Inc, New York.

Richard E. Brown, C.S., 2016. Becoming the utility of the future, risks and opportunities. IEEE Power Energy Mag. 14 (5), 57–65.

U.S.Energy_Informaion_Administration. 2016. Table 1.2. Summary Statistics for the United States, 2004–2014. Retrieved from Electricity: www.eia.gov.

2

TRANSMISSION GRID SMART TECHNOLOGIES

Abdel-A. Edris[1] and Brian W. D'Andrade[2]

[1]IEEE Fellow, San Jose, CA, United States [2]Exponent Inc., Washington, DC, United States

CHAPTER OUTLINE

2.1 Introduction

Electricity is fundamental to civilization as a means to convey both energy and information. The electric power system is a very complex system and consists of various large sources of power generations, vast transmission networks, and local distribution systems. Given the complexities, there are inefficiencies of the current system such as inadequate power flow control,

The Power Grid. DOI: http://dx.doi.org/10.1016/B978-0-12-805321-8.00002-1

inadequate reactive power management, power angle and voltage instabilities, inadequate planner/operator training, and inefficient use of collected data. Therefore, there is a need to address these inefficiencies by the development of new technologies for improving grid performance characteristics such as reliability and quality of power delivery.

Electric power systems, after several decades of slow development, are now experiencing tremendous changes due several factors including the need for large scale integration of renewable energy resources, aging assets, energy efficiency, and concerns of system vulnerability to major disturbances (e.g., cascading failure events and system-wide blackouts) (2013a; 2013b; Baldwin et al., 2004). Furthermore, the increasing complexity of these operations will ultimately require the introduction of more intelligence, communication, and control coordination into electric transmission grids for improved security, economy, and efficiency.

2.2 Smart Electric Transmission Grid: A Definition

Some key characteristics for the evolutionary transformation of the existing electric grid into a smart grid include the ability to rapidly detect, analyze, respond, and restore power after a failure event. The grid design should mitigate and be resilient to physical and cyber-attacks and provide power quality needed by 21st-century users. The grid should accommodate a variety of resources (including demand response, combined heat and power, wind, photovoltaic, and end-use efficiency), and the grid should fully enable and be supported by competitive markets (Gagwari and Bhatia, 2015; Moslehi and Kumar, 2010; Al-Haddad, 2010; Line et al., 2011).

2.3 Smart Grid Road Map

Transforming current electric grids into a smart grid may be hindered by the following: grid conditions and requirements, such as variable power generator supply, integration of distributed resources, and long distances between power resources and loads; operational factors, such as greater transmission network complexity, aging infrastructure and dwindling human capital, and high power delivery quality; external influences, such as legal and regulatory framework, integration of renewable energy resources, carbon dioxide reduction, and energy efficiency (2016; Farhangi, 2010).

The electric grid therefore has to undergo significant adjustments to overcome these challenges with the growing global demand of electricity. Any adjustments to an electrical grid must, however, meet the following four basic requirements: capacity (satisfying increasing demand for electrical energy), reliability (satisfying quality and reliability/availability of electricity), efficiency (from production and transport to consumption of electricity—energy has to be saved), and sustainability (integration of low carbon energy sources).

In order to implement the adjustments while meeting the above four requirements, the following characteristics should be met: ability to sense overloaded regions and to re-route power flows to reduce overload and prevent potential outage and cascading failures; real-time communication between consumer and utility, allowing the consumer to optimize energy usage based on environmental and/or price preferences; capability of auto-balancing supply and demand; carbon dioxide reduction; accommodation of any energy resource with a robust transformation of these energy resources into consumer end-use with minimal human intervention; and allowing unlimited use of renewable energy sources aimed at minimizing carbon footprints.

2.4 Understanding Transmission System Performance and Operation

2.4.1 Electric Power Transfer Fundamentals

Most of today's transmission grid is a complex network of transmission lines interconnecting all the generator stations and all the major loading points in the power system. Separate AC (Alternating Current) systems are often synchronously interconnected with AC transmission lines to form a power pool in which energy can be transported between the individual systems. In this arrangement, at a given time, some systems may be importing and others exporting power. Also, some systems only provide the service of transporting power through their transmission network to facilitate particular transactions. The power pools with individual systems can further be tied together with synchronous AC, or asynchronous DC (Direct Current) links to form a large power grid to provide highly

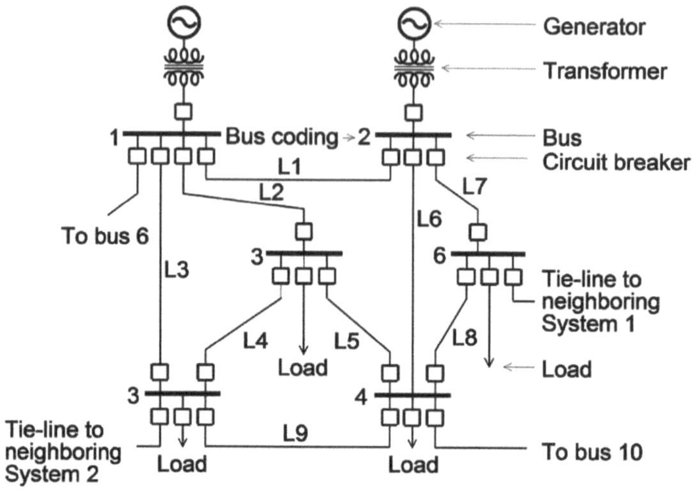

Figure 2.1 A simplified example schematic of an electric power system structure.

reliable electric power supply at minimum cost. The main characteristic of today's transmission system is an overall loop structure, as illustrated with a simple power system schematic in Fig. 2.1.

2.4.2 Real and Reactive Power Flow

Real and reactive power flows on large integrated power transmission systems are constrained by voltage and power stability criteria. In many cases, the limits dictated by stability criteria are far below the "inherent" thermal capacity of transmission lines. Even with the adoption of stability measures, power flow levels could still be below the thermal limit due to the "uncontrolled" flow levels as determined by Kirchhoff's circuit laws.

Based on Kirchhoff's laws, real and reactive power flows on a transmission line are determined by transmission line impedance, voltage magnitudes, and angle difference at the transmission line's end, as illustrated in Fig. 2.2. Real and reactive power flow, as determined by circuit laws, result in power levels where some transmission lines may operate near thermal limits while other transmission lines are not fully loaded. The result is unused transmission capacity of the transmission grid that lowers the overall efficiency of the grid.

Another issue of having uncontrolled power flow is "loop flow" or "parallel flow" where less than 100% of a contractual power flows over the designated path and the rest of the power flow strands over other transmission paths. As a result, the

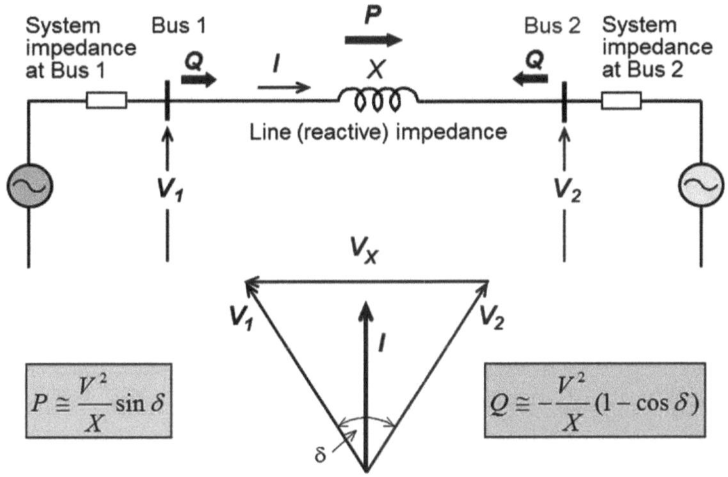

Figure 2.2 Real and reactive power flow on a transmission line.

uncontrolled power transfer is generally lower than the inherent thermal transmission capacity limit.

The challenge, therefore, is to be able to load all transmission lines up to their respective thermal capacity limits while maintaining the overall system reliability (i.e., optimizing the stability limit closer together with the thermal limit). Successfully overcoming this challenge will result in a "smart" utilized transmission grid. In some cases, the increased transmission capacity could be 30–60%, depending on the specifics of the considered transmission grid.

2.4.3 Controlling Power Flow Parameters

As shown in Fig. 2.3, shunt and series transmission compensations (Edris, 1990), as well as transmission angle control, can significantly change the power flow on a compensated transmission line. Thus, it is reasonable to expect that suitable and fast control techniques would be able to vary the power flow in the system to counteract grid contingencies and thereby enhance the grid stability limit, increase transmission capacity, and ultimately result in maximizing utilization of transmission asset of the controlled grid.

2.5 Role of Power Electronics in Smart Transmission Grids

The basic limitations of power flow are line impedance, angle and voltage stability, and thermal limits. These limitations

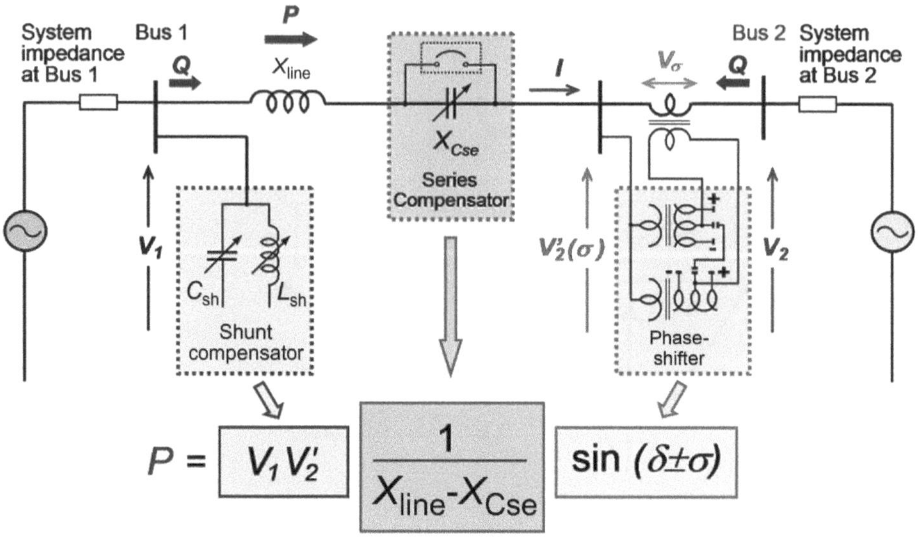

Figure 2.3 Shunt and series transmission compensations, as well as transmission angle control.

often result in underutilization of transmission assets in the sense that scheduled power transfer levels could be far below 20–60% lower than the inherent transmission capacity.

One solution to overcome these limitations would be adding new lines to the transmission network. Apart from the cost, such a solution under the prevailing environmental and regulatory constraints is not feasible. Furthermore, adding a new line, while increasing transmission capacity, doesn't provide additional means for controllability of the transmission network. There is, therefore, a need to introduce smart technology solutions and concepts to overcome limits and inflexibility of existing transmission grids.

2.5.1 Power Electronics-Based Transmission Controllers

Dynamic voltage variation and angle instability problems in AC transmission systems put limits on maximized utilization of the inherent capacity of transmission grids. Transmission controllers based on the use of power electronics enable a rapid, dynamic compensation and control of voltage and power flow; hence, they allow reliable increased utilization of the existing transmission system with no need for new transmission lines. Transmission technologies based on the use of power electronics are: flexible AC transmission system (FACTS) technology, high voltage DC (HVDC) transmission, line commutated converter (LCC) and/or voltage sourced converter technologies, and segmentation and grid shock absorber technology.

FACTS and HVDC technologies are "pillars" for enabling a smart transmission grid. These technologies provide the needed flexibility, controllability, reliability, and accessibility for transforming a transmission grid to become smart. FACTS and HVDC technologies provide increased transmission capacity; relieve transmission bottlenecks; improve flexibility and controllability of transmission grid; enable bulk power transmission in the gigawatt range over long transmission corridors; reduce carbon dioxide emissions; enable access of large wind, hydro, and solar power plants; and increase robustness and reliability of the transmission grid.

2.5.2 Functional Applications of FACTS Technology

FACTS technology is defined as, "alternating current transmission systems incorporating power electronic-based and other static controllers to enhance controllability and increase power transfer capability" (1997). The following are a few examples demonstrating the functional capability of implemented FACTS technology.

2.5.3 Marcy Substation

Marcy substation was a transmission bottleneck, limiting the power to be transferred from the northern and western area of the New York transmission system (Fardanesh et al., 2002). Relieving this bottleneck required a strong shunt dynamic voltage support at the Marcy substation. Fig. 2.4 shows the structure of the installed CSC (Convertible Static Compensator, 2003) at the Marcy substation.

The CSC consists of two 100 MVA voltage sourced converters, one shunt transformer with two secondary series transformers, and a set of motor operated switches. This structure provides a number of principal configuration control modes, +200 MVAR dynamic voltage support (STATCOM) mode at the Marcy substation, +100 MVAR shunt and 100 MVA series injection (unified power flow controller (UPFC)) mode, and 100 MVA series injection in one line and another 100 MVA series injection in the other line (interline power flow controller (IPFC)) mode.

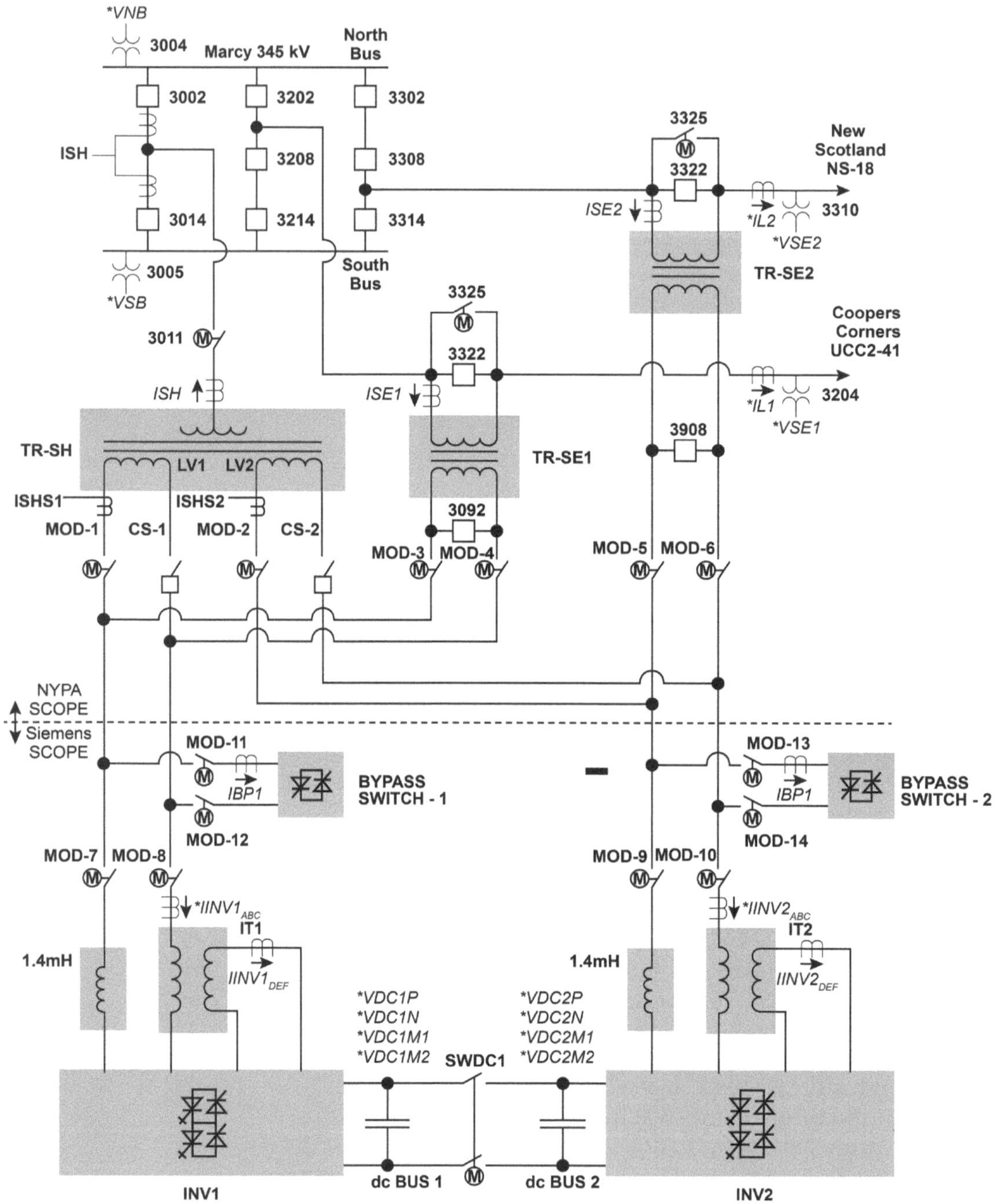

Notes: Various components (switches. PTs, CTs, etc.) which do not interface to the CSC Central Control are not shown.
*Asterisks indicate analog signals used for CSC control

Figure 2.4 Installed CSC at 345 kV Marcy Substation.

The benefits from installing the Marcy CSC included relief of the transmission bottleneck at Marcy substation and strong dynamic voltage support at Marcy, which resulted in an increase of transmission capacity by about 200 MW. This increase equates to enough power for approximately 200,000 homes with unprecedented smart controllability and flexibility in the transmission grid.

2.5.4 Inez Substation

American Electric Power (AEP) selected its Inez Substation in eastern Kentucky for the location of the world's first UPFC installation, which comprises two ±160 MVA voltage-sourced GTO-thyristor-based inverters. The installation was the first demonstration of this type of equipment connected in series with a transmission line (Renz et al., 1998; Schauder et al., 1998).

The benefits from installing the Inez UPFC included: increased power transfer capability of 100 MW; increased transmission capacity for years of load growth on AEP's transmission system and the reduction of real power losses by more than 24 MW, which is equivalent to a reduction of carbon dioxide emissions by about 85,000 tons per year.

2.5.5 Segmentation and Grid Shock Absorber

A large scale interconnected power system is complex due to a number of interacting factors that are not fully controlled or anticipated. Due to this complexity, major and widespread disturbances occur from time to time and result in major power outages and blackouts.

Hence, a "grid shock absorber" concept was introduced to mitigate disturbances (Clark et al., 2008). The impetus for the grid shock absorber concept is the need for a device to limit cascading faults. The idea of the grid shock absorber stems from the characteristic of a voltage sourced converter-based back-to-back (VSC BtB) device that connects electrically separated transmission segments.

As shown in Fig. 2.5, VSC BtB provides a four-quadrant real and reactive power control at its two ends; therefore, it has an inherent capability of providing voltage support at its point of interconnections. Furthermore, BtB acts as a firewall, stopping propagation of a disturbance from one segment to another. That is, it prevents cascading failures.

The concept was tested on the Eastern Interconnection (EI), which had experienced a major blackout in August 2003.

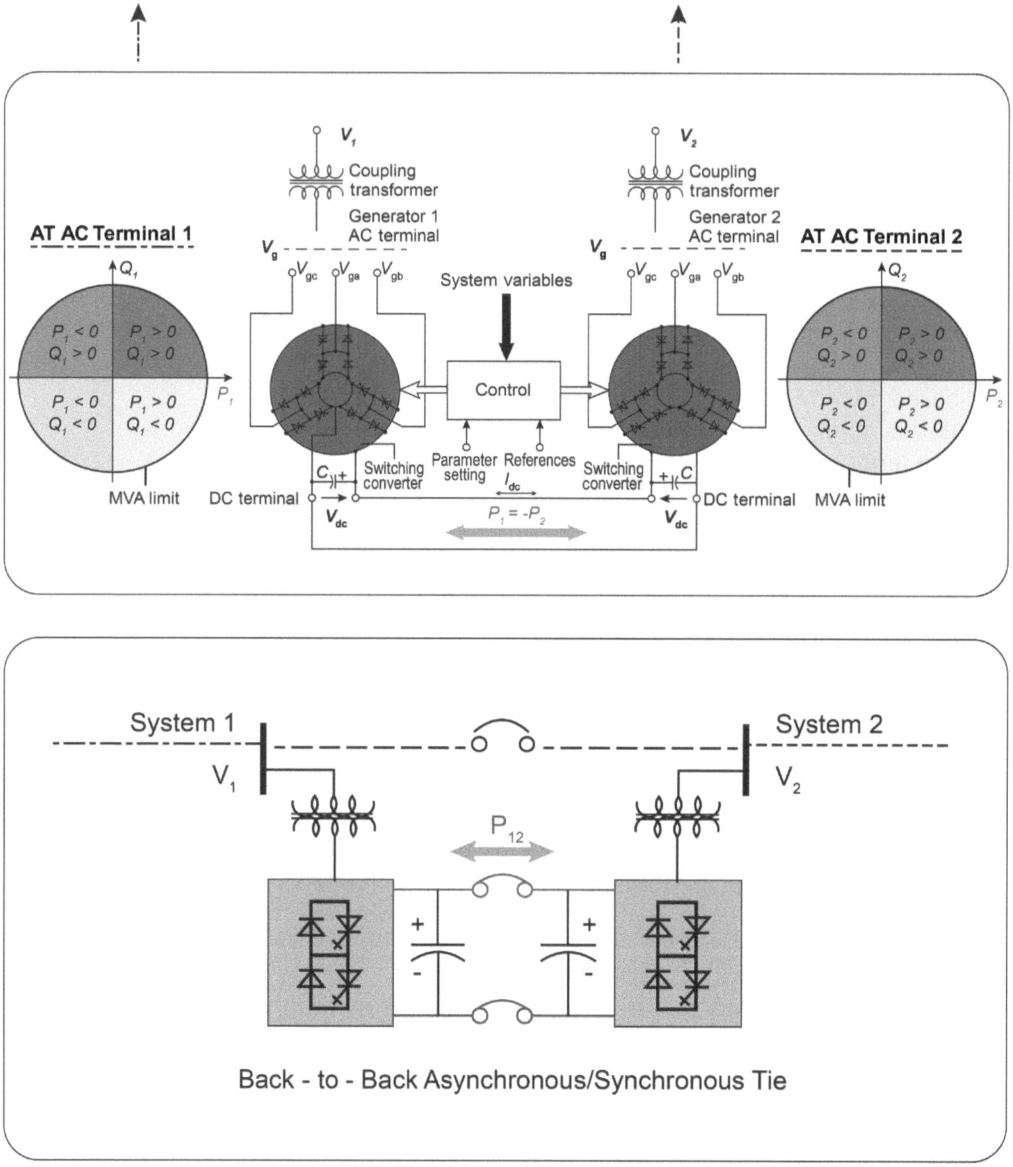

Figure 2.5 VSC back-to-back.

As shown in Fig. 2.6, EI was segmented into four segments. The AC interconnection between the respective segments was replaced by a VSC BtB, where the power transfer levels between the respective segments are kept the same as with AC interconnections.

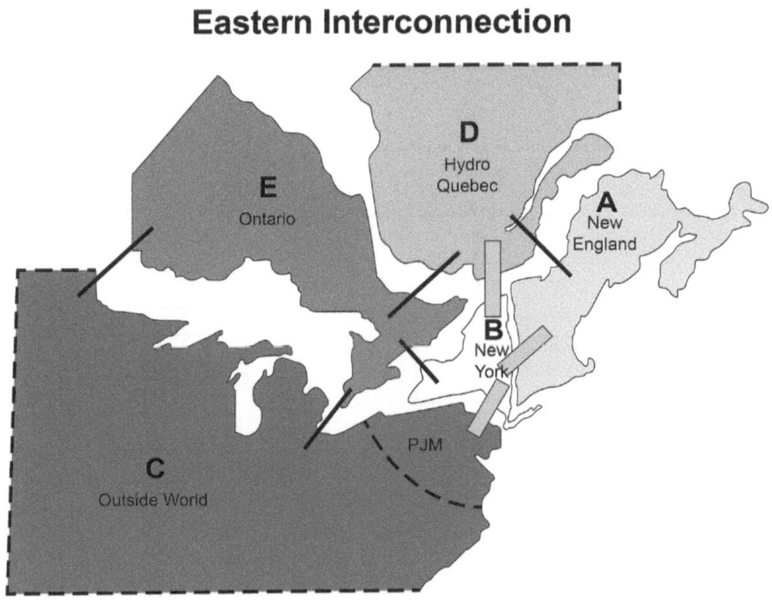

Figure 2.6 Segmentation and grid shock absorbers deployment on the eastern interconnection.

The simulated case is a loss of a major generation, 2700 MW in segment A, which is similar to what happened in 2003. As a result of this loss of generation, the frequency in segment A drops significantly and the compensation of the loss of generation is partially obtained from the spinning reserve in segment A, and the MW balance is transferred from and between the remaining segments controlled by a PI controller shown in Fig. 2.7, where the input signal is the frequency difference of the respective interconnected segments.

The results of this simulated case are shown in Fig. 2.8, comparing the case with AC interconnection and the case with the segmentation and grid shock absorbers. The results clearly showed that the concept of segmentation and grid shock absorber was a novel and smart solution for preventing cascading failures and voltage collapse.

Power electronics-based transmission controllers transform the transmission grid from being a passive network to a smart and active network resulting in increased transmission capacity, and improved flexibility and controllability. The ultimate achievements are an increase of power transfer capability up to the inherent thermal capacity, with the necessary enhancement of transient, dynamic, and voltage stability; and fully controlled power flow along defined transmission corridors, which, therefore, minimize or eliminate capacity limiting loop flows.

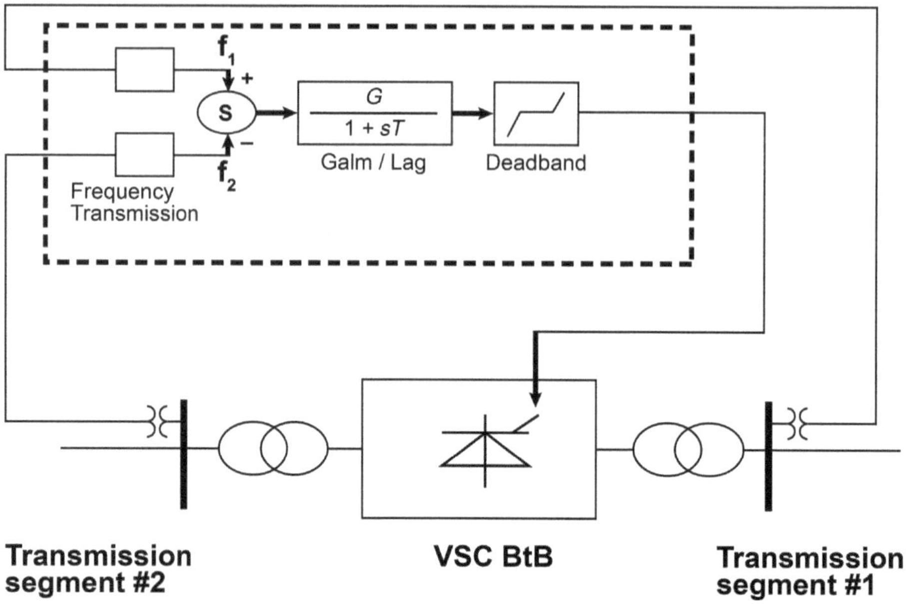

Figure 2.7 VSC BtB PI controller.

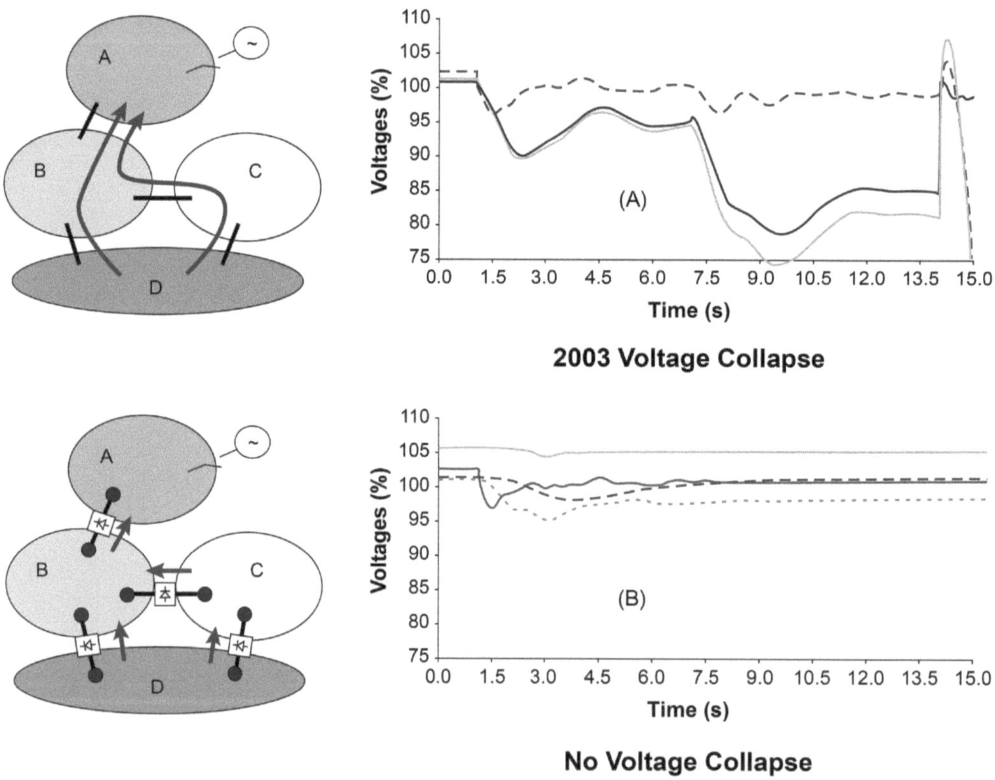

Figure 2.8 System responses with AC interconnection are shown in image top (A) and with segmentation and grid shock absorber are shown in image bottom (B).

2.6 Functional Applications of HVDC Transmission Technology

HVDC technology has unique functional capabilities in an electric transmission grid. It provides a controllable and flexible power link that may be used for interconnecting asynchronous grid areas, moving large amounts of power long distances, and/ or increasing transmission capacity and enhancing grid stability and reliability. Currently, there are two distinct HVDC technologies, one is based on the use of LCC and the other is based on the use of VSC. The former uses thyristor switching valves and the latter uses gate turn-off switching valves.

Fig. 2.9 shows a structural scheme of the LCC HVDC. The main characteristic of the LCC HVDC is a controllable link transferring real power (P) from one terminal to the other terminal. However, controlling real power transfer relies on the capability of the connected AC grids in providing adequate reactive power (Q) supporting the real power transfer. As a rule of thumb, each MW power transfer requires approximately 0.5 MVAR reactive power support. This means that reliable

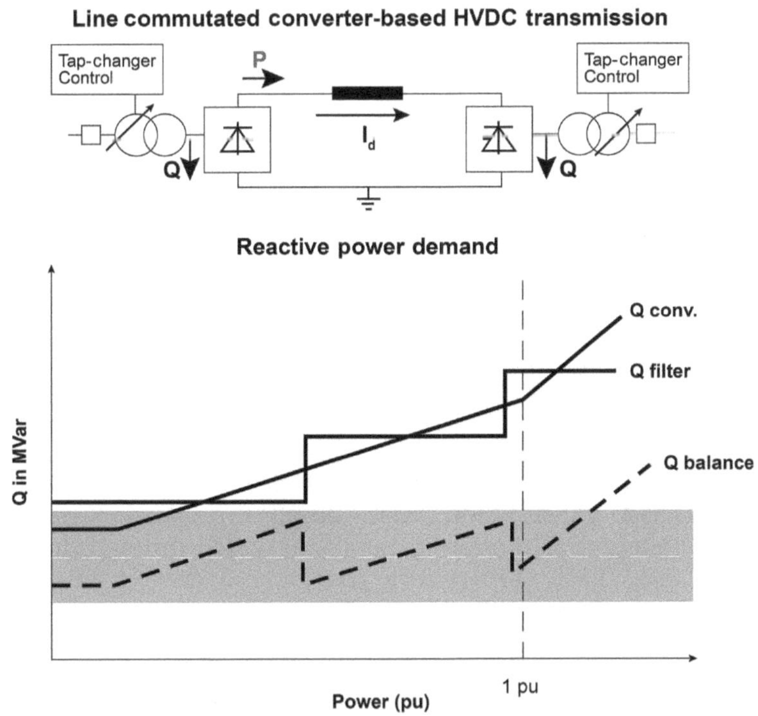

Figure 2.9 Schematic structure of line commutated converter-based HVDC.

operation of LCC HVDC is very much dependent on the strength of the connected AC grid areas as a ratio of the magnitude of the transferred real power. The measure of this strength is known as the short circuit ratio (SCR).Where:

$$\text{Short Circuit Ration(SCR)} = \frac{\text{System MVA(S)}}{\text{DC Power(Pd)}}$$

System MVA is the short circuit power at the point of connection of the converter terminal; Pd is the transferred DC power. The AC system is considered strong if SCR $>$ 3, the system is relatively weak 2$<$ SCR $<$ 3, while the system is considered weak if SCR $<$ 2.

A weak AC system may result in commutation failure and, therefore, unreliable DC power transfer. Strengthening a weak AC system is normally accomplished by the installation of synchronous compensator.

2.7 Monitoring

2.7.1 Dynamic Line Ratings: A Smart Technology for Increased Transmission Capacity

Dynamic line ratings (DLRs) are an inexpensive "smart" transmission technology that can result in increased transmission capacity from 10% to 20%, 80% to 90% of the time. Historically, DLR technology has been successfully applied at a number of power utilities across the United States with very beneficial results and impressive cost/benefit ratios.

The main idea of DLR is monitoring either weather conditions, wind speed and direction, and ambient temperature, or transmission line tension, or conductor temperature, or in some cases transmission line sagging of the critical span. The monitored data, together with the thermal model of the transmission line applied to the heat balance equation, yield a calculated characteristic temperature rise as function of time in response to step change in transmission line current.

DLR calculates dynamic thermal ratings (real-time ampacity) of overhead transmission lines, based on actual load and weather conditions that are generally accessed through the utility's SCADA/EMS (supervisory control and data acquisition/ electromagnetic susceptibility) system.

Traditional static ratings are overly conservative, since they are based on worst-case weather and load assumptions. Hence, the key benefits of DLRs are higher loading of overhead transmission lines using actual measured load and weather parameters; a better understanding of overhead transmission line thermal response, which results in increased reliability; and fuller utilization of transmission lines capacity, since DLR allows for informed decisions about capacity limits.

Dynamic thermal line ratings offer a way to increase transmission capacity under most weather and load conditions, while reducing the risk of failure under unfavorable conditions. DLR allows the transmission system operators to observe the true thermal limits of operation so they can seize opportunities to deliver more power during high demand and avoid unnecessary load shedding when contingencies arise. As a result, more power can be transferred with little additional investment in equipment.

The system operator typically sees circuit load (in MVA or amperes) and bus voltages. Real-time measurement and display of overhead line sag is far less common. Instead, operators limit overhead transmission line loading based on ratings developed from worst-case weather conditions. On the other hand, increasing the utilization of transmission assets calls for direct measurement of weather and transmission line parameters. System operators are accustomed to seeing ratings, so DLR calculation can process the real-time weather data being monitored and present the operator with the actual ampacity of the monitored transmission line.

While knowing the present thermal state (temperature, line sag, loss of insulation life, etc.) of overhead transmission lines is useful, operators want to see ratings and load in MVA or MW. Operators generally don't have the background to relate temperatures, sags, tensions, and other parameters to how much power a given transmission line can carry.

DLR allows the calculation and comparison of thermal ratings to electrical load using predicted models that may warn the operator of future problems. Monitors only report the measured temperature or other parameters at the present time and do not provide any information about the future.

The DLR system provides a solid basis for higher transmission line utilization. Fig. 2.10 illustrates the difference between dynamic ratings and traditional static (fixed) ratings. The rightmost curve is the actual rating distribution (dynamic ratings). This is a probability distribution of a line's thermal ratings calculated based upon a real-time monitoring. The dynamic ratings of the line typically vary over a range of more than 2:1. The very lowest ratings correspond to still air, maximum air temperatures, and full sun. A typical static thermal rating of about 700 amperes is shown at the left tail of the rating distribution.

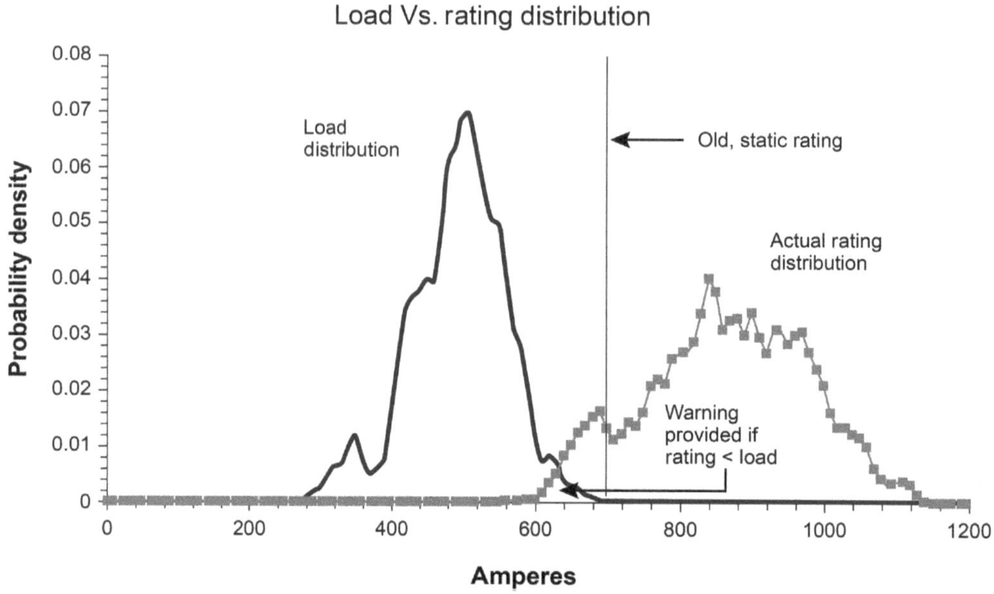

Figure 2.10 Monitoring overhead transmission line.

Clearly, the higher the static rating, the more frequently the actual real-time rating is less than the static rating. In general, the dynamic rating allows 10–20% increased transmission capacity 90–95% of the time. Furthermore, a dynamic rating could be lower than the static rating. This means that a dynamic rating can provide system operators with the information such as warnings to avoid overloading.

Monitoring and data integration play an important role in making a transmission grid a smart one. Dynamic thermal circuit ratings based on monitoring (e.g., weather condition, and/or tension of an overhead transmission line) is an example which might result in increased transmission capacity by 10–20% above the often-adopted static rating which is, usually, a conservative rating based on low wind speed and high ambient temperature.

2.8 Synchro-Phasor Measurement Technology

Another interesting monitoring technology for transmission grids is the use of synchro-phasors, where synchronized phasor data are collected from optimally located phasor measurement units on the transmission grid. These data could be used for wide area protection, control, and security assessment. Wide area monitoring, protection, and control are a well needed methodology to deal with the challenging complexity in managing the operation of large transmission grids. Deployment of synchro-phasor measurements in a transmission grid provides synchronized phasor data information, which can be integrated with EMS/SCADA for a wide assessment of transmission system performance (e.g., voltage stability, power system oscillations, frequency stability, etc.). Furthermore, the collected phasor data could be used for improving the operating performance of the control devices deployed in the transmission grid.

Fig. 2.11 shows a monitoring example, which can provide some of the diagnostic data assessing transmission system performance (e.g., voltage stability, local and inter-area power oscillations, frequency deviation, loss of synchronization, etc.).

2.9 Conclusion and Outlook

Electricity is fundamental to modern civilization. It conveys both energy and information, and it is by all measures is a very complex system, which consists of various large sources of power generations, vast transmission networks, and local distribution systems.

This chapter provides a brief description of transmission system performance and operation in terms of real and reactive power flow and transmission parameters that have a direct influence in controlling the flow of real and reactive power. Also, descriptions of the key characteristics considered to be some of the main drivers for the evolutionary transformation to smarten existing electric transmission grids were provided. Any adjustments to the electrical grid should meet four requirements: capacity (satisfying increasingly demand for electrical energy), reliability (satisfying quality and reliability/availability of electricity), efficiency (energy has to be saved from generation and transmission to consumption), and sustainability (integration of low carbon energy sources).

Some smart transmission technology options that have been shown to increase transmission capacity, flexibility, controllability, and performance reliability were reviewed. These technologies are based on the use of power electronics (FACTS and HVDC), real-time monitoring, and/or provide solutions to major transmission problems based on the root cause of these problems instead of mitigation of the effects.

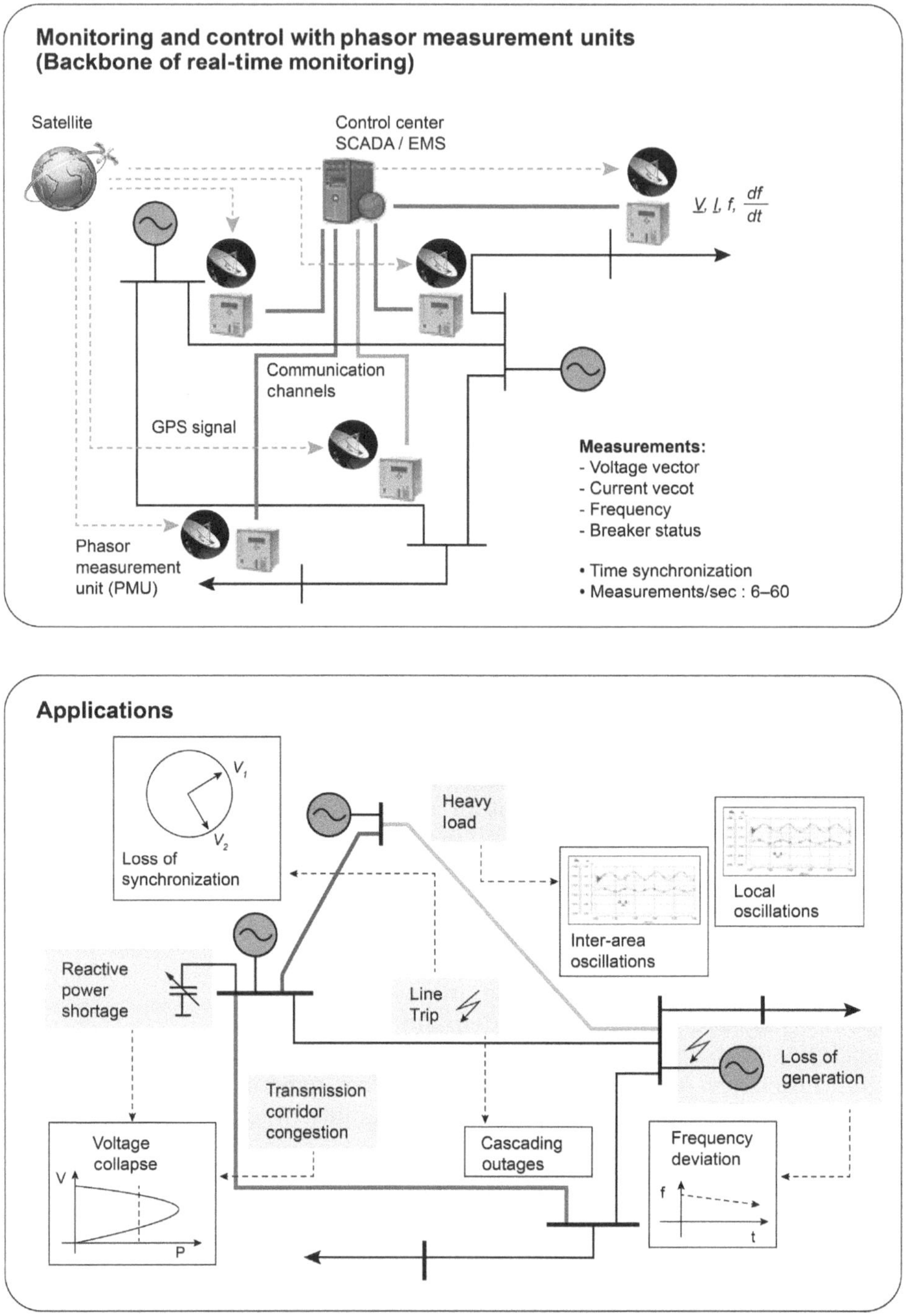

Figure 2.11 Monitoring and some diagnostic data assessing system performance.

References

2013 Report Card for America's Infrastructure, 2013a. Reston, VA, USA: American Soc. of Civil Eng., USA.

Aging Power Delivery Infrastructures, 2013b. Aging Power Delivery Infrastructures, second ed. CRC Press.

Al-Haddad, K., Power quality issues under constant penetration rate of renewable energy into the electric network. Power Electronics and Motion Control Conference (EPE/PEMC), 2010 14th International, 6–8 September 2010, 2010. S11-39-S11-49.

Baldwin, S., La Commare, K.H., Eto, J.H., 2004. Understanding the cost of power interruptions to U.S. electricity consumers. United States.

Clark, H., Edris, A.A., El-Gasseir, M., Epp, K., Isaacs, A., Woodford, D., 2008. Softening the blow of disturbances. IEEE Power Energy Magazine. 6, 30–41.

Convertible Static Compensator (CSC), 2003 for New York Power Authority: Final Report. EPRI, Palo Alto, CA and New York Power Authority, White Plains, NY.

Edris, A.A., 1990. Series compensation schemes reducing the potential of subsynchronous resonance. IEEE Trans. Power Syst. 5, 219–226.

Fardanesh, B., Shperling, B., Uzunovic, E., Zelingher, S., Gyugyi, L., Kovalsky, L., et al. NYPA Convertible Static Compensator, Validation of Controls and Steady State Characteristics. CIGRE Session 2002, Paris, France. 14–104.

Farhangi, H., 2010. The path of the smart grid. IEEE Power Energy Magazine. 8, 18–28.

Gagwari, D., Bhatia, R.S. A brief analysis of recent developments in smart grid. Signal Processing, Informatics, Communication and Energy Systems (SPICES), 2015 IEEE International Conference on, 19–21 February 2015, 2015. pp. 1–5.

IEEE Smart Grid Vision for Computing: 2030 and Beyond Roadmap, 2016. *IEEE Smart Grid Vision for Computing: 2030 and Beyond Roadmap*, 1–14.

Line, M.B., Tondel, I.A., Jaatun, M.G. Cyber security challenges in Smart Grids. Innovative Smart Grid Technologies (ISGT Europe), 2011 2nd IEEE PES International Conference and Exhibition on, 5–7 December 2011, 2011. pp. 1–8.

Moslehi, K., Kumar, R. Smart Grid—a reliability perspective. Innovative Smart Grid Technologies (ISGT), 2010, 19–21 January 2010, 2010. pp. 1–8.

Proposed terms and definitions for flexible AC transmission system (FACTS). 1997. IEEE Trans. Power Delivery. 12, 1848–1853.

Renz, B.A., Keri, A.J.F., Edris, A. World's first unified power flow controller on the AEP System. Cigre Session 1998, Paris, France. pp. 14–107.

Schauder, C., Stacey, E., Lund, M., Gyugyi, L., Kovalsky, L., Keri, A., et al., 1998. AEP UPFC project: installation, commissioning and operation of the ±160 MVA STATCOM (phase I). IEEE Trans. Power Delivery. 13, 1530–1535.

3

ADVANCES IN POWER CONVERTERS

Nareg Sinenian and Daniel Shai
Exponent Inc., New York, NY, United States

CHAPTER OUTLINE

Power converters used in the grid today are largely passive devices that convert transmission line voltages to distribution voltages or convert High Voltage Direct Current transmission lines to AC (Alternating Current). This chapter explores advances in power conversion technology that are an integral part of a smart grid. Power conversion technologies in use today and under development are discussed at a high level in Sections 3.1. The circuitry and systems comprising these power converters are discussed in Section 3.2. The final two sections discuss the applications of power converters, including their role in integrating distributed power sources and electric vehicles (EVs) with the grid.

The Power Grid. DOI: http://dx.doi.org/10.1016/B978-0-12-805321-8.00003-3

3.1 Introduction to Power Conversion

3.1.1 Power Conversion in the Grid

Power converters operate as part of the grid and are found in a number of points along the power grid, including generator plants, substations and distribution points. For example, the transmission substations frequently found in the grid link power plants with high-voltage transmission lines, and power substations provide residential and commercial access to the high-voltage transmission line. These substations link the high-voltage transmission lines to distribution lines which can be found on typical residential streets. The final step in power conversion takes place at the distribution poles. Here, one might typically find drums that contain transformers that link the distribution lines with the consumer.

Each of the substations described incorporate power converters that either step up or step down the voltage along the transmission path. Power efficiency dictates a need for high-voltage transmission, which is in part provided by the transmission substation. It converts the power plant output voltage to the several hundred kilovolt levels that are typical of a high-voltage transmission line. Power is transmitted through the high-voltage transmission line to a distribution substation where the voltage is reduced, typically on the order of 10 kV, so that it may safely be distributed throughout neighborhoods. Finally, at the link between distribution line and the consumer, the voltage is further reduced to typically several hundred volts before being transmitted into a consumer facility or residence. Thus, an important role of power converters is to provide voltage transformation.

Another role of grid power converters is to provide a degree of safety by isolating portions of the grid. Transformers inherently provide galvanic isolation (whereby current cannot directly flow from one part of the grid to another), which has some safety benefits. In addition, power converters are coupled with circuit breakers and earth switches that can switch one or more incoming or outgoing transmission lines onto a bus bar. A fault in one branch may therefore be isolated from the rest of the system. Switching also provides a means of connecting various power sources to the grid depending on demand.

Power converters may also provide power factor correction on the grid. The power is defined as the ratio of the real power to the apparent power (sometimes referred to as volt-ampere reactive, or VAR). Loads that are not purely resistive and have a reactive component introduce a phase difference between the voltage at the load and the current flowing through it. This is generally undesirable since excess current must be supplied by the grid to deliver the same amount of power to the load. Low power factor may also reduce the efficiency of the distribution network since the additional current will dissipate power in parasitic resistances along the transmission path. A power factor of unity is therefore desirable since it minimizes the amount of working current flowing through a load. Currently, power factor correction is achieved by adding capacitors downstream of grid power converters, e.g., downstream of the supply transformer at the entrance to a particular facility.

Another factor that affects power quality is the presence of line harmonics, which are distortions of the line waveform because of nonlinear loads or power generation sources. Harmonics are generally classified as either conducted harmonics or emitted harmonics. Conducted harmonics are conducted along power lines and may affect downstream electronics. At sufficiently high frequencies, the presence of harmonics in power conversion electronics can lead to harmonic distortion emission, causing interference with the surrounding environment. The IEEE and other standards and regulatory bodies each have standards on the amount of total harmonic distortion that is acceptable. Passive means of controlling line harmonics in conventional power converters include the filtering away of unwanted current and voltage perturbations.

Conventional power converters as described in this section largely comprise passive components such as voltage transformers, current transformers, coupling capacitors and safety circuitry, including for example lightning arrestors and circuit breakers. As such, power converters and substations on the grid are not capable of self-monitoring and control. For instance, substation operators control the earth switches in substations in response to changing load conditions. The load conditions are communicated to substation operators by load dispatchers, whose responsibility is to coordinate activities on power grid.

3.1.2 Smart Power Converters for the Smart Grid

The comparison between Smart Power converters and conventional power conversion on the electricity grid is akin to comparing control of motor vehicle traffic on the road prior to the automation of traffic lights. Imagine having to manually monitor the amount of traffic in each direction and to change lights accordingly. Statistics and trends of vehicle motion and traffic patterns would be virtually impossible to ascertain. Historically, and at present, power conversion systems on the grid operate in this relatively rudimentary way.

The smart power converter will be able to perform bidirectional power conversion, and to transform working voltages of either type of current flow (AC, Alternating Current or DC, Direct Current) to accommodate changing conditions. Moreover, the smart power converter will sense conditions, reconfigure itself accordingly and perhaps report sensed conditions electronically so that portions of the grid can be monitored from a centralized location. Essentially, the smart power converter combines the intelligence found in modern information technology with advanced power electronics topologies to satisfy these goals.

3.2 Power Electronics and Solid-State Transformers

3.2.1 Solid-State Transformers

Transformers are at the heart of the today's power grid. They are passive devices consisting of a number of coils that are coupled together using a high-permeability magnetic core. The core provides good coupling of magnetic flux between the coils that are wound around it, providing a degree of galvanic isolation while allowing the conversion of AC voltages. Consider the distribution transformer commonly found in residential neighborhoods, shown schematically in Fig. 3.1. It consists of a fixed number of turns wound around both the primary and secondary side of a core constructed of material that has a high magnetic permeability to provide good flux coupling between the two windings. The alternating source that drives the primary

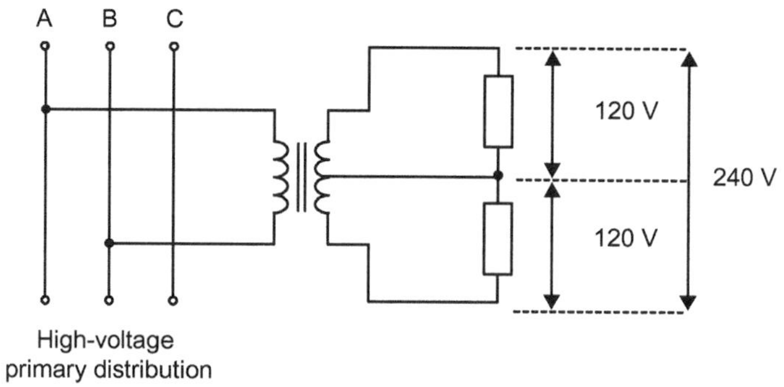

Figure 3.1 Typical distribution transformer found in neighborhoods in United States. Three-phase primary distribution is converted to split-phase 120 V as shown here and delivered to the home (Shoemaker & Mack, 2006).

winding generates a time-varying flux in the iron core, which is coupled to the secondary winding. This flux induces a voltage on the secondary winding and has the potential to drive a current through a connected load. The relationship between the AC voltage on the secondary side and that of the primary side is given by the ratio of turns on the secondary side to that of the primary side.

The distribution transformer is limited in that the nominal voltage transformation ratio (typically around 10 to 1 in the United States) cannot be modified without physically altering the windings. Similarly, the simple transformer is unable to readily regulate line voltage nor limit the amount of output current (e.g., for protection). Further, since the operating principles of the simple transformer require an AC voltage on the primary and secondary sides, it is difficult to integrate DC energy sources and energy storage devices (e.g., batteries) into the power grid.

In order to achieve the flexibility that a smart grid requires, transformers must be coupled with active devices and control logic in order to provide for desired voltage and power transformations. Such power electronics have a long-running history on a much smaller scale. For example, the AC to DC wall adapters that are ubiquitous in homes today (e.g., used in conjunction with laptop computers, mobile phones, etc.) utilize one or more transformers alongside power transistors and one or more integrated circuits that control the operation of the power transistors. Adapting the same technology for the grid, albeit with slightly different topologies, is advantageous for a number of reasons. Solid-state transformers (SST) are optimized to deliver an optimal target voltage at a desired frequency to a load that can be electronically modified or controlled to a significant extent. As a corollary, these transformers are capable of voltage regulation—the delivery of a steady voltage in the face of grid disruptions.

A single-stage SST analog of the distribution transformer is shown in Fig. 3.2. It consists of a standard transformer at its core,

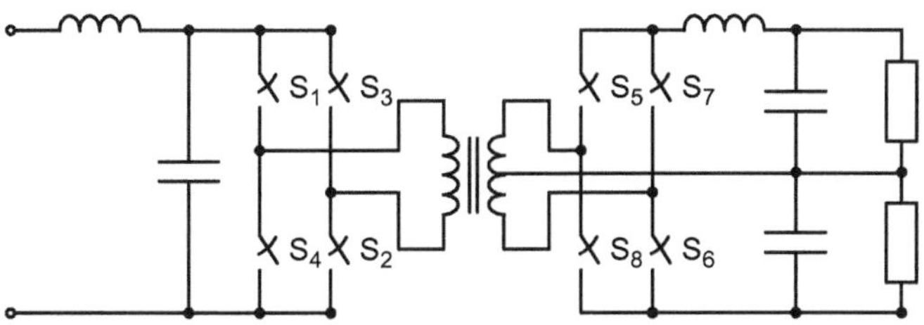

Figure 3.2 A full-bridge AC-AC converter for use in smart grid applications.

a number of switches, and filters on both the input and output of the transformer. The switches operate at a frequency that is higher than the input waveform (e.g., the line frequency) so as to modulate the AC source into a waveform that is then applied to the transformer. The second-order inductor and capacitor filters at the input to the converter minimize the amount of harmonics that arise from switching currents at the input. The filter formed by the split-capacitor network at the output similarly remove unwanted switching transients.

The converter's operation may be understood by first considering the special case where the output voltage equals the input voltage. During the first half of the switching cycle, switches S1 and S2 are switched on while switches S3 and S4 are switched off. In this case, the second half of the switching cycle the switch states are reversed. The switching scheme thus modulates the input waveform by inverting portions of the waveform at a rate equal to the switching frequency. The secondary side of the converter then recovers the original sinusoidal waveform by alternating between switching S5 and S6 in phase with S1 and S2 and switching S7 and S8 in phase with S3 and S4.

The voltage transformation ratio is altered by adjusting the relative phase between the switches on the secondary side relative to the primary side. If the switches on the secondary side are not exactly in phase with the primary side, then the original waveform will be modified as it is transmitted through the converter. For example, a small phase difference will introduce short periodic negative excursions that effectively reduce the average voltage. If the switching frequency is sufficiently high, these small excursions are readily averaged by the output filter to produce a clean sinusoid with a different voltage. Examples of the transformer waveform, output waveform, and switch timing are shown in Fig. 3.3 for two different phase angles. Sophisticated control systems that incorporate microcontrollers can be used to program voltage transformation ratios on the fly. Voltage regulation may also be achieved to some extent by monitoring the output voltage and making fine adjustments to the phase angles in real time. These are two of the important features common to smart transformers that are not readily found in the simple passive transformer.

The simple SST, known as a full-bridge AC-AC converter, demonstrates a number of important features typical of many SST topologies. First, it is often the case in such devices that the switching frequency needs to be many times the frequency of the input waveform. Higher switching frequencies in switched-mode power conversion reduce the filtering requirements,

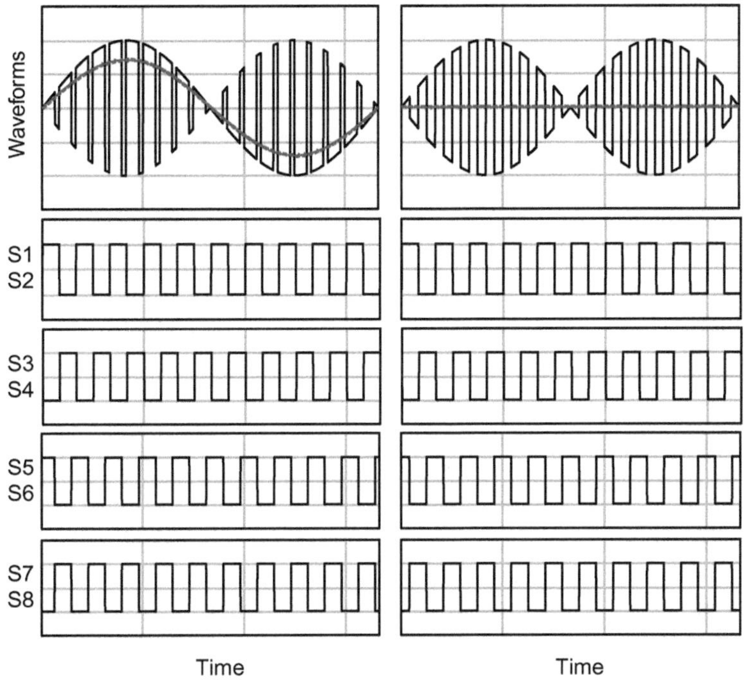

Figure 3.3 Transformer, output and switch waveforms for the full-bridge AC-AC converter operating at a relative switch phase angles of 30 degrees (left) and 60 degrees (right). The transformer waveform is overlaid on top of the output waveform (dashed).

allowing for relatively smaller inductors and capacitors, thereby reducing the cost and weight of power converters (Kassakian et al., 1991). Second, each of the switches used in the full bridge AC-AC converter need to block voltages that are on the order of transmission voltages (i.e., several kilovolts) when they are in the off state. As a result, the widespread implementation and success of SSTs requires switches that capable of operating at thousands of cycles per second, of blocking large voltages, and handling potentially high currents or power. One of the main challenges in recent history has been to develop such transistor switches, which are the subject of the next subsection.

One way to relax the high-voltage requirement on the transistor switches found in SSTs is to modularize them. The simple SST shown in Fig. 3.3 can be implemented as a combination of N modules as shown in Fig. 3.4A. Here, the input voltage is nominally equally divided between N modules. Since the output of each of the modules are combined in parallel, the output current requirement per module is lowered by a factor of N as well. There is a practical limit to the number of modules that can be combined since lowering the voltage and current requirements on the modules comes at the expense of more

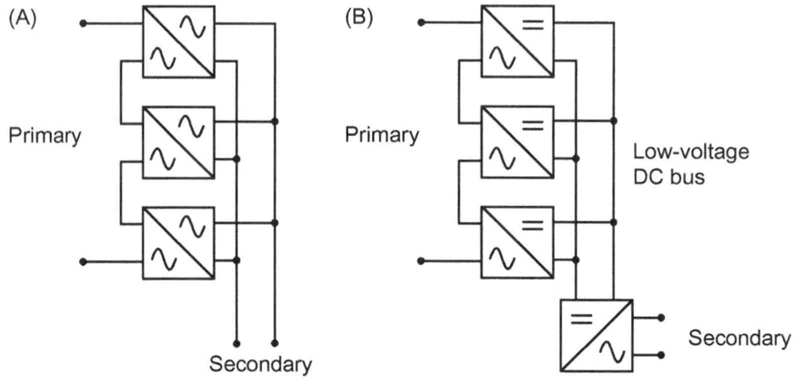

Figure 3.4 Architecture of modular single-stage and two-stage solid-state transformers. (A) Modular single-stage SST. (B) Modular two-stage SST.

transistors and associated hardware (e.g., thermal management). It is apparent that the number of components and hence the weight and size of an SST may be reduced as switches with higher voltage and current handling capabilities are developed.

The modular architecture can also be combined with an additional stage to form a modular two-stage SST as shown in Fig. 3.4B. In this configuration, the first stage of modular converters provides an intermediate DC output. This low-voltage DC bus is then connected to an AC inverter for grid interconnection. This configuration has a significant advantage in that it allows for distributed DC energy sources and energy storage devices to be readily incorporated into the grid. Power converter topologies associated with specific distributed renewable energy sources, including both grid-tie inverters and DC converters, are further discussed in Section 3.3.

A further benefit of incorporating active devices (e.g., transistors) into power converters is the ability to provide advanced power quality correction. In addition to basic passive filters, smart power converters may incorporate advanced control schemes for mitigating and suppressing harmonics. One such advanced technology is the use of active power filters. When using passive filters to remove unwanted distortions, the nature and magnitude of the distortion must be known in advance. In particular, the distortion must be well characterized so that an appropriate filter network can be designed. Once it is built into a conventional power converter, the filter may prove to be adequate but only for the distortions that it was intended to mitigate. Active power filters, however, are reconfigurable and as such adapt to a wider range of distortions and to variations in the nature of the distortion in real time. Conceptually, active filters can be understood as follows. The SST shown in Fig. 3.2

contains output filter capacitors that may be supplemented or replaced with power electronics. These electronics would present a low shunt impedance (to ground) for unwanted signals and a moderate impedance for desirable signals. Active power filters adjust the effective shunt impedance as desired by monitoring the frequency content of the current at the output of the converter in real time. The details of how active power filters are implemented and controlled in real time is a subject of ongoing research and several implementations currently exist (Marks, 2005; Li & Nejabatkhah, 2014).

3.2.2 Advances in Semiconductor Devices

Semiconductor devices play a significant role in the smart power converters we have considered. The advancement of semiconductors has been historically driven by microelectronics, where low power consumption, low voltage, small form-factor, and speed (e.g., high-frequency operation) are desirable attributes. Smart grid applications dictate some of the same requirements on semiconductors while introducing new ones. For instance, solid-state transformer applications require high-speed switching at frequencies that are not significantly different from the typical converters that are ubiquitous in laptop power adapters and in power electronics found within mobile devices. In addition, grid applications require that the active devices be capable of handling large currents and safely blocking high voltages (HVs) when they are switched off.

At high frequencies, switching losses can become appreciable and it is desirable to minimize these losses in order to maintain high efficiency. The power dissipated in an active device such as a transistor is the product of the instantaneous voltage across the device's terminals and the current flowing through the device (e.g., the collector-emitter current and voltage for the case of a bipolar transistor). For this reason, the active device must present a low resistance when it is conducting and prevent leakage current when it is off. The low resistance maintains a low terminal voltage during conduction and hence minimizes power dissipation.

The power handling capacity and operating frequency of power converters are constrained by the choice of semiconductor devices as shown in Fig. 3.5. Typical switching frequencies for solid-state transformers are in the range of 1–10 kHz, while the typical power handling capabilities of power converters are required to be on the order of 10 kVA (Falcones et al., 2010). The devices of interest and the focus of this section are therefore

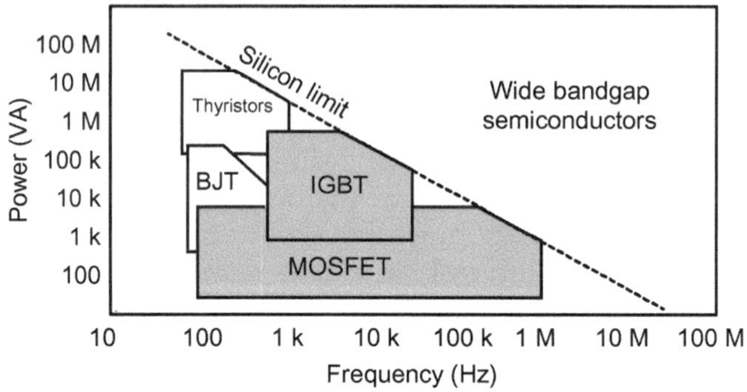

Figure 3.5 Power handling capacity and operating frequency of power converters that use various semiconductor devices. The operating limit for silicon-based devices is indicated by the dashed line (Wang, 2007).

IGBTs (Insulated Gate Bipolar Transistors) and MOSFETs. Recent advances in device structure combined with a move from silicon to wide bandgap semiconductors such as silicon carbide have enabled converter operation at higher power levels and frequencies.

There are a number of additional constraints placed upon semiconductor devices used in smart-grid applications. First and foremost, grid applications require blocking voltages that are about an order of magnitude larger (or more) than are currently found in typical power electronics in the home. One of the challenges is to scale the blocking voltage without compromising other performance metrics, such as high-frequency operation with low switching losses. The following subsections discuss recent advances in device technology to meet these needs.

3.2.2.1 Power MOSFETs

MOSFET technology dates back to the 1970s during which the first insulated-gate field-effect transistors were produced. The typical MOSFET consists of gate, source, drain, and body terminals along with regions of n-type and p-type semiconducting material as shown in Fig. 3.6. In regions of n-type silicon, there is an abundance of electrons in the conduction band of the silicon atoms that form the material, and these electrons are the majority charge carriers. In contrast, p-type silicon has a deficit of electrons in the valence band, also known as an abundance of holes, and the motion of these holes in the valence band is responsible for carrying charge.

In a typical planar MOSFET, the gate is insulated from the semiconductor using an oxide and the body terminal is

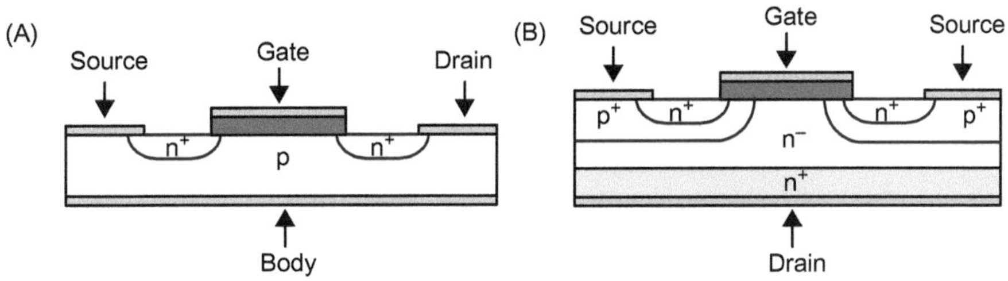

Figure 3.6 (A) Planar MOSFET structure (left) and (B) Vertical MOSFET structure (right).

internally connected to the source, leaving three exposed terminals. This has the effect of creating a diode between the source and the drain, which is able to block current flow from the drain to the source when the device is off. The MOSFET is operated as a switch by applying a positive voltage bias to the gate relative to the drain. This generates an electric field which attracts electrons from the p-type silicon toward the gate. These electrons fill the space between the two highly-doped n-type regions, forming an inversion layer and thereby providing a path for current flow from the drain to the source.

The planar MOSFET is not well-suited for switched-power applications for several reasons. The operating voltage of the device depends on the length of the channel (i.e., the width of the p-type region as shown in Fig. 3.6A), while the current rating of the device depends on the current-carrying cross-section and hence the width of the channel (i.e., the depth into the page as shown in the figure). For power applications, a planar MOSFET would need a significant spatial footprint to create the large cross-sectional area required to transmit large amounts of current. Vertical structures such as the one shown in Fig. 3.6B alleviate this problem and allow for higher currents. In this configuration, the current flows vertically thereby increasing the cross-sectional area through which current is transmitted while maintaining approximately the same package size. The voltage rating of the device depends on the thickness and doping concentration of the n-type layer, which can also affect the on-state resistance.

Two sources of inefficiency in power converters utilizing MOSFETs are switching losses and conduction losses. The switching losses in MOSFETs are understood by considering the gate capacitance. This capacitor must be charged in order to raise the gate voltage sufficiently high to turn on the transistor. Similarly, the charge must be removed in order to lower the voltage and turn off the transistor. If the transistor is used as a high-frequency switch, as in the case of the SSTs we have considered, the constant addition and removal of charge consumes a significant amount of power and generates heat that the bulk silicon must be able to handle. Switching losses are reduced by minimizing the amount of charge required to turn on the device. Conduction losses, by contrast, are reduced by decreasing the on-state resistance.

During the last several decades, improvements have been made to the basic MOSFET structure shown in Fig. 3.6 in order to improve performance in power applications. For instance, the introduction of lightly-doped drains enabled MOSFETs to block higher voltages when switched off (Saxena & Kumar, 2012). Geometric improvements, such as the introduction of trench-gates to reduce on-state resistance, have improved the current-handling ability of power MOSFETs (Shenai, 2013). Despite these improvements, a fundamental limit to MOSFET technology is the link between the on-state resistance and the breakdown voltage of the device. Specifically, the on-state resistance increases with the square of the breakdown voltage (Shenai, 2013). For this reason, it is difficult to obtain a high operational voltage while simultaneously carrying large currents. MOSFETs are therefore often utilized for medium-voltage applications, which include certain renewable sources and energy storage devices.

Ongoing research and development efforts in the context of smart grid applications center around the successful implementation of these devices using advanced semiconductors. The use of these wide band-gap semiconductors, including silicon carbide (SiC), is discussed in Section 3.2.2.3.

3.2.2.2 Insulated Gate Bipolar Transistors (IGBTs)

The history of IGBTs date back to the 1980s, when the first generation of devices were created. The IGBT is a three-terminal device comprising four layers of semiconducting material in an n-p-n-p arrangement as shown in Fig. 3.7. The regions beneath the gate are oxides that provide a layer of insulation between the gate and the device. The plus and minus modifiers next to the labeled regions denote heavily and lightly doped regions, respectively, which affect the conductivity of the region accordingly.

The IGBT is a voltage-controlled device that modulates the conductivity by injecting minority carriers from the p + region into the n- region. The amount of injection depends on the electric field between the gate and the collector, and hence the IGBT is said to be a voltage-controlled device. This is advantageous over conventional bipolar transistors in terms of

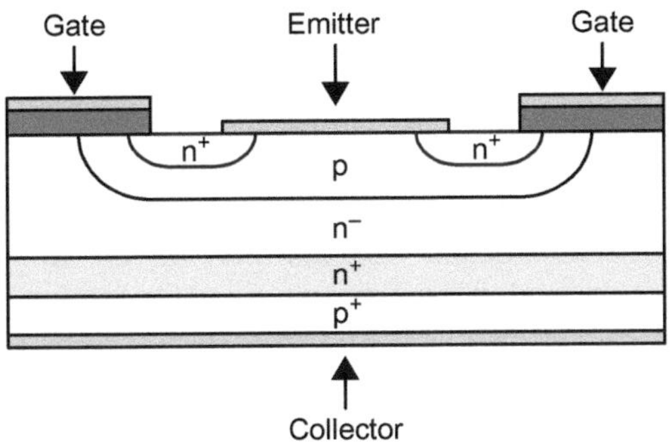

Figure 3.7 The structure of the gate is one of the key differences between a conventional bipolar transistor and an IGBT. The insulated gate allows voltage control of the device with minimal leakage current.

efficiency in power applications since in steady state no current has to be supplied to the gate terminal.

The first IGBT devices were prone to a number of problems, including thermal runaway and latch-up (i.e., where the device remains on as long as current is flowing, regardless of control voltage). Since then IGBTs have come a long way in terms of construction and performance improvements. The latch-up problem was addressed by carefully designing the geometry and the doping concentration of the of the various semiconducting layers. The advent of non-punch through (NPT) devices, or devices that lack the n^+ region shown in Fig. 3.7, have minimized tail losses, or the losses associated with having a decaying current for a short period of time after the device is turned off (März et al., 1998).

NPT-based devices also improved conduction losses associated with the on-state resistance of the devices (the emitter-collector voltage was minimized) and made it easier to use multiple IGBTs in parallel in order to increase the current handling capacity of a system. Paralleling IGBTs was problematic early on because of the negative temperature coefficient that could lead to thermal runway. A negative coefficient meant that these transistors would conduct more current as they would heat up, which in turn meant that they would conduct even more current leading to runaway. NPT technology solved these problems, allowing increased adoption of IGBTs for high power applications.

A number of additional improvements were made over the years in efficient in packaging of IGBTs into high heat dissipation modules (Kobayashi et al., 2007), and the introduction of trench geometry in gates, which allowed simultaneous low on-state resistance and a high breakdown voltage (Frank, 2008; Chou, 2008). Ongoing research and development is focused on producing IGBTs on advanced semiconductor technologies, such as silicon carbide (Wang & Cooper, 2010).

3.2.2.3 Wide Bandgap MOSFETs and IGBTs

The advances and optimizations in MOSFET and IGBT design have resulted in measurable gains in power applications. However, the use of plain silicon as a base material for these devices has fundamental limits. As discussed earlier, it is desirable in power electronics applications to be able to operate transistor switches at high frequencies. Operation at high frequencies has fundamental limits due to the heat that is generated as a result of switching losses.

Another requirement is the ability for these devices to operate at higher voltages. The ability for a transistor to block HVs when it is in the off state depends on several factors, including the intrinsic breakdown voltage of the bulk material. The breakdown electric field for silicon is approximately 300 kV/cm, which leads to MOSFET devices that operate at a few hundred volts and IGBTs that operate at around kilovolt levels. Geometric optimizations (e.g., modifying the dimensions of the p-well) can increase the length over which fields penetrate and therefore increase the operating voltage. This comes at the expense of a higher on-state resistance and increased heat production.

Recent advances have therefore focused on materials-based improvements, such as the use of wide band-gap semiconductors. The band-gap refers to the amount of energy that must be supplied to a medium in order to generate carriers (the energy required to move an electron from the valence band to the conduction band). Such semiconductors have both electrical and thermal properties that are superior to plain silicon, including higher breakdown electric fields and thermal conductivity. One such material is silicon carbide (SiC), which has a seven to eight times higher breakdown electric field and three times the thermal conductivity of plain silicon as indicated in Table 3.1 (Tolbert et al., 2005).

Each of the material properties associated with these semiconductors lead to advantageous device characteristics which in turn translate to overall power converter gains. A larger thermal conductivity means that the device more readily conducts heat to its surroundings. For a given operating temperature, this relaxes the cooling requirements imposed on the device and hence the converter as well (e.g., less heat sink mass is required in the case of passive cooling). Higher saturated electron drift velocities mean that the electrons are more mobile for a given electric field and can traverse the length of the device faster. This allows for the transistor and hence the converter to be

Table 3.1 Material Properties of Selected Wide Band-Gap Semiconductors as Compared to Silicon, Including Two Forms of Silicon Carbide and Gallium Nitride (Tolbert et al., 2005)

Property	Silicon	6H-SiC	4H-SiC	GaN
Bandgap (eV)	1.12	3.03	3.26	3.45
Breakdown field (kV/cm)	300	2500	2200	2000
Thermal conductivity (W/cm-K)	1.5	4.9	4.9	1.3
Saturated electron velocity ($\times 10^5$ m/s)	1	2	2	2.2

switched at a higher frequency, thereby reducing the converter's filtering requirements. In addition to providing for a higher breakdown voltage, a higher bandgap also allows for high-temperature operation. This is because as the temperature of a semiconductor is increased, energy is provided to the valence electrons that are then able to overcome the bandgap. This results in a loss of semiconducting properties and the material behaves more like a metal. For this reason, wide bandgap semiconductors can operate at higher temperatures while retaining their desirable electrical characteristics.

Currently, research and development of both SiC- and GaN-based devices is ongoing. The first SiC power MOSFET was introduced in 2011, with a blocking voltage of 1.2 kV and an on-state resistance of 80 mΩ (Cree, 2011). The production of SiC-IGBTs soon followed with 12 kV devices.

3.3 Grid Integration of Distributed Power Sources

One of the main benefits of a smart grid is the ability to incorporate a diversity of distributed power sources, including renewable energy sources and energy storage devices. Until now, integration of distributed power sources is challenging in part due to voltage and current compatibility. For instance, DC power sources are not easily integrated in today's AC grid using conventional grid technologies. Furthermore, the monitoring of distributed power sources proves to be a challenge in today's grid since there currently exists no network (e.g., a "power Internet") to control or to make adjustments to these sources. The following sections discuss a number of distributed power sources and various converter topologies that can be used to readily integrate them into the smart grid.

3.3.1 Integration of Renewables

A number of emerging renewable energy sources are under consideration for large-scale implementation. Two of the fastest growing are perhaps solar photovoltaics (PV) and wind power technologies. In 1995, approximately one GW of wind-generated electricity was consumed by the world. Fifteen years later in 2010, that number grew to 100 GW, catching up to nuclear installations. Solar PV has similarly grown at an exponential rate totaling installations of 50 GW in 2010 (Jäger et al., 2014). Given these trends, there is a significant amount of research and development in power conversion and grid integration of these technologies.

3.3.1.1 Power Conversion for PV

Photovoltaic power sources comprise a number of smaller building blocks. The smallest building block, the solar cell, has an output voltage of on the order of one volt. Several such cells are combined to form a PV module. These modules in turn may be combined into a panel, which may be further combined in series to form a string of panels. Several such strings are referred to as a PV array, which produces useful electricity for consumption. A single PV module can have an output voltage on the order of 50 VDC and can source up to 10 A. The output voltage of PV arrays and the requirements on power conversion therefore depend on how these PV modules are combined and connected to the grid.

An important parameter of PV modules that constrains interconnection and hence overall system design is the maximum power point (MPP), which specifies the operating voltage and current that maximizes module power output and hence system efficiency. Fig. 3.8 shows current–voltage characteristics of a typical PV module. Typical modules are able to output constant current up to a particular voltage at which point the current output capability beings to drop, resulting in a maximum power output near the critical voltage. The curves shown in Fig. 3.8 can vary from module to module due to both intrinsic and environmental parameters, including partial shading (e.g., due to clouds overhead), inclination (if the panels are not at all

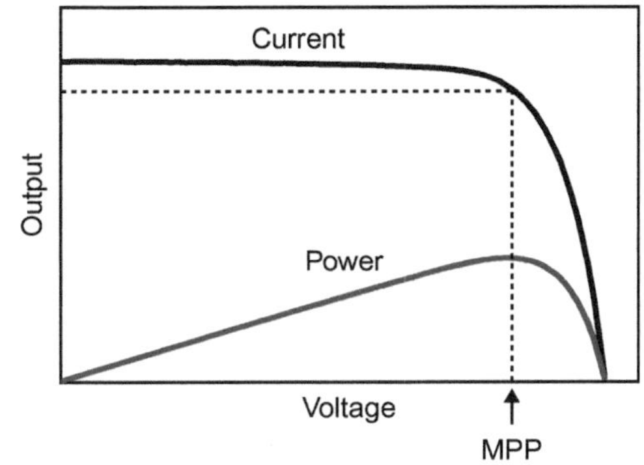

Figure 3.8 The current and power output of a solar PV modules as a function of operating voltage. The current remains constant over a range of operating voltages before it decreases, leading to a maximum power output point.

mounted at the same angle), and due to manufacturing tolerances. The efficiency of a module then depends on how much current is drawn from that module, and this is prescribed by both the power converter topology (i.e., the inverter design) and the system topology (i.e., the arrangement and interconnection of modules in an array with one or more inverters).

In order to provide for efficient operation, power converters implement various methods of detecting the operating point relative to the MPP and adjusting the operating point accordingly, a feature known as maximum power point tracking (MPPT). A number of simple methods of determining the MPP exist (Freeman, 2010; Jäger et al., 2014). These include, e.g., programming a power converter controller with a particular voltage that corresponds to the MPP. A similar alternative involves measuring the open-circuit voltage of the module and using it to scale a known nominal curve similar to the one shown in Fig. 3.8 to set the voltage corresponding to the MPP. The latter approach addresses to some extent variations due to manufacturing variability and environmental conditions. These simple methods are not direct measurements of the MPP and hence make some assumptions regarding the behavior of the module. For this reason, these techniques (along with other similar techniques) are termed indirect MPPT. There is a second class of direct MPPT which is somewhat more difficult to implement yet more accurate.

One common direct MPPT approach is to measure the current and voltage of the module in order to obtain the total instantaneous power. The operating point is then perturbed by a small amount so as to change the power output level. If output power increases with a positive voltage perturbation, then the MPP is at a higher voltage and the perturbation must take place again. Alternatively, if output power decreases with a

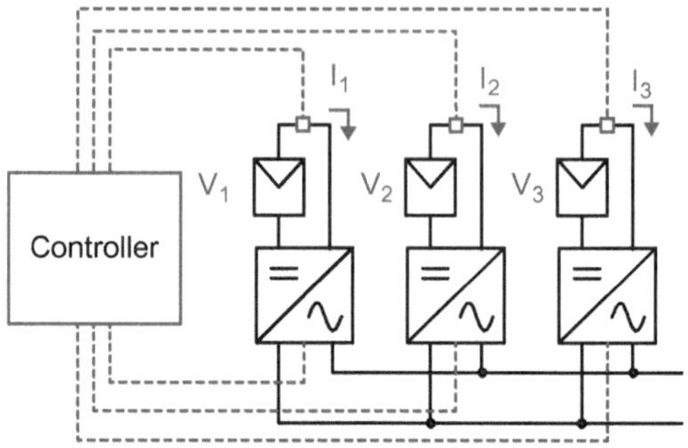

Figure 3.9 A system of solar PV modules each with a power converter that includes maximum power point tracking. The output voltage and current of each module is sensed and the power converter is operated to maximize the power output of each module.

positive voltage perturbation, then the current operating point is to the right of the MPP shown in Fig. 3.8. This iterative process is common known as the Perturb & Observe algorithm (Jäger et al., 2014). An alternative process, known as the incremental conductance method, leverages the fact that the slope of output power with respect to voltage is zero at the MPP. This method can be implemented as a feedback loop within a power converter such that a control variable in the converter's circuitry, such as a reference voltage or a duty cycle is varied (thereby changing the operating point) until the aforementioned condition is satisfied (Xiao et al., 2007).

For example, consider a system of PV modules each with a DC-AC power converter as shown in Fig. 3.9 (the converters may also be DC-DC converters if an intermediate DC bus is used before AC conversion). Sensing circuitry is incorporated into the system to individually monitor the voltage and current output of each PV module. An algorithm such as the incremental conductance method can then be used to adjust the operating point of each converter individually, thereby maximizing the power output of each module while maintain a stable output voltage on the AC bus (i.e., grid). Alternative implementations have been recently developed that sense the current after the inversion takes place, instead of calculating the module power as shown schematically in Fig. 3.9 (Bae et al., 2007). Such advances simplify the control electronics, thereby reducing both cost and size.

The system configuration shown in Fig. 3.9 is sometimes called the module inverter configuration. Several other system topologies are possible and each system design requires a different power conversion arrangement. Another common configuration is the string inverter configuration shown in Fig. 3.10. This scheme involves a number of PV modules stringed together in series to achieve a particular output voltage.

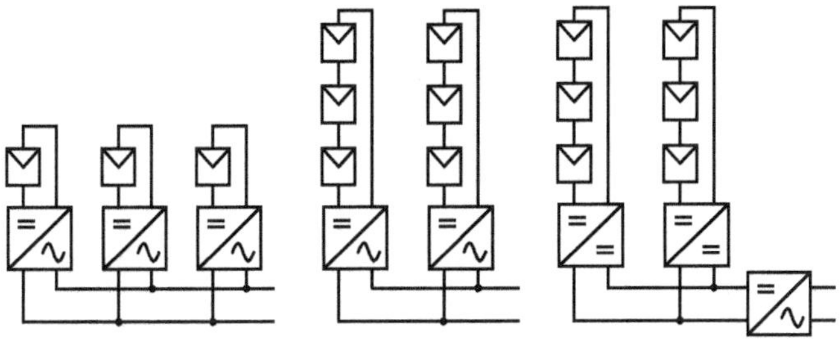

Figure 3.10 Common architectures of solar PV systems (Alahmad et al., 2011).

Multiple stringed panels may be then combined in parallel to provide a greater amount of current and hence power. The multi-string inverter is a variant of this scheme whereby the strings connect to an intermediate DC bus which is then inverted and tied to the grid.

The string configurations have both advantages and disadvantages over the module inverter configuration. Since only a single inverter is required for a number of panels, the cost and overall size of the system is minimized. This comes at the cost of having more demanding constraints on each of the converters. Specifically, each string comprises a number of modules in a series, which increases the overall voltage that the converter must able to handle. Since the current flowing through the string remains comparable to the module inverter configuration, the power converter must handle a higher power as well. In terms of efficiency, the string configurations are not ideal since the current flowing through each string of modules must be identical. Since the converter cannot independently choose the operating point, some modules may operate near peak efficiency while others do not.

Consider the case where shade is cast upon a single module within a string. Even if the current–voltage curves (e.g., Fig. 3.8) are well-matched across modules, the uneven irradiance can result in a shift in the curves and lead to a different MPP. In this situation, it is not possible to operate all modules within the string at peak efficiency. Moreover, for such partial shading scenarios, the power output as a function of voltage can exhibit multiple peaks. For example, there may be a second smaller peak in addition to the main peak due to partial shading of one module in a three-module string.

Advanced adaptive converter architectures are being developed to address these issues (Alahmad et al., 2011; Jäger et al., 2014). These adaptive converters consist of several DC-AC inverters intended for operation with strings of modules and several microinverters with lower power ratings that are intended for operation with a single module. A controller with a matrix of switches is able to connect a module either to a string or isolate the module, providing it with a microinverter. Such adaptive architectures are able to balance conversion efficiency against cost and size. There are sufficient microinverters to handle partial shading conditions (e.g., a few modules in an array), as opposed to the module inverter configuration discussed earlier. The challenge in implementing such systems is the complex control algorithms and decision-making that is involved, and research is ongoing.

3.3.1.2 Power Conversion for Wind

The adoption of wind electricity generation has grown exponentially since the early 1990s, with a total capacity of over 100 GW in 2010 (Jäger et al., 2014; Hansen et al., 2004). The adoption of smart-grid technologies, including advanced power conversion architectures and techniques have enabled more efficient extraction of energy, with recent designs that are capable of producing on the order of 10 MW. During the last decade, there has been a trend away from fixed-speed wind turbines in favor of variable-speed generators. Variable-speed turbines are desirable because they are able to optimize the output power for a range of wind speeds. The amount of power a turbine extracts for a given wind speed and turbine blade geometry (i.e., pitch angle of the blades into the wind) is a function of the ratio of the blade tip speed to the wind speed, otherwise known as the tip speed ratio. For a given wind speed, the pitch angle and the rotor speed determine whether the maximum power is extracted from the wind (Muljadi & Butterfield, 1999). Fixed-speed turbines have long used pitch angle control to adjust the amount of power extracted from the wind. This aerodynamic method of control requires little power electronics and is largely meant to limit power at high winds so as to not damage the mechanical components.

Power electronics is playing an increasing role in variable-speed turbines, since a variable rotor speed can be used to readily alter the tip speed ratio and provides for more efficient operation. A typical variable-speed turbine consists of a rotor with attached blades which are connected to a generator. A gearbox may optionally be used to convert the low-speed high-torque output of the generator to higher rotational speed if required by the generator. The generator is then connected to the grid using power conversion and control electronics. Variable-speed turbines typically incorporate induction generators that consists of a wound rotor and a stator. Useful power is produced when the rotor operates at super-synchronous frequency. The synchronous frequency is determined by the frequency of the AC drive that is applied to the stator. The stator is largely inductive and hence draws reactive power from the grid or another source. The rotation of the turbine and hence the rotor windings then induces currents in the stator and rotor, thereby delivering actual power to the grid.

One topology that incorporates power electronics for controlling power output of wind turbines while allowing for variable-speed operation is shown in Fig. 3.11A. Aerodynamic pitch control is used to limit the power output at high speeds, while the operating speed and power output at low and medium output levels are altered by controlling the amount of current flowing in the rotor. Rotor current flow is controlled using an AC drive provided by an AC-DC converter tied to a DC-AC converter which is in turn fed by the grid. This topology is known as the doubly-fed generator or a partial-scale power converter and accounts for a substantial fraction of variable-speed turbines in operation (Blaabjerg et al., 2006; Blaabjerg et al., 2011). In this architecture, the power converter is

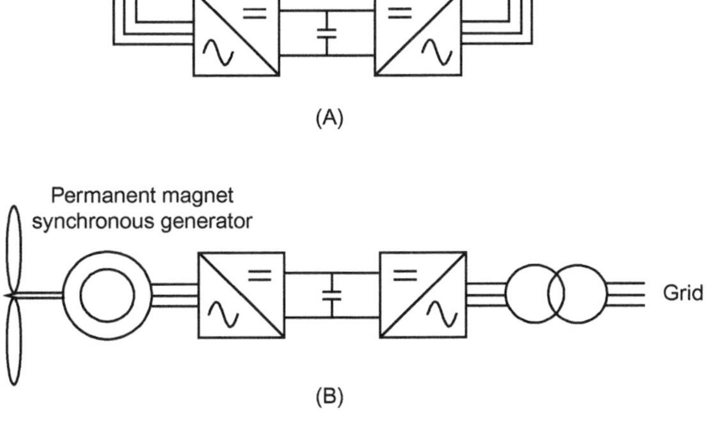

Figure 3.11 Power conversion architectures for wind turbine generators, including (A) Partial-scale power conversion and (B) Full-scale power conversion.

essentially a frequency converter that controls the rotor frequency and hence the turbine speed. It is known as a partial-scale converter since the power handling capability of the converter need only be a fraction of the overall turbine power output. This is because the converter drives only the rotor windings, while a substantial amount of power is transferred directly from the stator to the grid. The natural progression, then, is to incorporate full-scale power conversion whereby the full power output of the turbine is converted to the desired output voltage and frequency. This topology may utilize the asynchronous induction generator discussed earlier or a synchronous generator utilizing a wound stator or permanent magnets as shown in Fig. 3.11B. The higher power levels required of full-scale power conversion come at a higher cost, but the current trend is to use PSMG-based full-scale systems (Blaabjerg et al., 2011).

Induction generators and synchronous generators utilizing wound stators must be initially powered by the grid to operate. Since the windings are largely inductive, these generators draw reactive power from the grid. The advantage of using power converters (such as the architecture shown in Fig. 3.11A) is that the reactive power and hence the power factor may be controlled. In particular, control is achieved by the AC-DC converters, which can provide power factor correction during the back-to-back AC-DC conversion. Such generators also require the use of additional power electronics for fault ride-through requirements. Ride-through refers to the turbine's ability to continue producing power on the grid even when the grid experiences an interruption (and therefore cannot supply reactive power to the generator's stator). The advantage of PSMG-based full-scale power conversion systems over the use of induction motors (partial-scale) and synchronous motors (full-scale) is that no reactive power is required to drive the stator, and hence such systems are more readily able to ride through grid faults.

The implementation of the AC-DC converters used in these power conversion architectures may be classified as either direct or indirect. The back-to-back power converters shown in Fig. 3.11 are indirect since there exists a DC link between the two converters. The DC link provides the advantage of decoupling the generator from the grid and aids in low-voltage grid faults (e.g., assists ride-through). This comes at the expense of a potentially large DC bus capacitor that must be charged (Teodorescu et al., 2011). Direct converter topologies eliminate the latter and its associated benefits, but come with the advantage of lower thermal loading of components and less switching

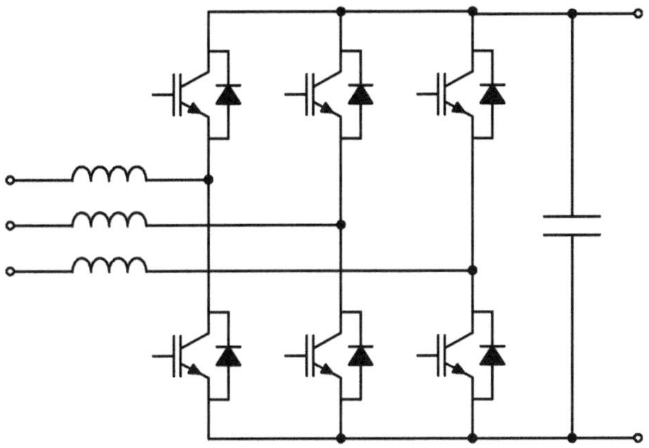

Figure 3.12 A bi-directional three-phase AC-DC converter comprising diodes, transistors and inductors that can be placed back-to-back to form an AC-AC converter for use in wind turbine power conversion.

losses. An example of one commonly used indirect converter is shown in Fig. 3.12. Only half of the symmetric converter is shown which consists of six switches with associated diodes, three inductors (one for each phase), and a DC bus capacitor as discussed earlier. Additional filter capacitors and inductors on both the generator and grid sides are required (not shown) in order to limit the harmonic content on their respective sides. Conversion from AC to DC utilizes the diodes, which rectify the three-phase AC to DC with ripples that are filtered by the bus capacitor. Conversely, DC conversion back to AC makes use of the switches to generate square waves 120° apart, which are filtered by the inductors to produce a sine wave output. A symmetric topology therefore allows for bidirectional power conversion. Alternatively, diodes may be used on one side and transistors on the opposite side if one-way conversion is desired.

Wind turbines are often integrated into a power generation plant, as in the case of both onshore and offshore wind farms. Doubly-fed induction generators may be simply coupled on the grid AC side using a transformer. Systems that use full-scale power conversion can be similarly connected, or alternatively connected onto an intermediate DC bus. In this scenario, each turbine utilizes an AC-DC converter on the generator side to connect to a high-voltage DC transmission system. For offshore systems, this is ideal since high-voltage DC transmission may be used to efficiently transmit power over long distances to an onshore DC-AC converter that ties the farm to the grid. A number of wind farm topologies are possible and offer tradeoffs between individual speed control, redundancy, cost, and other factors (Blaabjerg et al., 2006).

3.3.2 Energy Storage Systems

The renewable energy sources discussed in the preceding sections have the common feature that power output levels are not constant throughout the day. Moreover, peak power output may not necessarily coincide with peak power consumption. In order to fully benefit from renewables, the grids of the future must incorporate a mechanism for storing the energy harvested using solar and wind technologies.

Grid-tied energy storage systems (ESS) can utilize a variety of energy storage mediums, including for example batteries with lead-acid, lithium-ion and sodium-sulfur chemistries These systems not only provide a means to store energy for later use, but also offer other benefits such as load and frequency regulation (US Dept. of Energy, 2013; Luo et al., 2015). Systems utilizing lithium-ion and sodium-sulfur chemistries, e.g., are expected to deliver up to 10 MW of power with capacities approaching tens of MWh (IEC, 2012). There are various tradeoffs between different chemistries that make each one suitable for storage of electrical energy over different timescales, power levels, and energy densities.

Storage of energy in these systems therefore amounts to charging the battery with the appropriate charging current profile, which requires rectification of the AC grid supply and appropriate control. The energy may then be supplied to the grid at a later time as needed by inverting the stored DC supply and feeding it to the grid. Grid-tie power converters for ESS must therefore provide bidirectional AC-DC power conversion.

Advanced power conversion architectures provide for a means to integrate the ESS as part of a renewable energy source (as opposed to being tied to the grid in isolation). For example, a solar photovoltaic installation comprising PV modules alongside an ESS may utilize converter architecture of the type shown in Fig. 3.13, which includes an intermediate DC bus with ports for the renewable and the ESS systems and bidirectional conversion between the bus and the grid (Clark, 2012). In this multiport architecture, the controls system monitors several variables, including grid demand, battery capacity, solar power production, and adjusts power flow accordingly. If excess power is produced relative to demand, the converters are configured to divert power to the storage system. Alternatively, the converters can be configured to deliver power to the grid either from the storage system or the renewable source should demand require it.

The DC-DC converter blocks can be specific to the type of renewable that is to be combined with an ESS. For example,

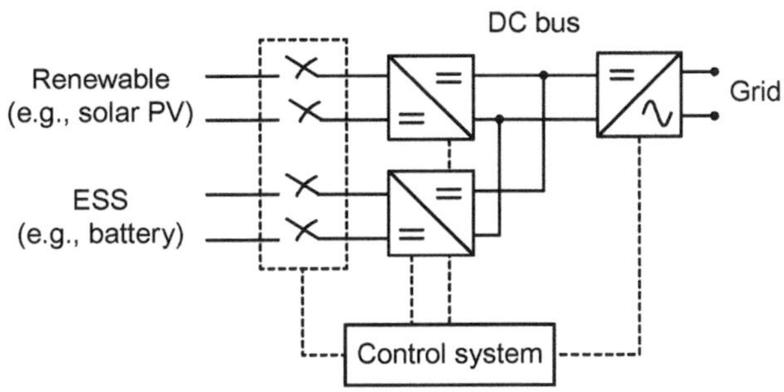

Figure 3.13 Power converter architecture for integrating a renewable energy source with an energy storage system (ESS). The converter consists of several DC-DC converter components connected to an inverter over a DC bus. A control system determines the direction of power flow depending on power production and grid demand (Clark, 2012).

DC-DC converters that implement MPPT would be used if a solar PV was to be combined with an ESS. The grid-tie inverter and DC-DC converter block used to incorporate the storage system are necessarily bidirectional so that energy can be transferred to and from the battery.

The bidirectional AC-DC converter can be broadly classified into single-phase and three-phase systems. Single-phase systems convert between DC and single-phase AC, whereas three-phase systems convert to and from three-phase AC. We have already seen one example of a three-phase converter in 3.3.1.2 for wind-turbine applications. Single-phase converters are appropriate for low power systems; they are simpler to implement and easier to control than three-phase counterparts. Recent advances in converter topologies have reduced the number of components required to implement these converters.

One such example is the single-phase AC-DC shown in Fig. 3.14 (Qian, 2011). It comprises four transistor switches, each in parallel with a diode, along with an input inductor and an output capacitor. During AC-DC conversion, the diodes rectify the grid voltage, which is filtered by the bus capacitor to provide clean DC power to the ESS while the transistors remain in the off state. During the positive half of the switching cycles, diodes D1 and D2 are active, applying a high potential to the positive terminal of the ESS and therefore delivering power. Similarly, during the negative half of the switching cycles, diodes D3 and D4 are active. For inverter operation, the transistors S1 and S2 are turned on so that the ESS is connected to the grid during the positive half-cycle, thereby charging the grid through the inductor. During the negative half-cycle, the polarized battery cannot readily provide energy to the grid. For this

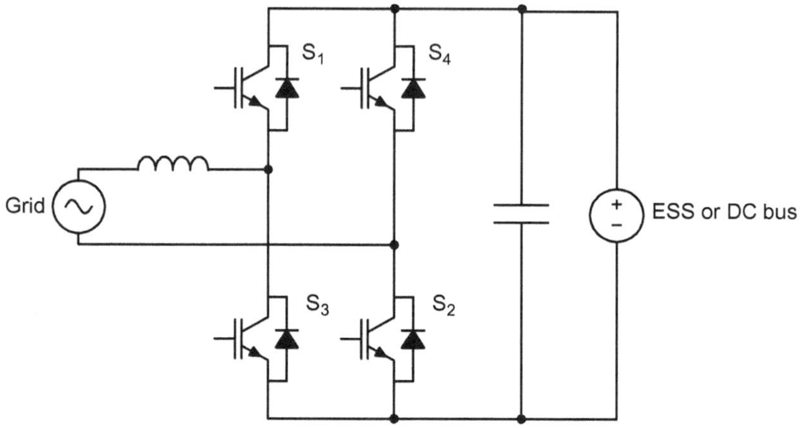

Figure 3.14 Single-phase bi-directional AC-DC converter (Qian, 2011).

reason, only S2 is turned on and D4 is activated so that only the charged inductor connects to the grid, providing it with energy.

Recent research and development efforts have focused on more sophisticated topologies that provide additional benefits. For example, advanced topologies are under research and development that alleviate or completely remove the need for large bus capacitors and allow both MOSFETs and IGBTs to be used in the implementation (Qian, 2011). Numerous other topologies are under consideration, including converters with three-phase grid connections and transformer isolation. The choice of topology is a tradeoff between switching losses in the converter, acceptable output voltage ripple (e.g., the amount of filtering), harmonics that are generated on the grid, the DC bus voltage, and much more (Trintis, 2011).

The example demonstrates the basic principles behind bi-directional AC-DC conversion. Sophisticated control mechanisms and additional circuitry are required for proper operation. For example, operation in rectification mode requires careful control of the charging profile by limiting the amount of current reaching the storage system. In fact, one of the factors influencing efficient grid integration of battery-based ESS is the power converter's ability to estimate the remaining battery capacity or the state of charge (SOC). Accurate SOC estimation is important since it is the metric used by the converter to determine whether the ESS is able to handle additional energy storage. Battery management systems are electronic systems that employ various methods to estimate charge, including coulomb counting, monitoring of battery voltage, and more recently predictive algorithms (Rahimi-Eichi & Chow, 2012).

Coulomb counting involves measuring the amount of current flowing into or out of the battery. If the flows are measured in real time, an estimate of the capacity at any one time can be attained. This method is simple to implement yet inaccurate since the amount of energy supplied to charge a battery might not be entirely available upon discharge due to inefficiencies. Battery voltage monitoring is another method, whereby the voltage of a cell is accurately measured and compared against model voltage versus discharge curves. This technique is also inaccurate since environmental variables such as temperature can affect the shape of these curves. A proper calibration against these factors is therefore required. Recent advances have improved upon these simple techniques by improving existing models and developing advanced computation methods that can be implemented in embedded controllers (Kim et al., 2014).

3.4 Integration of EVs

EVs are increasing in popularity because of their fuel efficiency and role in decreasing emissions. At present, EVs constitute a small fraction of total power consumption on the grid but the number of EVs on the market is expected to grow. Annual sales of plug-in vehicles (including EVs and plug-in hybrid vehicles) have increased from 50,000 in 2012 to 100,000 annually in 2013–15 (Inside EVs, 2016), and the EVs Initiative (Electric Vehicles Initiative, 2013) has stated a goal of having 20 million vehicles on the global market by 2020. The combination of EVs and the smart grid represents the confluence of two rapidly evolving areas. As a consequence, these technologies are developing together and to some extent can accommodate each other. Advanced power converters enable this accommodation, by providing a multitude of charging options that work in synergy with the variable pricing for consumption and even rebates for power donation that the new grid infrastructure will be able to accommodate (Morgan, 2012). Concurrently, advances in semiconductor and power system design are making EVs more efficient and are therefore likely to increase their popularity and thus their role in the power industry of the future.

3.4.1 EV Power Demands and Consumption

Efficient energy storage, conversion and management are paramount within fully electric vehicles as there is no alternative

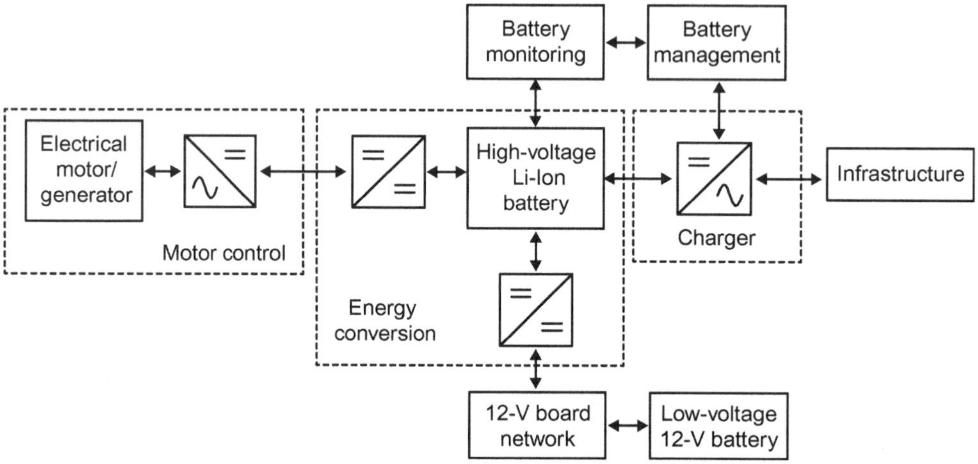

Figure 3.15 Schematic overview of vehicle power system outlining varied power requirements (Texas Instruments, 2013).

power source, compared to, e.g., a hybrid vehicle, which has both an electric motor and an internal combustion engine. EVs use a battery array as their energy supply and require multiple power conversion stages for various functions of the vehicle, each of which is illustrated schematically in Fig. 3.15. The battery when fully charged has an energy capacity typically between 15 and 30 kWh and operates at a voltage of several hundred volts. AC grid voltage must be rectified and converted to an appropriate value using an AC/DC converter, which can either reside within the vehicle or be integrated into a DC charging station. Energy stored in the battery is converted to traction power for use by an AC motor with the aid of an onboard DC/AC converter, typically consisting of a DC/DC converter and a separate DC/AC inverter. Finally, power from the HV battery must be converted onboard to charge the 12 V vehicle battery that powers many of the standard peripherals in the vehicle. The main power conversion steps of rectification and inversion will be discussed in more detail in the following subsections.

3.4.2 Fast and Smart Chargers

One the largest grid adaptations to support the proliferation of EVs is the incorporation of charging infrastructure. Charging at home overnight has long been an effective option because of the relatively long periods that a car is spent unused along with the relatively low power load on the grid during the overnight hours. However, charging in the field is equally necessary, and because importance is placed on minimizing charging times in the field, this option has significantly higher power requirements, as explained next.

Although the detailed individual power requirements of EVs may vary, all EV chargers require a similar set of power electronics components. AC power from the grid is rectified into a DC source, typically using a diode bridge. The DC power is then converted to an appropriate voltage level using a DC/DC converter, similar to the full bridge converter described in Section 3.2.1, which is controlled to generate an appropriate charging profile required for the onboard battery.

The majority of EVs use lithium-ion (Li-ion) batteries because of their high energy density, a feature that is desirable for weight considerations, although some vehicles use other battery chemistries such as nickel metal hydride (NiMH) (Young et al., 2013). Many batteries typically charge in a dual mode, first using a constant current mode until a large fraction of their rated voltage is reached, at which point the charger switches to a constant voltage mode while the current is tapered off to bring the battery to full capacity. Regardless of the specific charging profile used, variable charging behavior necessarily requires a smart charger comprising power, monitoring, and control electronics that can communicate with the vehicle's onboard battery management system.

Despite EVs being a relatively young technology, standards have been ratified to specify certain aspects of EV charging behavior. Within the United States, the de facto standard for EV charging is SAE J1772 "SAE Electric Vehicle and Plug-in Hybrid Electric Vehicle Conductive Charge Coupler" and is published by the Society of Automotive Engineers. SAE J1772 specifies three types of EV chargers that are summarized in Table 3.2.

Table 3.2 Typical Power Characteristics of Various EV Charger Classes as Specified by SAE (SAE, 2011)

Charger Type	Voltage	Power (kW)
AC Level 1	120 V	1.9
AC Level 2	208/240 V	19.2
AC Level 3 (TBD)	TBD	>20
DC Level 1 (not finalized)	200–450 V	36
DC Level 2 (not finalized)	200–450 V	90
DC Level 3 (TBD)	200–600 V	240

AC Level 1 chargers run on a standard 120 V AC outlet in the United States and consequently require little changes in infrastructure, but the relatively low power delivery capacity of these chargers results in a lengthy charging time, and thus AC Level 1 chargers are suitable for extended charges such as in the overnight hours. AC Level 2 chargers use either 208 or 240 V split-phase power similar to large household appliances such as a clothes dryer. As of the January 2010 revision of SAE J1772, AC Level 2 chargers are limited to 80 A and therefore can supply up to 19.2 kW. Both AC Level 1 and AC Level 2 chargers require the EV to have an onboard rectifier/converter to charge the battery, and the necessary size and weight requirements for designing onboard chargers limits their power handling capacity. A Level 3 AC charger, which will deliver more than 20 kW of power, has also been proposed.

DC chargers have also been proposed by SAE and are included in Table 3.2, but have not yet been standardized. With these chargers, power conversion is done off board and DC power is delivered to the EV by bypassing the onboard converter. Since the AC/DC conversion is done off board, size and weight considerations for the rectifier are less stringent and consequently allow the converter to deliver more power. Because of the higher power delivery capabilities, these chargers cater toward EVs that need to charge in the field. In addition to the proposed SAE charger, multiple other versions of the DC charger exist, such as the Tesla Supercharger (Tesla Motors, n.d.; CHAdeMO, n.d.)

Fast charging times are desirable in the field and this leads to relatively large power delivery requirements. For instance, a 20 kWh EV battery would take more than 11 hours to charge using an AC Level 1 charger, and more than an hour to charge using AC Level 2. For a 50 kW DC Eaton charger (Eaton Power Corporation, 2013), the same battery could be charged in approximately 24 minutes, making on-the-go charging accessible for EV owners.

3.4.3 EV Inverters

Energy stored in the HV battery needs to be subsequently converted to AC power to drive the vehicle's traction motor. In some ways this is a more mature technology, having existed for over a decade in hybrid EVs. Compared to the AC/DC converters discussed earlier, the DC/AC inverter has more stringent power requirements because it needs to supply the immediate needs of the electric motor in real-time. Furthermore, a strong

Table 3.3 Combined (City + Highway) Fuel Economies for Selected EVs (United States Department of Energy, 2016)

Vehicle	Fuel Economy (kWh/100 mi)
2015 BYD e6	54
2016 Mercedes-Benz B250e	40
2016 Tesla Model S	33
2016 Ford Focus	32
2015 Nissan Leaf	30
2014 BMW i3	27

emphasis is placed on efficiency since the total capacity of the vehicle battery can be a limiting factor in vehicle usage. Finally, the inverter needs to function in reverse mode, providing rectification to transmit energy back to the battery when regenerative braking is used.

Given the emphasis on battery capacity and efficiency, it is worth investigating the typical power needs of an EV during use. This depends on many factors, such as speed, wind and road conditions, aerodynamics, and auxiliary systems usage. The Unites States Department of Energy, however, assigns a fuel economy to EVs using standardized methods in the same way it does for conventional gasoline powered vehicles. Fuel economy values for several representative EVs produced between 2014 and 2016 are included in Table 3.3.

As an estimate for typical EV power requirements, considering a highway speed of 65 miles per hour and an average fuel economy of 36 kWh/100 miles, an EV traction motor would require an average power of 23.4 kW provided by the vehicle inverter. Power requirements of tens of kW, combined with the HVs present in the power system influence the choice of components used in the inverter, as discussed next.

EV inverters are similar to the bidirectional three-phase converter illustrated in Fig. 3.12. The inverter accepts HV DC input and through a series of appropriately timed switches and converts power to three-phase AC that is then transmitted to the traction motor. Because of the large power requirements discussed earlier, IGBTs are typically used as switches in the inverter. IGBTs are robust enough to handle the high-voltage switching requirements for automotive applications (Hauenstein, 2010). Control electronics can adjust the switching frequency of the IGBTs thereby changing the AC power supplied to the motor. For a power traction motor, switching frequencies are typically in the range of 5–10 kHz.

In the case where regenerative braking is used, power from the traction motor is transmitted in the reverse direction through the power conversion step to charge the HV battery. In this case, the IGBTs can be opened and the six diodes function as a full bridge rectifier. Rectified power can then be transmitted through the DC/DC converter operating in a buck capacity in order to charge the HV battery. It has also been suggested that this pathway could also be used to charge the battery using an AC grid power source (Johansen, 2013). In this scenario a connection would be made on the traction motor side (the AC side of the vehicle's power conversion system), and the vehicle's onboard rectifier and DC/DC converter would be used in reverse to convert and transmit power to the battery.

While some have proposed more complex inverter designs, such as utilization of multilevel inverters generate output that better approximates a sine wave (Tobert et al., 1998) (Josefsson, 2015), many of the advances in EV inverter technology have focused primarily on increasing thermal cooling capacity and decreasing loss (Kimura et al., 2014). More efficient cooling designs have led to a significant increase in inverter power density, allowing smaller and lighter converters to be implemented, as well as new circuit designs that reduce transients to enable faster switching. Furthermore, advances in semiconductor design, such as the transition from NPT to PT IGBTs discussed earlier, as well as new materials choices, such as the use of SiC, have also led to more efficient semiconductor structures with better conductivity and lower losses (Ide et al., 2011).

3.4.4 Role of EVs in the Smart Grid

One of the tenets of the smart grid is an increasing resilience and adaptability to changing loading requirements. Without these enhancements and adaptability, the proliferation of EVs could represent a significant consumption problem in areas where peak power draws are already approaching limiting values. But through the new capabilities present in the smart grid, incentives can be provided to shift vehicle charging behavior to off-peak hours, thereby flattening out the daily electricity consumption curve and reducing costs to both vehicle owners and electricity providers.

For example, the relatively low power transfer of Level 1 and Level 2 charging stations discussed earlier naturally incentivize vehicle owners to charge at home overnight when the vehicle is likely unused. Further incentives could be provided through use adaptability and communications capabilities of the smart grid to institute lower fees during overnight hours and higher fees during the day when grid stress has increased.

For those who need to charge their vehicle during the day, significantly higher energy transfer rates, such as those available in fast DC charging stations, are desirable to minimize the inconvenience of stopping and waiting for an EV to charge. Higher power delivery in turn requires a revised power infrastructure. During the day when power loads across the grid are highest, the additional capacity to charge a large fleet of EVs may not necessarily be available, which has led some to propose alternate options for fast DC charging stations, such as onsite energy storage in battery banks, or onsite battery exchange.

Another interesting prospect is utilization of EVs as power sources, whereby excess power could be restored back to the grid during peak consumption hours for increased compensation. Such vehicle-to-grid power donation represents a new layer of adaptability of the grid in times of stress. The variable electric marketplace that the smart grid will make possible has strong synergies with the mobile and adaptable behavior of EVs, and it is likely that these two industries will reinforce each other and grow together in the coming years.

3.5 Future Trends

Grid-scale power converters are becoming increasingly more flexible, intelligent, and robust so that they can support the smart grid of tomorrow. As smart power converters become more ubiquitous, they will enable wider adoption of renewable energy technologies, ESS, and integrate with EVs.

The ability of power converters to provide bi-directional power conversion while retaining a degree of configurability to meet consumer demand is enabled by the use of active solid-state devices in an otherwise passive grid. Improvements to the underlying technologies of semiconductor devices and in particular switching devices will lower the cost, size, and weight of power converters leading to wider adoption. Advances are continuously being made to both the geometry and structure of semiconductor devices and more recently the choice of materials used to manufacture these devices. Recent advances in

SiC-based devices has enabled high-voltage and high-temperature operation of devices that were previously limited.

Advancements and improvements in power converter topologies are also being implemented. Power converters for use with renewable technologies are implemented using more sophisticated topologies that maximize the power output and hence efficiency of stochastic renewable energy sources such as wind and solar. Ongoing research in power conversion incorporates not only these advanced circuit topologies, but also sophisticated control algorithms that improve the converter's ability to monitor and optimize the operating point as necessary. Some examples of these were already discussed in the context of solar power conversion and ESS.

The flexibility afforded by tomorrow's power converters will make it easier to incorporate distributed power sources into the power grid, and the bi-directional flow of power supported by advanced power converters will allow homes with PV to become distributed sources of power as well, delivering excess energy back to the grid. The adoption of EVs together with consumer PV may lead to a greater role for power converters: the smart home.

References

Alahmad, M., Amer Chaaban, M., Lau, S.K., 2011. An Adaptive Photovoltaic-Inverter Topology. IEEE, Anaheim.

Bae, H.-S., Park, J.-H., Cho, B.-H., Yu, G.-J., 2007. New MPPT control strategy for two-stage grid-connected photovoltaic power conditioning system. J. Power Electron. 7 (2), 174–180.

Blaabjerg, F., Chen, Z.T.R., Iov, F., 2006. Power Electronics in Wind Turbine Systems. IEEE, Shanghai.

Blaabjerg, F., Liserre, M., Ma, K., 2011. Power Electronics Converters for Wind Turbine Systems. IEEE.

CHAdeMO Association, n.d. [Online] Available from: <http://www.chademo.com/> (accessed 29.08.16).

Chou, W., 2008. Choose your IGBTs correctly for solar inverter applications. Power Electron. Technol. August, 20–23.

Clark, K., 2012. Technology Roadmap Workshop for Increased Power Electronic Grid Applications and Devices. NIST, Gaithersberg.

Cree, 2011. Cree launches industry's first commercial silicon carbide power MOSFET; destined to replace silicon devices in high-voltage (≥1200-V) power electronics. [Online] Available from: <http://www.cree.com/news-and-events/cree-news/press-releases/2011/january/110117-mosfet> (accessed 23.02.16.).

Eaton Power Corporation, 2013. DC quick charger for electric vehicles. Technical Data TD0EV00004E.

Electric Vehicles Initiative, 2013. Global EV outlook: understanding the electric vehicle landscape to 2020. [Online] Available from: <https://www.iea.org/

publications/freepublications/publication/GlobalEVOutlook_2013.pdf> (accessed 06.03.16).
Falcones, S., Mao, X., Ayyanar, R., 2010. Topology Comparison for Solid State Transformer Implementation. IEEE, Minneapolis.
Frank, W., 2008. Power electronics for solar power inverter systems. [Online] Available from: <http://www.electronicproducts.com/Power_Products/Invertors/Power_electronics_for_solar_power_inverter_systems.aspx> (accessed 07.03.16.).
Freeman, D., 2010. Introduction to photovoltaic systems maximum power point tracking. Texas Instruments Application Report SLVA446.
Hansen, A.D., Iov, F., Blaabjerg, F., Hansen, L.H., 2004. Review of contemporary wind turbine concepts and their market penetration. Wind Eng. 28 (3), 247–263.
Hauenstein, H., 2010. High-voltage IGBTs move into automotive apps. Electronic Design Europe.
Ide, K., Nakatsu, K., Hombo, H., Maki, K., 2011. Power electronics component technologies for green mobility. Hitachi Rev. 60 (6), 310–315.
Inside EVs, 2016. Monthly Plug-In Sales Scorecard. [Online] Available from: <http://insideevs.com/monthly-plug-in-sales-scorecard/> (accessed 07.03.16).
IEC, 2012. Grid integration of large-capacity Renewable Energy sources and use of large-capacity Electrical Energy Storage, Geneva: s.n.
Jäger, K., et al., 2014. Solar Energy: Fundamentals, Technology, and Systems. Delft University of Technology, Delft.
Johansen, J., 2013. Fast-Charging Electric Vehicles Using AC. University of Denmark.
Josefsson, O., 2015. Investigation of a Multilevel Inverter for Electric Vehicle Applications. Chalmers University of Technology.
Kassakian, J.G., Schlecht, F., Verghese, G., 1991. Principles of Power Electronics. first ed. Prentice Hall, Reading.
Kim, T., et al., 2014. State of Charge Estimation Based on a Real- time Battery Model and Iterative Smooth Variable Structure Filter. IEEE, Kuala Lumpur, pp. 132–137.
Kimura, T., et al., 2014. High-power-density inverter technology for hybrid and electric vehicle applications. Hitachi Rev. 63 (2), 41–47.
Kobayashi, Y., Maruyama, R., Mochizuki, E., 2007. New Concept IGBT-PIM Using Advanced Technologies. Fuji Electric Review, 53(2), pp. 69–72.
Li, Y., Nejabatkhah, F., 2014. Overview of control, integration and energy management of microgrids. Journal of Modern Power Systems and Clean Energy. 2 (3), 212–222.
Luo, X., Wang, J., Dooner, M., Clarke, J., 2015. Overview of current development in electrical energy storage technologies and the application potential in power system operation. Appl. Energy. 137, 511–536.
März, M., Knapp, A., Billmann, M., 1998. High-speed 600V IGBT in NPT technology. Power Semiconductors.
Marks, T. G. a J., 2005. Control Techniques for Active Power Filters. IEEE, pp. 369–381.
Morgan, T., 2012. Smart Grids and Electric Vehicles: Made for Each Other? International Transport Forum.
Muljadi, E., Butterfield, C., 1999. Pitch-Controlled Variable-Speed Wind Turbine Generation. IEEE, Phoenix.
Qian, H., 2011. A High-Efficiency Grid-Tie Battery Energy Storage System. Virginia Polytechnic Institute, Blacksburg.
Rahimi-Eichi, H., Chow, M.-Y., 2012. Adaptive Parameter Identification and State-of-Charge Estimation of Lithium-Ion Batteries. IEEE, Montreal.

SAE, 2011. SAE Charging Configurations and Ratings Terminology ver. 031611. [Online] Available from: <http://www.sae.org/smartgrid/chargingspeeds.pdf> (accessed 29.08.16).

Saxena, R.S., Kumar, M.J., 2012. Trench Gate Power MOSFET: Recent Advances and Innovations. Nova Science Publishers, New York.

Schweber, B., 2011. SiC power MOSFET pushes past Si-based MOSFETs/IGBTs with lower losses, lower capacitance, easier drive. [Online] Available from: www.eetimes.com (accessed 28.02.16.).

Shenai, K., 2013. Switching megawatts with power transistors. The Electrochemical Society Interface, Spring, pp. 47–53.

Shoemaker, T., Mack, J.E., 2006. Lineman and Cableman's Handbook. eleventh ed. McGraw-Hill.

Teodorescu, R., Liserre, M., Rodríguez, P., 2011. Grid Converters for Photovoltaic and Wind Power Systems. John Wiley & Sons, Ltd.

Tesla Motors, [No date]. Supercharger. [Online] Available from: <https://www.tesla.com/supercharger> (accessed 07.03.16.).

Texas Instruments, 2013. Hybrid and electric vehicle solutions guide. Texas Instruments Document SZZA058C.

Tobert, L.M., Peng, F.Z., Habetler, T.G., 1998. Multilevel inverters for electric vehicle applications. Power Electron. Transp.79–84.

Tolbert, L.M., et al., 2005. Power Electronics For Distributed Energy Systems And Transmission And Distribution Applications. Oak Ridge National Laboratory, Oak Ridge.

Trintis, I., 2011. Grid converters for stationary battery energy storage systems, Aalborg University.

United States Department of Energy, 2016. [Online] Available from: <https://www.fueleconomy.gov/> (accessed 07.03.16.).

US Dept. of Energy, 2013. Grid energy storage.

Wang, H., 2007. Investigation of Power Semiconductor Devices for High Frequency High Density Power Converters. Virginia Polytechnic Institute and State University, Blacksburg.

Wang, X., Cooper, J.A., 2010. High-voltage n-channel IGBTs on free-standing 4H-SiC epilayers. IEEE Trans. Electron Devices. 57 (2), 511–515.

Xiao, W., Dunford, W.G., Palmer, P.R., Capel, A., 2007. Regulation of photovoltaic voltage. IEEE Trans. Ind. Electron. 54 (3), 1365–1374.

Young, K., Wang, L.Y., Strunz, K., Wang, C., 2013. Electric vehicle battery technologies. In: Garcia-Valle, R., Peças Lopes, J.A. (Eds.), Electric Vehicle Integration into Modern Power Networks, Power Electronics and Power Systems. Springer, New York.

4

SMART GRID ENERGY STORAGE

Balasubramanian Pinnangudi[1], Michelle Kuykendal[1] and Shoham Bhadra[2]

[1]Exponent Inc., Phoenix, AZ, United States [2]Exponent Inc., Natick, MA, United States

CHAPTER OUTLINE

The Power Grid. DOI: http://dx.doi.org/10.1016/B978-0-12-805321-8.00004-5

4.1 Introduction

4.1.1 Energy Storage Systems (ESS)—A Key Enabler to Smart Grids

By some estimates, the United States (US) is projected to consume 4000–5000 tera-watt-hours of electricity by 2050 (Fig. 4.1).

The increased demand will put tremendous stress on the generation, transmission, and distribution infrastructure, with the consequence that the aging electric grid is likely to encounter difficulties in providing these quantities of electricity at a level of reliability that is expected of it in the US economy (Abraham, 2002). Modernizing the electric grid is a clear technological priority to meet the ever-growing energy needs of the nation and it should also address climate change by decarbonizing the electric grid through the integration of renewables (Fig. 4.2).

The challenges faced in the implementation of grid expansion include balancing economic and commercial viability, resiliency, cyber-security, impact on carbon emissions, and environmental sustainability. Energy storage technologies play a significant role in meeting these challenges and are a key enabler of grid modernization, addressing the electric grid's pressing needs by improving the operational capabilities of the grid as well as deferring and/or reducing infrastructure investments while maintaining a robust power delivery system (Gyuk et al., 2013; Bussar et al., 2013).

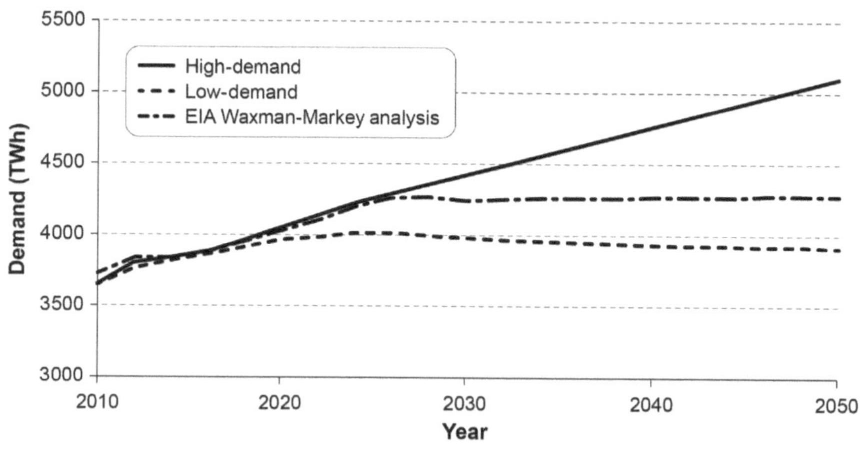

Figure 4.1 Demand trajectory for the low-demand and high-demand baselines through 2050 (Hostick, 2012).

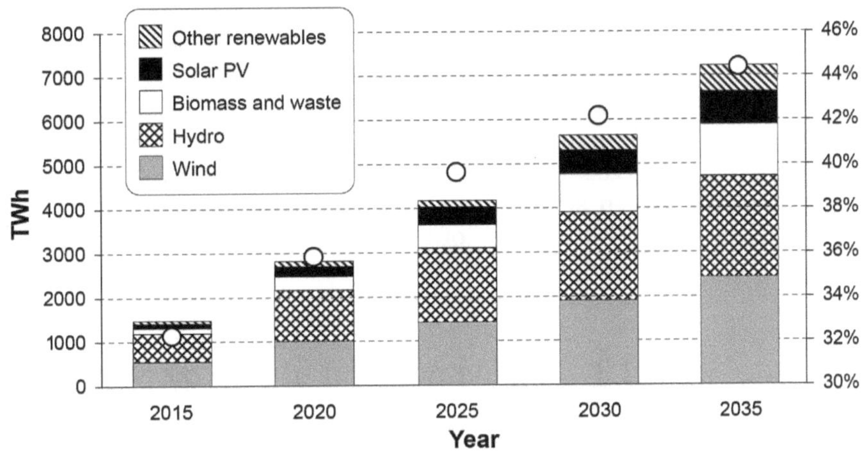

Figure 4.2 Global renewables-based electricity generation relative to 2009 by technology (IEC Market Strategy Board, 2012).

4.1.2 Market Demand—Need for ESS

Electricity differs from other markets in two aspects, which generate the market demand for ESS (Roberts and Sandberg, 2011; IEC Market Strategy Board, 2011) and are described below:

1. Electricity generation and consumption must be balanced at all times. The load demand continuously and randomly fluctuates, and power generation must be adjusted accordingly to meet instantaneous changes in the load. An imbalance between the supply and demand in an electric grid will be detrimental to network stability and power quality even if it does not result in an unsatisfied demand. If the demand is greater than generation, the frequency drops and vice versa. Small deviations from the nominal frequency (60 Hz in the United States) can adversely affect sensitive end-user equipment and devices.
2. Power sources may be located anywhere on the grid: close to load centers, dispersed across the network, or in remote offshore/desert locations. Power generation usually occurs far from where the electricity is consumed. The generators and the consumers are connected through transmission and distribution networks to form a power system. In a deregulated environment, the transmission network is freely available to all of the market participants that may want to use it. This could result in a situation where the actual flow of electricity over the transmission line or a piece of equipment are

constrained below the desired levels and is referred to as transmission congestion. Constraints may be imposed either by the physical or electrical capacity of the line/equipment, or by operational directives that are enforced to protect the security and reliability of the electric grid. Transmission congestion can result in higher electricity costs for consumers. In severe conditions, it can impair grid stability by limiting the number of generation sources available and rendering a geographical area prone to unplanned, expensive outages.

4.1.3 Implementation of Energy Storage—Challenges

A changing energy landscape with a move to fuel-free power in the form of wind and solar energies is great for environmental sustainability. However, it makes the reliable delivery of power when and where it is needed a greater challenge due to fast changing power flows caused by their intermittent availability and highly variable output. Unlike traditional centralized generation plants, renewable energy sources are remote, distributed, and volatile, and their integration into the existing grid requires fundamental changes to the architecture and controllability of the grid (Roberts and Sandberg, 2011). This calls for smart and efficient power transmission/distribution networks and energy storage to provide a balance between generation and consumption, and to maintain grid stability. Storage is critical to smart grid technology due to its role in complementing renewable energy sources. With the right amount and type of storage technology that is optimally controlled and broadly distributed in the network, renewable energy can be transformed into a dispatchable generation source. The economics of energy storage will be a major driver in determining how quickly distributed energy storage solutions are implemented in electric grids. As renewable energy technologies mature and their large-scale implementation becomes affordable, the desire to use inexpensive and abundant alternative energy will increase. An increased awareness about climate change and a goal toward reducing carbon emissions are going to further encourage the development of both renewables and storage (Fig. 4.3).

To realize their full benefits grid-wide, ESS must be managed in harmony with other power generation, loads, and ESS in the network. The smaller and distributed ESS can be grouped together into virtual power plants that can be managed at the grid level in conjunction with traditional power plants. With an

Figure 4.3 Energy storage can be integrated seamlessly into the existing infrastructure in order to support the power grid.

increase in the percentage of renewables in the electric grid, forecasting the generation capacity and load demand becomes essential. Accurate forecasting using advanced mathematical techniques to model the constraints of energy storage, transmission/distribution network, market, etc., allows for the optimal use of energy storage across the grid. An understanding of the different elements of the power landscape, their dependencies and interdependencies, and how the different pieces fit together is the key to defining an effective energy storage strategy for smart grids (ABB, 2012; Modern Grid Strategy Team, 2009).

4.1.4 Storage Technologies—Metrics

For widespread deployment of grid-scale energy storage, and to provide maximum benefit for the power industry and its customers, storage technologies must meet a set of economic, performance, and system-design targets to offer the right combination of performance and cost effectiveness (Eyer and Corey, 2010). While the metrics and targets vary depending on the specific energy storage technology and the location of the application, they can be used as a guideline to assess the individual technologies.

These metrics include:

System economics: As would be the case in any industry, system economics are a critical metric in the electric power industry. Consumers expect to receive a safe and reliable power service at an affordable price, which makes the lifecycle cost of storage technologies very important. The energy storage technology must be competitive with the other currently available technologies for a given application and provide benefits sufficient to offset their capital, operating, and lifetime costs. The overall costs associated with the storage technologies must continue to decline for their broad implementation in the grid.

Technical performance: Technical performance and system economics are interrelated factors. While being an affordable option, technologies must be able to meet the performance needs of the intended application; otherwise it is unlikely that they are considered as an alternative. The performance needs are application specific. For example, metrics for battery ESS would typically include power rating, capacity, efficiency, rate of charge/discharge, response time, cycle life, etc.

System design: A range of technologies and competencies are required to build an efficient, reliable, and durable grid ESS. These include but are not limited to the storage media (battery, flywheel, regenerative fuel cell, capacitor, super-capacitor, etc.), power electronic conversion, system- and grid-level control, and smart forecasting. Energy storage technologies based on direct current requires a conversion step using power electronics to connect to the Alternating Current (AC) grids used for power transmission and distribution, while energy storage technologies based on AC rely on power electronics to integrate them optimally to the AC grid. The power electronic systems provide a stable interface between the energy storage media and the AC grid. System design has a significant impact on the overall system economics and technical performance, and vice versa. Additionally, the system design must meet the safety standards of the electric power industry; it needs to manage the health and safety risks to utility workers and the surrounding community.

ESS can be broadly characterized based on the following parameters:

Power and energy rating: Power is the rate at which energy can be supplied to the load. A storage system's power rating represents the system's maximum power output under normal operating conditions. Some types of storage systems (like sodium–sulfur batteries) can also discharge at a relatively higher rate for short periods of time often referred to as the emergency rating. This feature can be valuable when additional power is required infrequently to meet urgent needs for a short duration. However, it must be noted that the storage efficiency and equipment damage will decrease and increase respectively with higher discharge rates as compared to the nominal rate of discharge (Lichtner et al., 2010).

Discharge duration: Discharge duration is the time period that an ESS can discharge at its rated output power in a single charge/discharge cycle. It is an important criterion to determine the technical and economic viability of an ESS for a given application (Lichtner et al., 2010). Different energy storage technologies can be compared based on their rated power, energy content, and discharge time as shown in Fig. 4.4.

Power and energy density: Power density can be defined as the rated output power per unit mass or volume of the storage system. The energy density can be similarly defined as the amount of stored energy per unit mass or volume of the storage device. These criteria become important in situations where there are limitations on space and/or weight, such as heavily populated urban areas (Lichtner et al., 2010). Different energy storage technologies can be compared based on their power and energy density as shown in Fig. 4.5.

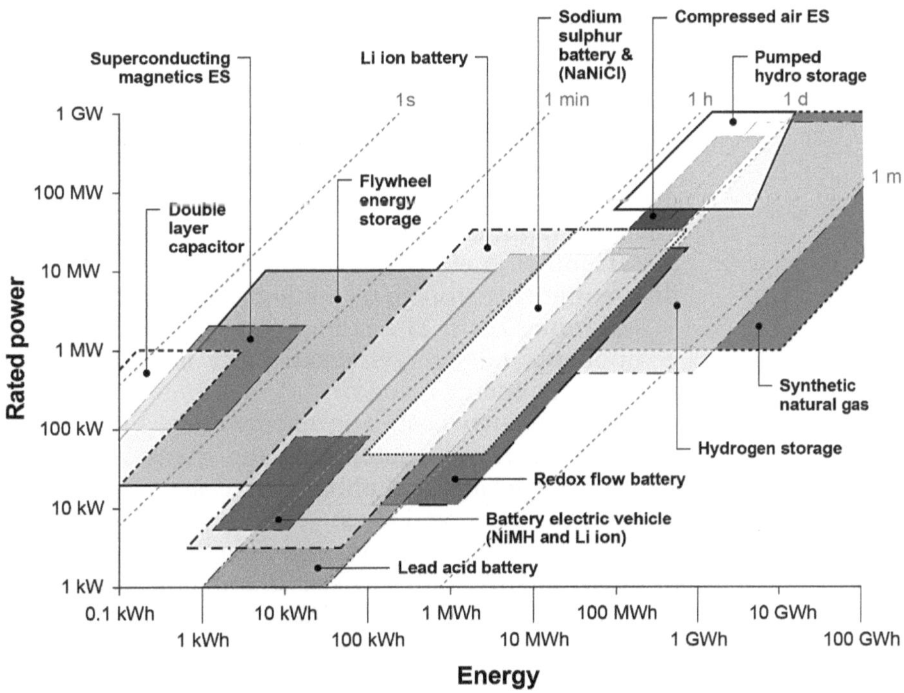

Figure 4.4 Comparison of rated power, energy content and discharge time of different energy storage technologies (IEC Market Strategy Board, 2011).

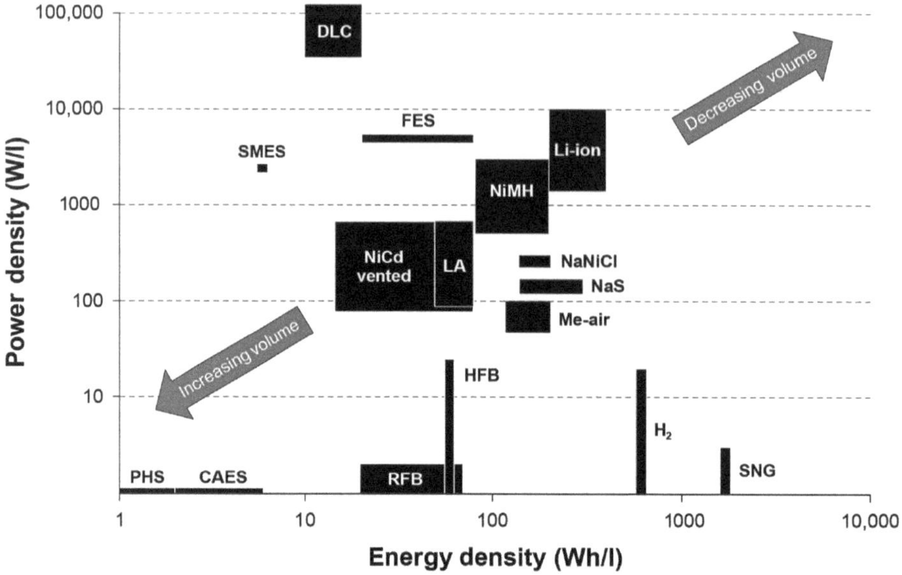

Figure 4.5 Comparison of volumetric power and energy density of different energy storage technologies (IEC Market Strategy Board, 2011).

Round-trip efficiency: Round-trip efficiency is the amount of energy that can be retrieved from the storage system divided by the electrical energy input. It is a measure of the energy lost during one charge/discharge cycle of the storage device and, hence, determines the overall usefulness of the storage technology (Lichtner et al., 2010). A comparison of cycle efficiencies of different energy storage technologies is shown in Fig. 4.6.

Operating costs: The storage operating cost consists of two components (a) energy-related costs and (b) operating costs that are not energy related.

a. Energy-related cost: The energy-related cost consists of all costs incurred in purchasing the energy used to charge the storage device including the costs associated with the round-trip energy losses.

b. Nonenergy-related cost: The elements of nonenergy-related costs include labor costs, maintenance costs, equipment costs, and decommissioning and disposal costs.

1. Labor costs: Labor is required for the operation of large storage facilities and/or blocks of aggregated storage capacity, whereas little to no labor is required for small/distributed systems that are designed for autonomous operation. Depending on the storage type and system

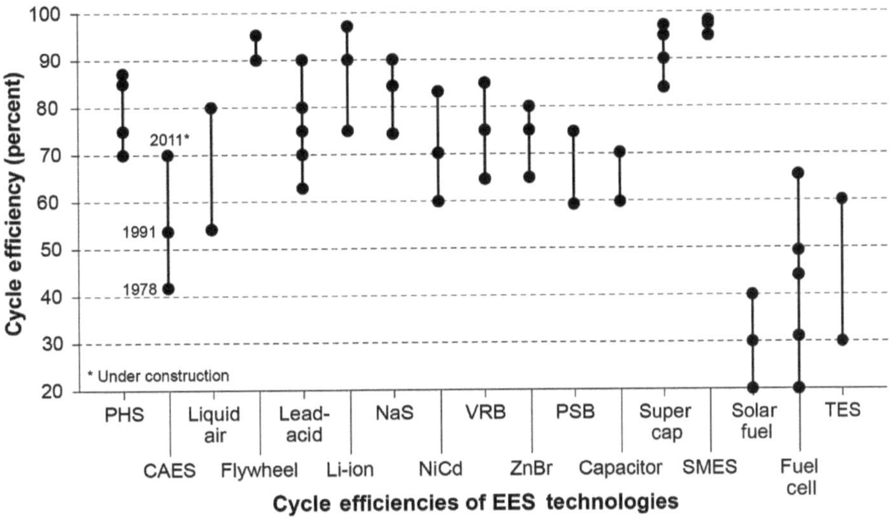

Figure 4.6 Comparison of cycle efficiencies of different energy storage technologies (Luo et al., 2015).

size there could be a wide range of labor costs that are possible.

2. Maintenance costs: Maintenance costs are incurred during normal scheduled and unscheduled repairs, and replacements for equipment, buildings, or other storage plant-related infrastructures. Maintenance costs can be fixed (the cost is the same irrespective of how much the storage is used) or variable (the cost depends on the frequency and duration of storage use) and are highly circumstantial.
3. Replacement costs: Equipment and subsystems within a storage system age over their expected lifetime. Eventually, they wear out and have to be replaced; the associated costs for replacement are termed "replacements costs." Similar to maintenance costs, replacement costs are highly variable and are dependent on the technology/circumstances. The variable costs should be minimized and that is the key to cost-effective storage technology.
4. Decommissioning and disposal costs: There will be nontrivial decommissioning costs associated with almost any storage system which should be included in the total cost to own and operate them. For example, when decommissioning a vanadium redox system, the solid ion exchange cell membranes should be disposed of in the same manner as any corrosive material. The liquid electrolyte

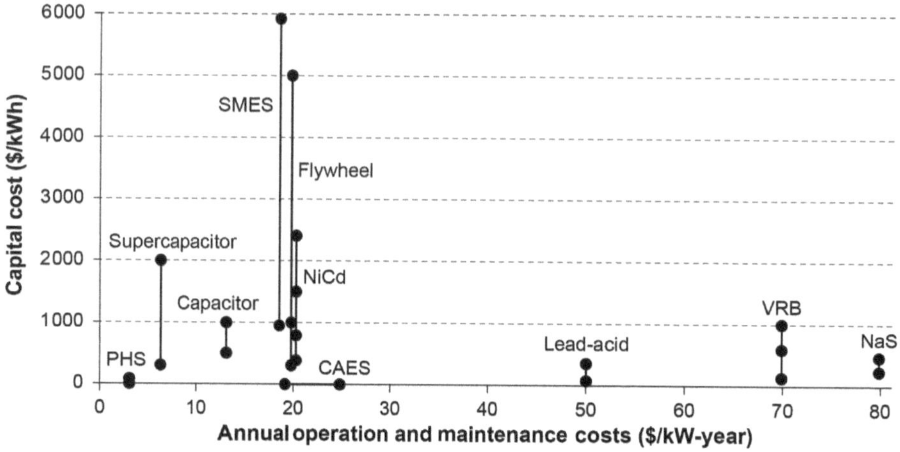

Figure 4.7 Comparison of capital, operating and maintenance costs of different energy storage technologies (Luo et al., 2015).

should be recycled; if not, vanadium should be extracted before further processing of the liquid and subsequent disposal (Lichtner et al., 2010). The capital, operating, and maintenance costs associated with the different energy storage technologies are shown in Fig. 4.7.

Lifetime discharges: Energy storage media degrade with each charge/discharge cycle. The rate of degradation depends on the type of storage technology, operating conditions, etc. For some storage technologies—especially batteries—the extent of discharge will also have an effect on their useful life; deep discharge, where most or all of the energy is drained, is more damaging when compared to shallow discharge. Vendors may be able to manufacture batteries with superior performance, increased cycle life and tolerance to deep discharges. Regardless, the storage medium needs to be replaced after their expected useful life and the costs to replace them should be included in the variable operating costs of the storage system (Lichtner et al., 2010).

Response time: Response time is the amount of time required for an ESS to be brought online and discharge energy. The transmission and distribution requirements change instantaneously in an electric grid and the storage system should respond to these changes quickly. Most types of storage have a response time in the order of several seconds or less; compressed air energy storage (CAES) and pumped hydroelectric storage (pumped hydro) tend to have slower response times but are quick enough for the several applications. In contrast, the response times of engines/combustion turbines to changes in demand are in the order of several seconds to minutes, or even hours in the case of coal and nuclear energy. Response time tends to be representative of the ramp rate; the ramp rate is defined as the rate at which the output power of the storage system can change (Lichtner et al., 2010).

Charge rate: Charge rate is the rate at which the storage media can be charged. If the storage device cannot re-charge quickly enough, it may not have sufficient energy to meet the requirements of a specific application. The charge rate can be similar, less or more than the rate of discharge and is dependent on the equipment capacity, condition, and underlying physics/chemistry of the storage technology (Lichtner et al., 2010).

Energy retention and standby losses: Energy retention is the amount of time that a storage device can retain its charge. Storage devices have a tendency to self-discharge or otherwise dissipate energy while not in use. The energy lost is referred to as standby losses. Each chemistry is different in terms of the underlying chemical reactions and the rate of self-discharge. Technologies that are based on chemical media are prone to self-discharge while the ones that use mechanical means to storage energy are prone to energy dissipation. In pumped hydro, energy losses occur in the form of water evaporation and CAES may lose energy due to air escaping from the reservoir. Standby losses are less of a concern when the ESS is used frequently than when there are extended periods of inactivity (Lichtner et al., 2010).

Transportability: Some locational benefits only last for 1 or 2 years. Transportability can be especially useful if the storage can be relocated to where it is most needed and/or where the benefits are most significant. Transportability can ultimately improve the cost to benefit ratio of ESS (Lichtner et al., 2010).

Modularity: Modularity enables design flexibility and allows for easy scaling. The system can be adapted to suit changing customer needs, load growth, etc., by increasing/decreasing the capacity when required. It is an attractive feature especially to the utilities to manage uncertainties and risks associated with transmission and distribution investments (Lichtner et al., 2010).

4.1.5 Chapter Outline

Section 4.2 of this chapter will provide an overview of the different energy storage technologies including both deployed and emerging technologies. The various applications of ESS in an electric grid, challenges to their widespread deployment, trends/future outlook, and market/regulatory trends are discussed in Sections 4.3, 4.4, 4.5, and 4.6 of the chapter, respectively.

4.2 Energy Storage Technologies

4.2.1 Overview of Technologies

Interest in developing energy storage technology is growing every day. Numerous energy storage technologies presently span the development lifecycle, from early research to widespread deployment. The need for energy storage that is integrated into the power grid has become obvious to stabilize power delivery during unpredictable, high-demand times, both within a single day and across months.

As shown in Fig. 4.8, pumped hydro technology leads the field in deployment, holding approximately 95–99% share of the market (Gyuk et at., 2013). However, pumped hydro installations are limited by geographical accessibility and are large scale, which makes installations, such as the one shown in Fig. 4.9, inherently expensive and not universally realizable.

There are numerous complementary technologies that span the full spectrum, from mature technologies to those that are only just emerging and are not yet widely deployed and in commercial use. Energy storage technologies include pumped hydro, CAES, flywheels, superconducting magnetic energy storage (SMES), electrochemical capacitors (EC), hydrogen electrolyzers coupled with fuel cells, synthetic natural gas (SNG)

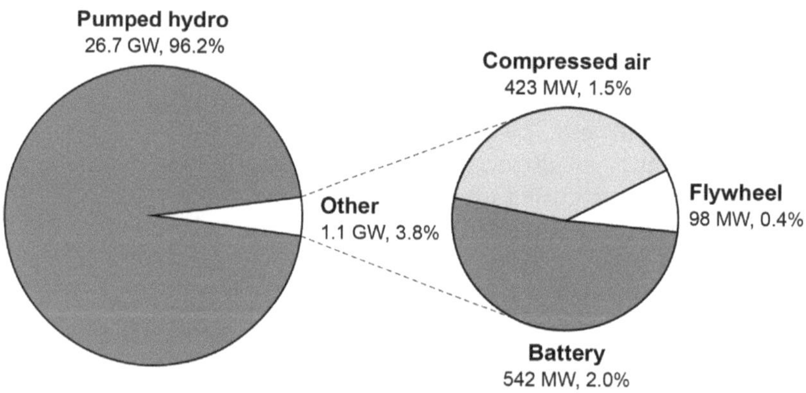

Figure 4.8 Market share of energy storage technologies as of 2013 (Gyuk et al., 2013).

Figure 4.9 Geographical limitations of pumped hydroelectric energy storage.

and numerous battery technologies, including lead-acid, lithium-ion (Li-ion), sodium batteries (e.g., sodium sulfur (NaS)), nickel batteries (e.g., nickel cadmium (NiCd), and nickel metal hydride (NiMH)) and flow batteries (e.g., vanadium redox). There are new developments being made every day to enhance more mature technologies, thus making them more flexible, and to bring down costs of emerging technologies, making future deployment more feasible.

4.2.2 Pumped Hydro Energy Storage

Pumped hydro energy storage has existed since the late 19th century and comprises the use of two dammed water reservoirs in which water is pumped from the lower reservoir to the upper reservoir during off-peak, low-cost hours. The water is then allowed to flow back down to the lower reservoir powering a turbine coupled with a generator to produce energy on demand. Pumped hydro is a highly matured technology and currently has the most widespread deployment among energy storage technologies. An example of a pumped hydro installation is shown in Fig. 4.10.

Figure 4.10 Pumped hydroelectric power.

Advantages of pumped hydro are that it is a highly developed, mature technology. Its high energy storage capacity makes it ideal for energy management. However, it has a slow response time making it ill-suited for peak power regulation. Pumped hydro is currently the most cost effective form of energy storage once a plant is up and running, but the initial investment cost is extremely high and installations are only feasible in specific geographically remote locations. The geographical limitations also require that the travel distance to the grid can be significant, and the environmental impacts cannot be ignored. Efficiencies can range from 76% to 85% and advantages of pumped hydro plants are that they have a long lifetime (approximately 50–60 years) and unlimited cycle stability (Gyuk et al., 2013). Even in mature technologies such as pumped hydro, new research developments are continuously pursued, including the use of variable speed pumping to enhance its ability to be integrated with other technologies.

4.2.3 Compressed Air Energy Storage

CAES is another well-developed mature technology offering capacities reaching upwards of 400 MW with discharge durations in the tens of hours (Pasadena Water and Power, 2014). While CAES does not see the same widespread deployment as pumped hydro, it is similarly an energy management technology, storing high energy at a relatively low cost. In a CAES system, air is either stored in an underground cavern, like what is shown in Fig. 4.11, or above ground in vessels. During off-peak,

Figure 4.11 Compressed Air Energy Storage (CAES).

low-cost hours, air is pumped into storage and then during peak, high-cost hours, energy is extracted from the pressurized air by mixing the compressed air with natural gas, burning it and allowing it to expand in a modified gas turbine. CAES systems have a large capacity and high reliability, and they can start without extraneous power. Like pumped hydro, CAES systems also have geographical requirements which limit their deployment and they also have a relatively high environmental impact. A disadvantage of CAES systems is their lower efficiency, around 50%, due to round trip conversion of energy into pressurized gas and then back out again.

Another advantage of CAES plants is their ability to allow successful integration of significant amounts of renewable resources while enhancing the transmission system and providing grid stability during intermittent operations. CAES technology can also provide essential grid services, such as energy regulation, capacity reserve, and reactive power for voltage support. These plants have the size and flexibility to enable energy management applications to maintain and improve the grid system reliability and relieve transmission congestion. Energy storage like CAES enhances the grid by making the grid more efficient, especially by facilitating the integration of renewable energy generation technologies, which will assist in achieving the full potential of renewables and will provide an industry model for a grid-enabled diversified energy portfolio.

4.2.4 Flywheel

Another mechanical technology is the flywheel, which was introduced in the 1970s, an example of which is shown in

Figure 4.12 Flywheel.

Fig. 4.12. Advancements have been made in flywheel technology, allowing advanced designs to spin in the tens of thousands of revolutions per minute. Compared to other mechanical technologies, including pumped hydro and CAES, the flywheel is a fast-responding, short-duration discharge technology used primarily for frequency regulation. A flywheel stores rotational energy in an accelerated rotor. When there is a surplus of electrical energy, the flywheel is accelerated, and when energy needs to be extracted from the flywheel, it is permitted to slow. Flywheels respond rapidly to frequency changes, which make them great for stabilizing frequency when integrated into the power grid.

The primary advantages of flywheels are their rapid response, high peak power output and long cycle life. Additionally, they have high cycle stability, long lifetime, little maintenance, high power density, and use environmentally inert material. However, compared to other technologies, flywheels have a relatively low energy density and are not feasible for longer term energy storage. Additionally, flywheels have a low current efficiency and high level of self-discharge because of air resistance and bearing losses.

4.2.5 Lead–Acid Batteries

Lead–acid batteries are a mature technology because they have existed since the late 19th century. They have a low cost for their high performance, they are easily recycled, and they are simple to charge. A cartoon depiction of a lead–acid battery is shown in Fig. 4.13.They have a relatively high efficiency of

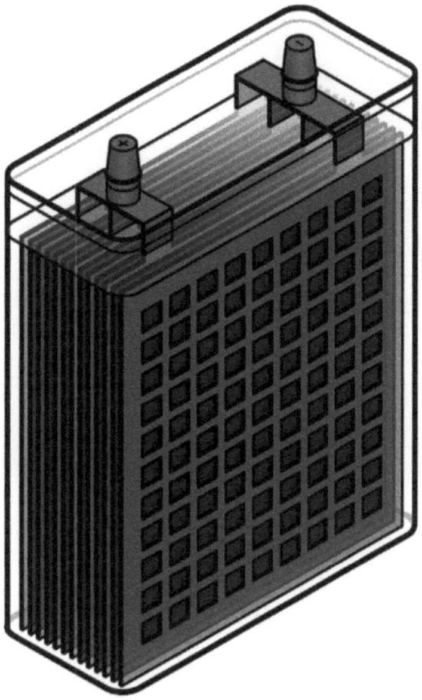

Figure 4.13 Lead–acid battery.

approximately 80–90% (IEC Market Strategy Board, 2011). However, their usable capacity decreases when high power is discharged. For this reason, lead–acid batteries have a limited depth of discharge. They have a typical cycle life of approximately 1500 cycles at 80% depth of discharge (IEC Market Strategy Board, 2011).

There are different types of lead–acid batteries, including vented designs and maintenance-free valve-regulated lead–acid batteries. Lead–acid batteries have seen relatively wide deployment due to their low cost and relatively long service life, upwards of 15 years, and there are advancements still being made to lead–acid battery design, which could enable them to become a more feasible option for energy management. One disadvantage of lead–acid batteries is their dependence on hazardous and restricted lead. Additionally, they have a relatively lower volumetric energy density, which makes their deployment somewhat impractical for energy management applications, and so they are primarily used for power applications.

4.2.6 Lithium-ion Batteries

Li-ion technology has become widely deployed in the portable consumer electronics world since the 1990s and it has

Figure 4.14 Lithium-ion battery.

more recently become the go-to energy storage method for electric vehicles due to its extremely high energy density-to-weight ratio. Li-ion batteries, shown in Fig. 4.14, can utilize a variety of electrode and electrolyte chemistries, but they all contain a cathode comprising a lithiated metal oxide, and a carbon-based anode. To charge the battery, Li-ions travel from the cathode to the anode through electrolyte and reversibly deposit onto the carbon-based anode such that they are able to travel back to the cathode during discharge.

Continuing advancements in Li-ion battery construction and larger-scale manufacturing make it feasible for larger-scale power grid applications. However, manufacturing costs are still extremely high compared to other technologies and essential circuitry must be included to safely manage operation due to the highly restricted charge and discharge requirements. Li-ion battery state of charge must be precisely managed in order to keep them safe and to prevent over-discharging. Li-ion batteries have a higher nominal cell voltage over other battery technologies (3.7 V).

Versatility in electrode design and scalability also enables Li-ion batteries to be catered to nearly any power and discharge requirements with discharge duration capability from seconds to weeks. Enabled by their versatile design capabilities, Li-ion batteries can have drastically differing discharge durations. Li-ion batteries are an excellent technology for power quality and frequency regulation and they also offer great potential for energy management. While roundtrip efficiency is high (near or above 95%), cycle life is adequate for grid applications (~5000 cycles) (IEC Market Strategy Board, 2011). While costs are still currently high, continuing technological and manufacturing improvements, including recycling, will enable Li-ion batteries to become ever more beneficial to grid applications.

4.2.7 Sodium Sulfur Batteries

NaS batteries are a unique technology that is well suited for both power and energy management grid applications, including the integration of renewables into the grid. NaS batteries comprise molten sulfur at the positive electrode and molten sodium at the negative electrode, which requires that they are kept between 300°C and 350°C. The active materials are separated by a solid electrolyte, which permits the movement of sodium ions during discharge from the negative electrode to the sulfur at the positive electrode. NaS batteries have a cycle life upwards of 4500 cycles, and a discharge time around 6–7 hours (IEC Market Strategy Board, 2011). Their high volumetric energy density makes them candidates for longer timescale applications, such as grid stabilization, and their fast response (on the order of milliseconds) makes them good also for power quality and frequency regulation. One of the disadvantages of NaS technology is their high cost. They require an external heat source because they cannot be allowed to cool or they may otherwise undergo damage, which requires backup heat sources as well in an installation. Another challenge is the required sealing to prevent sodium from being exposed to moisture and to prevent corrosion.

4.2.8 Nickel-Based Batteries (NiCd and NiMH)

Nickel-based rechargeable battery technology has achieved substantial market penetration for numerous applications, due especially to their great performance at very low temperatures. NiCd is a well-established technology that was introduced in the early 20th century. However, advancements in both power and energy density came with the development of NiMH technology as a potential replacement for NiCd, which contains highly toxic cadmium, and while NiMH technology costs on the order of Li-ion, it is a safer battery. Due to their wide-operating temperature range, high cycle life, and reliability within relatively extreme environments, nickel-based batteries will remain irreplaceable in the foreseeable future (Office of Basic Energy Sciences, 2007).

4.2.9 Flow Batteries (Vanadium Redox)

A flow battery is one in which two liquids, separated by a membrane and circulated in order to enable ion exchange between them. By isolating the liquid electrolytes, flow batteries offer a long cycle life and can make use of their complete discharge range. Additionally, they have a low internal discharge rate. However, compared to other battery technologies, flow batteries have a lower energy density and their physical constraints make it more challenging for them to compete with other more flexible battery technologies. There are numerous flow battery constructions, many of which are still early in the development stages. One such example is the vanadium redox battery (VRB) that utilizes vanadium-based electrolytes. One advantage is the flexibility in capacity introduced by this technology that depends on the physical size of the electrolyte tanks in that capacity can be increased with increased tank size. VRB flow batteries offer lifetimes upwards of 10,000 cycles (IEC Market Strategy Board, 2011).

4.2.10 Superconducting Magnetic Energy Storage

One emerging technology is SMES, which requires extremely low temperatures (single digit Kelvin) to surpass the critical temperature for superconductivity in a coil. Energy is stored in the induced magnetic field as Direct Current (DC) current flows, essentially resistance free, through the coil. Energy is then extracted when the coil is discharged. Advantages of this developing technology include a high cycle life and an extremely high round trip efficiency (>95%) that challenges most other technologies (IEC Market Strategy Board, 2011). SMES is designed for short duration discharges of high peak power, which makes it well suited for power quality and frequency regulation. However, like many other cutting edge technologies, it is currently cost prohibitive to manufacture, especially because of the high-cost refrigeration units and power consumption required to maintain the units in single Kelvin operating temperatures (Gyuk et al., 2013). Additionally, SMES faces competition from other technologies which rival SMES peak power but do not require superconductivity, nor the energy cost of extreme cooling. Like other technologies that are well suited for power management, SMES is not well suited for longer discharge duration, energy management applications.

4.2.11 Electrochemical Capacitors—Power

While energy storage technologies utilize a vast variety of storage media, ECs store energy directly as charge, which enables them to have high efficiency and cycle life (Office of Basic Energy Sciences, 2007). EC technology has been known for decades but has been utilized primarily in small-package consumer and power electronics (IEC Market Strategy Board, 2011), and EC technology is still trying to push its way into larger-capacity applications, e.g., in electric vehicles and grid energy storage. Energy can be discharged quickly from ECs because there is no energy conversion required, which makes ECs excellent candidates for power applications, such as frequency regulation and voltage stabilization (Gyuk et al., 2013). Like other emerging technologies, however, EC remains expensive and will require further advancements in manufacturing technology to reduce costs within feasible limits.

4.3 Energy Storage Applications

4.3.1 Overview of Energy Storage Applications

The vast diversity of technologies offers great opportunity to cater energy storage to a variety of applications and their requirements, including discharge response time, discharge duration, and total capacity. Because of the very different construction and performance of each of these technologies, they are each suited for different grid applications, which can be broadly divided into power applications requiring very fast response times and energy applications requiring higher total energy capacities. Power management applications include frequency regulation, power quality improvements, and time shifting, e.g., to address voltage faults. These applications require fast response, short duration discharges, and high power. Energy management applications include grid stabilization, backup-energy reserves, and integration of renewables into the grid. These applications require high-energy capacity to support longer duration discharges, which energy management technologies can sustain by storing energy during off-peak hours and supplying energy during peak demand (Fig. 4.15).

Established large-scale technologies, such as CAES and pumped hydro, are capable of long discharge times and incredibly high energy capacity. Pumped hydro installations have been scaled upwards of 4000 MW with discharge durations lasting on the order of hours to days (Gyuk et al., 2013), which makes

Figure 4.15 Energy storage systems can be used for power management and energy management applications for the electric grid, solving many of the issues associated with long-distance power transmission.

these technologies ideal for grid energy management. In contrast to the capabilities of energy management-focused technologies, are power management–focused technologies, such as flywheels and SMES. They are ideal for short discharge durations on the order of milliseconds to hours, making them optimal for improving grid power quality, frequency regulation, and time shifting. Several different electrochemical battery technologies, such as Li-ion and sodium sulfur batteries, cross into both energy and power management application facilitated by the flexibility in their design and performance, which includes variable discharge durations. The complexity of the smart grid, as shown in Fig. 4.16, underlines the benefits of energy storage integration.

4.3.2 Energy Management Applications

4.3.2.1 Grid Stabilization

As the power grid becomes more distributed, the maintenance of a stable power grid becomes increasingly more concerning. Highly variable loads have long been a power grid challenge, especially near the end of lengthy transmission lines.

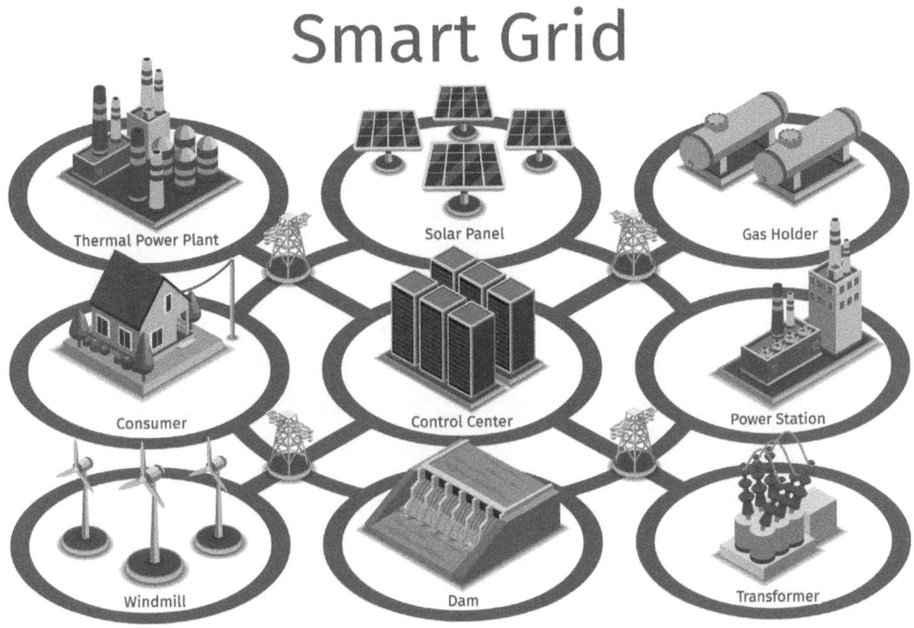

Figure 4.16 The smart grid integrates both variable loads and time-dependent supplies.

However, a highly variable power supply is a newer concept. Large capacity energy storage technologies, like pumped hydro and CAES, may be able to support the grid in longer duration power supply lags in order to maintain a stable grid.

4.3.2.2 Renewables Integration

The massive push to make major investments, both at the government and consumer level, in renewable energy generation has created a more diverse power grid with a need for more sophisticated power supply schemes. The ever-changing economic landscape and worldwide environmental outlook has prompted the further development of numerous renewable technologies and installations. However, integration of these variable and unpredictable power sources into the power grid introduces added complexity that has not previously been required. End users must now be able to both take and return power to the grid, and energy storage technologies can add grid flexibility to make renewables integration, such as the solar installation shown in Fig. 4.17, seamless.

Figure 4.17 Integrating renewable technologies, like solar cells, into the power grid is one of the ways energy storage technologies can add grid flexibility.

4.3.2.3 Backup Energy Reserves

Due to the very different energy demands at different times of the day it is important to store energy, when possible, during the off-peak hours when energy costs are lower. Energy storage, such as pumped hydro and CAES, can be replenished during low-demand times and can be stored for significantly longer durations than other energy storage technologies. This enables backup energy reserves to be utilized in the most demanding of times when the grid is stressed, such as in the summer months.

4.3.3 Power Management Applications

4.3.3.1 Power Quality

To ensure efficient grid operation, it is essential that under changing-power demands across hours and physical locations, grid power is supplied at as near to ideal voltage and frequency as possible. Numerous energy storage technologies can be rapidly turned on and off to effectively balance both real and apparent power, especially to safeguard against abrupt reactive load changes such that they do not drastically alter the power supply.

4.3.3.2 Frequency Regulation

Frequency regulation and needed stabilization requires fast-responding energy storage technologies to counteract any sudden loss of power generation. In the event of a short-term upset in the power grid, fast responding technologies, like flywheels and batteries, can react to maintaining grid power frequency.

4.3.3.3 Time Shifting

Similar to its larger capacity counterpart, the building of backup energy reserves, time shifting is the utilization of energy storage technologies to store any—and ideally all—excess energy generated at any time of the day, especially from renewable resources such as wind and solar power. Since solar power, e.g., is generated only during daylight hours, if there is any excess production, it will be lost if not for energy storage technologies. Often the excess power generation is minimal compared to what is possible with backup energy reserves like pumped hydro, and so technologies like battery technologies, which are shorter duration discharge, higher power outputs can be ideal to store any excess energy generated even during peak hours in order to deliver it back to the grid even within the same day.

4.3.3.4 Load Following

Especially in distributed grid architectures, in the event that a local load cannot be maintained by the present power supply, energy storage technologies must be fast acting to respond to sudden load changes. In this application, energy storage technologies must be able to operate at variable capacity in order to precisely follow load demands without high losses. During peak demand hours, generation (or supply) must increase output to accommodate demand, and during off-peak hours, supply must variably decrease output to respond accordingly to demand.

4.3.3.5 Peak Shaving

Similar to other grid applications, peak shaving is the utilization of energy storage to consistently reduce the costly energy requirements on the grid during peak demand hours. Effective peak shaving requires technologies with somewhat longer discharge durations that may be able to support multiple-hour discharges with relatively large energy requirements.

4.3.3.6 Transient Stability

When the power grid is subjected to unexpected changes, such as sudden large load variations or down generators or power lines, grid stability can be jeopardized due to the inability of the grid to respond quickly enough to maintain power quality. The presence of grid-integrated fast-responding energy storage technologies can absorb the burden of transients.

4.4 Key Challenges to Widespread Deployment

Despite the advances in ESS as a means of supporting a smart grid, widespread deployment has not yet been achieved. The reasons for this lack of deployment are numerous and have been studied by numerous organizations, including the Department of Energy (Gyuk et al., 2013). The main challenges that have been identified for widespread deployment are cost competiveness, reliability and safety, regulatory hurdles, and limited industry acceptance.

4.4.1 Cost Competitiveness

Before ESS can become a viable support for a smart grid, it must attain cost competiveness with current grid technologies. According to the US Energy Information Administration, the average price of electricity to ultimate customers across all usage sectors in April 2016 was 9.81 cents/kWh (US Energy Information Administration, 2016). This ranged from 6.39 cents/kWh for industrial customers to 12.43 cents/kWh for residential customers. Meanwhile, a November 2015 study on levelized cost of storage by Lazard shows that the price of energy storage scales strongly with system size (Lazard, 2015). A 5 kW Li-ion-based storage system would cost roughly $1.03/kWh, whereas a 100 MW Li-ion based storage system would cost just 18.8 cents/kWh. This is still roughly double the cost of current grid electricity, but efforts are underway to decrease these costs to bring energy storage technology to grid parity. This includes research into new storage technologies (like flow batteries), discovery of new battery chemistries, improvements to current technologies, improvements in cost effectiveness of nonstorage elements in ESS, and value-adding through "stacked" services (Gyuk et al., 2013).

The strong energy storage research community has pursued new and improved technologies with greater and greater fervor as public interest in ESS increases. Beyond the lab, the establishment of economies of scale has resulted in dramatic cost reduction of ESS. For example, the cost of Li-ion battery packs used in electric vehicles, shown in Fig. 4.18, and home energy storage applications has decreased by 14% ± 6% from 2007 to 2014, with further decreases expected as larger facilities come online in the future (Nykvist & Nilsson, 2015).

There are many kinds of ESS, and the maturity of these systems varies widely. As stated by the US Department of Energy in their overview of development and deployment of codes and standards, some ESS, such as pumped hydro, have been well-established for some time with few major changes in recent times and are considered safe and reliable, limiting the need for continuous updates to codes and standards. Others, like batteries and fuel cells, are evolving at a rapid pace, and the standards have not managed to keep pace given the slow pace at which standards are written and adopted, which is on average at least 6 years (Conover, 2014; Fig. 4.19).

Strong reliability and safety standards can improve perception and comfort with ESS, and can result in quicker adoption. As such, a unified and concerted effort to keep codes and standards in step with advances in ESS technology is paramount. The US Department of Energy has hosted workshops to bring researchers and industry contacts together (US Department of

Figure 4.18 The popularity of electric vehicles has helped to drive down the cost of lithium-ion batteries.

Figure 4.19 Strong reliability and safety standards can help to prevent failures like the October 2012 blackout in New York City following Hurricane Sandy.

Energy Office of Electricity Delivery and Energy Reliability, 2015). Large professional organizations, such as the Institute of Electrical and Electronics Engineers (IEEE), also host standards meetings with relative frequency (IEEE-PES ESSB, 2016). A step forward for this process would be a governing organization to continuously sample and inform the field on the state of ESS codes and standards, identifying the gaps between current and future codes and standards. This sort of organization, coupled with a feedback process from members, would allow codes and standards to close the identified gaps between current ESS and the latest research and developments, while continuing to identify any incongruities that arise. Undoubtedly, as ESS gain traction and get deployed in more populated areas, the codes and standards will be forced to evolve faster, as there will be increased pressure to ensure the safety and reliability of the system relative to a more remote deployment area.

4.4.2 Other Challenges

4.4.2.1 Regulatory Hurdles

As interest in ESS grows, the need to address regulatory issues has become more and more pressing, with greater penetration levels requiring more significant regulatory changes (Stanfield and Vanega, 2015). One of the major issues hindering the incorporation of ESS into regulatory frameworks arises from the dual nature of ESS as both an electricity consumer and supplier (Deloitte, 2015). Focusing on ESS as a supplier results in two kinds of uncertainties, jurisdictional uncertainties and cost recovery uncertainties (Stein, 2014). As explained by Prof. Amy L. Stein, jurisdictional uncertainty exists in determining whether ESS constitute wholesale or retail power, with the former requiring federal regulation and the latter requiring state regulation; cost recovery uncertainty is more involved. It is based on the labeling of ESS assets as generation, transmission, or distribution facilities, each of which have different capacities for cost recovery from the initial capital investment. Some forms of ESS can only achieve cost-effectiveness when classified within multiple asset categories and being compensated accordingly, entering an area that regulatory frameworks have not yet defined (Stein, 2014). The current regulations are misaligned, and this prevents ESS from being utilized in areas where they could be most useful.

According to the Interstate Renewable Energy Council (IREC), the clearest path forward for proper regulatory reform that can accommodate ESS involves four key actions (Stanfield and Vanega, 2015):

1. Demonstrating interest in storage
2. Clarifying existing rules as they apply to storage
3. Stimulating the storage market
4. Including storage in broader context when planning for the future

These actions can help to improve investment in ESS, by placing them onto an equal playing field with current generation, transmission, and distribution methods. Once the regulatory landscape becomes more favorable, widespread deployment of ESS becomes an easier task.

4.4.2.2 Limited Industry Acceptance

As can be expected with any disruptive technology, the electric power industry has been slow to accept ESS for a number of reasons. Much of this stems from limited experience and knowledge of how ESS can be deployed and what technological improvements have been made. For example, there is little understanding of how to standardize interfaces and controls for ESS with respect to current technologies (Ecofys, 2014). The task of retooling entrenched systems or the effort necessary to develop new technologies that are compatible with existing systems is a large barrier to acceptance. These existing technology biases have wide-reaching consequences, as they prevent further investment in ESS research, and many innovations are unable to make the jump from lab-scale to utility-scale.

The US Department of Energy believes that the way to move past this barrier lies in the development of better planning tools that can harness the benefits of ESS, and the development of algorithms to properly employ these resources in such a way as to improve profitability (Gyuk et al., 2013). Ultimately, this requires coordination between industry, regulators, and researchers.

4.5 Trends and Future Outlook

4.5.1 Introduction

Energy storage has already been commercially deployed for some applications and is poised to become an important element in the electricity market of the future. Grid-scale ESS are uniquely suited to address the needs of existing and emerging electric utility-related opportunities and challenges as discussed in the earlier sections of this chapter. The major drivers that determine the future of ESS include, but are not limited to, the following (Hostick et al., 2012; IEC Market Strategy Board, 2011; Lichtner et al., 2010; IEC Market Strategy Board, 2012):

- Growth in demand for electricity
- Carbon mitigation, high energy efficiency, and increased use of renewables
- Need for an improved power quality and reliability
- Economics of storage—technologies that are cost effective and competitive
- Policies, regulations, and legislations that would provide incentives for storage
- Research and development that leads to innovation in storage technologies
- Sophistication and savvy of load and distributed resource aggregators
- Long-term energy security
- Proliferation of plug-in electric and hybrid electric vehicles
- Increased use of modular distributed energy resources
- An increasingly smart electricity grid will enable effective integration and dispatch of renewables and distributed resources

The storage opportunity involves numerous stakeholders. Understanding their interests and relationships are critical since the benefits do not all accrue to the same stakeholder. Storage projects may require a significant amount of coordination and cooperation between the diverse stakeholders. Also, the benefits need to be aggregated into attractive value propositions because in many situations, two or more benefits are required to offset the total investment cost (Lichtner et al., 2010).

There are a few key barriers to widespread deployment of energy storage. It is critical that we understand these barriers before evaluating the prospects for energy storage in general and specific storage technology, application, value proposition, or project. Some notable challenges include, but are not limited to, the following (Gyuk et al., 2013; Lichtner et al., 2010):

- Cost-competitive storage system—currently, storage has a relatively higher $/kW installed
- Limited familiarity, knowledge, and experience base; standards and evaluation methodologies/tools need to be developed and codified
- Need for an equitable regulatory environment
- Industry acceptance
- Limited risk/reward sharing mechanisms
- Financing of new technology is challenging
- Coordinating among numerous stakeholders may be expensive and time-consuming

4.5.2 Technology Trends

The grid energy storage market is strong and is set for further growth. A study performed by Navigant Research indicates that the global market for utility-scale energy storage is expected to grow from $675 million annually in 2016 to $15.6 billion annually in 2024. As of 2013, pumped hydropower was dominant as the storage technology of choice with a market share, shown in Fig. 4.8, of about 96% of the deployed and anticipated energy storage projects in the United States, while CAES, batteries, and flywheels account for the remaining 4%. Similar trends have been observed internationally (Baldwin et al., 2015).

It should be noted that no energy storage technology scores high in all dimensions (Simbolotti and Kempener, 2012). Each storage technology has its own performance characteristics that make it better suited for a specific grid application when compared to others (Gyuk et al., 2013). The storage technology of choice depends on the system capacity, application, marginal cost of peak electricity, etc. A variety of complementing technologies will be required to fully address the needs of an electric grid with large-scale integration of renewable energy.

Pumped hydropower, electrochemical batteries, and hydrogen/SNG seem to emerge as the technologies of the future. Pumped storage is a mature technology with a proven track record and is the cheapest option for large-scale energy storage (>50 MW). The lack of suitable installation sites and local environmental impact are the barriers to their further expansion. Battery systems have emerged as a feasible alternative for applications less than 50 MW. However, their widespread deployment has been challenged by significant uncertainty over the usable life, performance, safety, and cost. Hydrogen/SNG storage technology has the potential, due to their high energy density and quick response time, but they must be further researched and developed to address concerns related to cost, safety, reliability, efficiency, and lack of physical infrastructure (IEC Market Strategy Board, 2011; Deloitte, 2015). Technology advances, combined with careful planning, improvements in design, collaborative field demonstrations, standardization, and equitable regulatory environment will result in lower storage technology cost, higher services revenue, increased user confidence, and, eventually, large scale deployments as shown in Fig. 4.20.

Renewable energy is critical to transforming electric grids to meet the challenges of the future. As the market for renewable

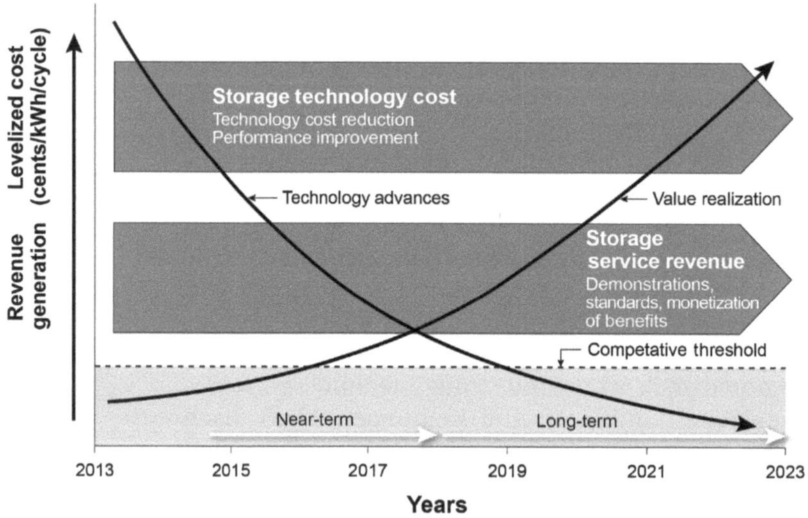

Figure 4.20 The future of storage technology (Gyuk et al., 2013).

energy grows, the need for ESS will increase. To ensure smooth connection of ESS, potential issues related to their largescale integration to the electric grid, coupled with renewable energy, should be thoroughly investigated and addressed. Also, necessary technical specifications and regulatory frameworks need to be in place. Lifecycle cost of storage technologies is critical to their widespread adoption. Significant research and development efforts need to be undertaken both at a component and at a system level to make the storage technologies more economically viable. Interoperable components and systems, sophisticated communication and automation technologies, and control algorithms must be ensured for successful project deployments (IEC Market Strategy Board, 2011).

4.6 Market and Regulatory Trends

Despite the previously discussed difficulties, many efforts are underway to improve the deployment of ESS in the United States. These efforts, at both the state and national level, are focused on improving and enacting policy in such a way that investment in ESS can be less risky, with opportunities for incentives for adoption.

4.6.1 State Level

Given the localized nature of grid energy storage, many of the relevant market and regulatory trends are occurring at the state level. Currently, 41 states have announced, begun construction on, or are currently deploying some form of grid energy storage (Sandia Corporation, 2016). The operational deployments total 24.1 GW, of which roughly 93% consists of pumped-hydropower storage. This shows a 2.7 GW increase in total capacity since 2014, and a 2% increase in non-pumped-hydropower storage technologies (NREL, 2016). With advances in both renewable generation and battery technologies, it is expected that new ESS installations will focus on battery-based energy storage, along with other advanced energy storage methods. This is evidenced by data showing that of the new ESS deployments that are under construction or have been announced, 848 MW use electrochemical (battery), electromechanical (flywheels), or thermal methods, making up nearly 15% of the planned 5.4 GW of energy storage. The breakdown of operational and planned storage capacity in the United States as of 2016 can be seen in Fig. 4.21.

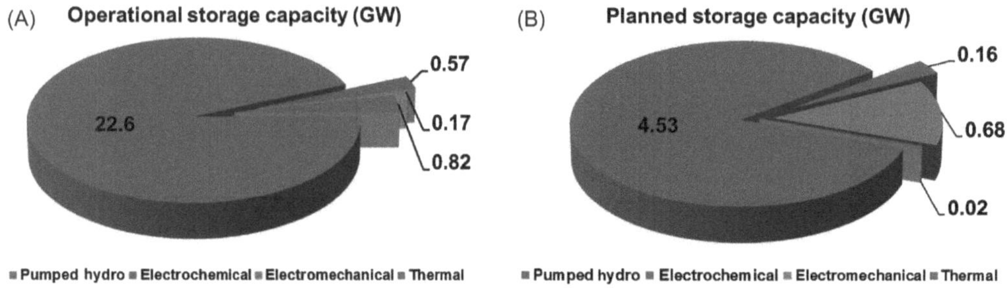

Figure 4.21 Grid energy storage capacity in the United States as of 2016. (A) Operational storage capacity. (B) Planned storage capacity (Sandia Corporation, 2016).

While it is clear that state utilities and consumers are moving toward adoption of ESS, there is still much regulatory work to be done to bolster the deployment. Since 2010, roughly 14 bills related to energy storage were introduced by at least 10 different states (NREL, 2014; Mullendore, 2016; Bovarnick, 2015). The bills tended to focus on establishing financial incentives for ESS deployment, policy planning, ESS technology potential investigations, and other methods for encouraging ESS adoption. Table 4.1 details some of the activities that have been undertaken or are underway in these states.

Thus, it is clear that the state-level regulatory practices are focused on stimulating the energy storage market through incentives.

4.6.2 National Level

Efforts at the national level have focused on creating a more even playing field for emerging energy technologies, including ESS, by introducing fair compensation schemes to incentivize investment in ESS by regional transmission operators (RTOs) and independent system operators (ISOs). The major initiatives have been through Federal Energy Regulatory Commission (FERC) orders, specifically FERC Order 719, FERC Order 745, FERC Order 755, FERC Order 784, and FERC Order 792. Each order is outlined in Table 4.2.

There have been some difficulties in implementing these orders. For example, ISO New England has yet to enact FERC Order 755, and has stated that they will implement it by 2018 (ISO New England, 2016). This is, in large part, due to legal challenges to the FERC's ability to enact and enforce these orders, with a case being brought before the US Supreme Court

Table 4.1 State-Level Market and Regulatory Policy Activities

State	Activity	Year(s)	Description
Arizona	S.B. 1465	2015	Provides disclosure requirements to be included in agreements for the sale or lease of a distributed energy generating system, including ESS (Arizona State Senate, 2015).
California	A.B. 2514	2010	Requires the California Public Utilities Commission (PUC) to establish targets for and pursue better ESS deployment at the grid scale (Skinner, 2010).
	Mandatory procurement requirements for Investor-Owned Utilities (IOUs)	2013	Requires IOUs to procure ESS equal to 1% of their peak capacity by 2020 (California Public Utilities Commission, 2013).
	CAISO roadmap	2014	Identifies actions to aid in the creation of a marketplace for ESS (California ISO, 2014).
	Rulemaking 1211005	2012–2016	Suggests and clarifies policies for fee reductions/ exemptions for ESS (California Public Utilities Comission, 2012).
Connecticut	S.B. 1078	2015	Provides regulatory tools for procuring affordable and reliable electricity with provisions for the procurement of ESS (Connecticut State Senate, 2015a,b).
	S.B. 1502	2015	Connecticut state budget, including provisions for the soliciting and evaluation of proposals related to ESS (Connecticut State Senate, 2015a,b).
Florida	SunSmart E-Shelter program	2010–present	Builds emergency shelters at Florida schools with solar + storage systems (Florida Solar Energy Center, 2014).
Hawaii	H.B. 2618	2014	Would have established tax credits (either a 20% investment tax credit or an 8 cents/kWh utilization tax credit) for the use of a grid-connected ESS with capacity >1 MW. Was not signed into law (Hawaii State Legislature, 2014).
	S.B. 2932	2014	Would have established an energy storage portfolio standard for Hawaiian electric utilities. Was not signed into law (Hawaii State Senate, 2014).
	Hawaiian electric company request for proposals	2013	Solicits proposals for a 60–200 MW ESS on the island of O'ahu (Hawaiian Electric, 2013).

(*Continued*)

Table 4.1 (Continued)

State	Activity	Year(s)	Description
Massachusetts	Energy storage initiative	2015	Attract ESS manufacturers, accelerate development, expand markets, and enact regulatory reform for better ESS deployment. $10 million budget allocation (Massachusetts Office of Energy and Environmental Affairs, 2015).
Minnesota	H.F. 3	2015	Omnibus employment, economic development, jobs, and energy bill, including a provision that utilities shall identify investments that further grid modernization, including deployment of ESS (Minnesota State House of Representatives, 2015).
New Jersey	Energy Resilience Bank	2014	Allocates $200 million to build distributed energy resources to prevent future outages (like those experienced during Hurricane Sandy) (New Jersey Board of Public Utilities, 2014).
	Renewable electric storage program	2016	Provides financial incentives for electric ESS integrated with renewable energy generation projects at non-residential sites. Current budget is $6 million (New Jersey's Clean Energy Program, 2016).
New Mexico	Senate joint memorial 43/ house joint memorial 10	2013	Requests the identification of legislative and regulatory incentives to encourage development of renewable energy storage (New Mexico State Legislature, 2013).
	Energy policy & implementation plan	2015	Outlines a 12-objective framework, including objectives to identify the feasibility of new ESS (New Mexico Energy, Minerals, and Natural Resources Department, 2015).
New York	NY battery and storage consortium	2010	Coalition of energy storage companies, universities, and regulators meant to accelerate commercial deployment of ESS in New York (New York State's Clean Energy Standard, 2010).
	"Reforming the Energy Vision" initiative	2014	Directs the New York Public Service Commission to pursue regulatory, market, and technology reform to implement more renewables, including ESS (New York Department of Public Service, 2016).

(*Continued*)

Table 4.1 (Continued)

State	Activity	Year(s)	Description
Oregon	H.B. 2193-B	2015	Requires Oregon electric companies with >25,000 customers to procure between 5 MWh and 1% of peak capacity in energy storage by 2020 (Oregon State House, 2015).
Texas	S.B. 943/PUC Project #39657	2011	Affords ESS projects the right to interconnect, the right to obtain transmission service, and the right to sell at wholesale (Texas State Senate, 2011).
	PUC Project #39917	2012	Defines the energy used while charging an ESS as a wholesale resource (Public Utility Commission of Texas, 2012).
Vermont	H.B. 40	2015	Creates a renewable energy standard with requirements that start in 2017, including the establishment of infrastructure for ESS for renewable energy storage (Vermont State House, 2015).
Washington	H.B. 1289	2013	Allows qualifying utilities to count electricity dispatched from an ESS charged by a renewable source at 2.5X the output (Washington State House, 2013a,b).
	H.B. 1296	2013	Requires an electric utility to include ESS assessments in its integrated resource plan (Washington State House, 2013a,b).
	H.B. 1115	2015	Establishes a $13 million grant to advance clean and renewable energy technologies, including ESS (Washington State House, 2015).

(SCOTUSblog, 2016). The US Supreme Court upheld the FERC's jurisdiction in a 6–2 decision, showing that support for energy storage extends beyond just regulatory bodies.

Furthermore, strides have also been made on the legislative front. In May 2015, Sen. Martin Heinrich (D-NM) introduced US Senate bill S.1434—the Energy Storage and Deployment Act of 2015, which set a mandate for electric power suppliers that sell at least 5 GWh of electricity annually to have energy storage equal to 1% of their average peak power demand by 2021, and 2% by 2025 (US Senate, 2015). Then, in May 2016, Rep. Michael M. Honda (D-CA-17) introduced US House of Representatives bill H.R.5350—the Energy Storage Act of 2016, which introduced

Table 4.2 FERC Orders Pertaining to Energy Storage

FERC Order Number	Year	Purpose
719	2008	Requires that RTOs (regional transmission operators) and ISOs (independent system operators) accept bids from demand response resources to participate in providing energy and ancillary services, while also requiring a 5-minute calculate of prices for energy and ancillary services. Helped establish ESS as a viable demand response resource (Federal Energy Regulatory Commission [FERC], 2008).
745	2011	Requires RTOs and ISOs to pay market price to demand response resources, putting it on equal footing with generation resources in wholesale markets. Incentivizes ESS by establishing fair compensation (FERC, 2011a,b).
755	2011	Requires RTOs and ISOs to pay for performance by compensating demand response resources more for their services in frequency regulation. Incentives ESS by offering better compensation (FERC, 2011a,b).
784	2013	Opens ancillary services market to third-parties. Expands FERC Order 755 pay-for-performance requirements for RTOs and ISOs to take into account speed and accuracy of frequency regulation resources. Specifically mentions ESS with regards to accounting and reporting requirements (FERC, 2013a,b).
792	2013	Expanded scope of Small Generation Interconnection Agreements and Procedures to include ESS. This allows ESS to receive the same rates and terms for interconnection with the grid (FERC, 2013a,b).

investment tax credits for ESS installed at homes and business (US House, 2016). In July 2016, Sen. Heinrich introduced a similar bill in the US Senate (US Senate, 2016). Additionally, in October 2015, a bipartisan Battery Energy Storage Caucus was launched to investigate how ESS deployment could be accelerated (PR Newswire, 2015).

Further supporting much of the legislative work being undertaken to improve ESS deployment in the United States, the White House has also announced executive actions and held a Summit on Scaling Renewable Energy and Storage with Smart Markets to bring together regulators, operators, and the private-sector to push for commitments to further ESS deployment. The announced actions and commitments are expected to result in at least 1.3 GW of energy storage procurement or deployment in the next 5 years. The executive actions focus on conducting research on the impact of incorporating ESS into the grid, increasing the federal government's storage capacity, bolstering the storage capacity of the US armed forces, promoting microgrids in rural communities, and enhancing data access and sharing (The White House, 2016). It is clear from the variety and level of initiatives being undertaken that the US regulatory environment is moving actively toward a position of improving and aiding ESS deployment by cutting down market barriers.

References

ABB, 2012. Energy storage: keeping smart grids in balance. [Online] Available from: <https://library.e.abb.com/public/59a2be960fdb777a48257a680045-c04a/ABB%20Energy%20Storage_Nov2012.pdf> (accessed 29.11.15.).

Abraham, S., 2002. National Transmission Grid Study. US Department of Energy, Washington, DC.

Arizona State Senate, 2015. Arizona senate bill 1465. [Online] Available from: <http://www.azleg.gov//FormatDocument.asp?inDoc = /legtext/52leg/1r/bills/sb1465o.asp&Session_ID = 114> (accessed 18.08.16.).

Baldwin, S., et al., 2015. Enabling Modernization of the Electric Power System. US Department of Energy, Washington, DC.

Bovarnick, B., 2015. Integrated energy storage in the United States. [Online] Available from: <https://www.ny-best.org/sites/default/files/resources/NRGstorage-report.pdf> (accessed 18.08.16.).

Bussar, R., et al., 2013. Battery Energy Storage for Smart Grid Applications. EUROBAT, the Association of European Automotive and Industrial Battery Manufacturers, Rome.

California ISO, 2014. Advancing and maximizing the value of energy storage technology—a California roadmap. [Online] Available from: <https://www.caiso.com/Documents/Advancing-MaximizingValueofEnergyStorageTechnology_California Roadmap.pdf> (accessed 18.08.16.).

California Public Utilities Comission, 2012. Rulemaking 1211005. [Online] Available from: <https://apps.cpuc.ca.gov/apex/f?p = 401:56:17132411274950::NO:RP,57,RIR:P5_PROCEEDING_SELECT:R1211005> (accessed 18.08.16.).

California Public Utilities Commission, 2013. Proposed decision on rulemaking 10-12-007. [Online] Available from: <http://docs.cpuc.ca.gov/PublishedDocs/Published/G000/M078/K929/78929853.pdf> (accessed 18.08.16.).

Connecticut State Senate, 2015a. Connecticut senate bill 1078. [Online] Available from: <https://www.cga.ct.gov/asp/cgabillstatus/cgabillstatus.asp?selBillType = Bill&bill_num = 1078&which_year = 2015> (accessed 18.08.16.).

Connecticut State Senate, 2015b. Connecticut senate bill 1502. [Online] Available from: <https://www.cga.ct.gov/asp/CGABillStatus/cgabillstatus.asp?selBillType = Bill&bill_num = 1502&which_year = 2015> (accessed 18.08.16.).

Conover, D., 2014. Overview of development and deployment of codes, standards and regulations affecting energy storage system safety in the United States. [Online] Available from: <http://energy.gov/sites/prod/files/2014/10/f18/Overview%20of%20Development%20and%20Deployment%20of%20Codes%2C%20Standards%20and%20Regulations%20Affecting%20ESS%20Safety%20in%20the%20US%20August%202014_1.pdf> (accessed 05.07.16.).

Dehler, J., 2015. Self-consumption of electricity from renewable sources. [Online] Available from: <http://www.insightenergy.org/system/publications/files/000/000/016/original/RREB6_self-consumption_renewable_electricity_final.pdf?1435062407> (accessed 05.07.15.).

Deloitte, 2015. Energy storage: tracking the technologies that will transform the power sector. [Online] Available from: <http://www2.deloitte.com/content/dam/Deloitte/us/Documents/energy-resources/us-er-energy-storage-tracking-technologies-transform-power-sector.pdf> (accessed 05.07.16.).

Ecofys, 2014. Energy storage opportunities and challenges. [Online] Available from: <http://www.ecofys.com/files/files/ecofys-2014-energy-storage-white-paper.pdf> (accessed 05.07.16.).

Eyer, J., Corey, G., 2010. Energy Storage for the Electricity Grid: Benefits and Market Potential Assessment Guide. Sandia National Laboratories, operated for the United States Department of Energy by Sandia Corporation, Albuquerque, Livermore.

Federal Energy Regulatory Commission [FERC], 2008. Order No.719. [Online] Available from: <http://www.ferc.gov/whats-new/comm-meet/2008/101608/E-1.pdf> (accessed 18.08.16.).

FERC, 2011a. Order No. 745. [Online] Available from: <http://www.ferc.gov/EventCalendar/Files/20110315105757-RM10-17-000.pdf> (accessed 18.08.16.).

FERC, 2011b. Order No. 755. [Online] Available from: <http://www.ferc.gov/whats-new/comm-meet/2011/102011/E-28.pdf> (accessed 18.08.16.).

FERC, 2013a. Order No. 784. [Online] Available from: <http://www.ferc.gov/whats-new/comm-meet/2013/071813/E-22.pdf> (accessed 18.08.16.).

FERC, 2013b. Order No. 792. [Online] Available from: <http://www.ferc.gov/whats-new/comm-meet/2013/112113/E-1.pdf> (accessed 18.08.16.).

Florida Solar Energy Center, 2014. SunSmart E-Shelters Program. [Online] Available from: <http://www.fsec.ucf.edu/En/education/sunsmart/index.html> (accessed 18.08.16.).

Gyuk, I., et al., 2013. Grid Energy Storage. US Department of Energy, Washington, DC. Available from: <http://energy.gov/sites/prod/files/2014/09/f18/Grid%20Energy%20Storage%20December%202013.pdf>. (accessed 05.07.16.).

Hawaii State Legislature, 2014. Hawaii house bill 2618. [Online] Available from: <http://www.capitol.hawaii.gov/session2014/bills/HB2618_SD1_.PDF> (accessed 18.08.16.).

Hawaii State Senate, 2014. Hawaii senate bill 2932. [Online] Available from: <http://www.capitol.hawaii.gov/Archives/measure_indiv_Archives.aspx?billtype = SB&billnumber = 2932&year = 2014> (accessed 18.08.16.).

Hawaiian Electric, 2013. Request for proposals No. 072114-01. [Online] Available from: <https://www.hawaiianelectric.com/Documents/clean_energy_hawaii/rfp/rfp_document_072114_01.pdf> (accessed 18.08.16.).

Hostick, D., et al., 2012. Renewable Electricity Futures Study Volume 3: End-Use Electricity Demand. National Renewable Energy Laboratory, Golden.

IEC Market Strategy Board, 2011. Electrical Energy Storage. International Electrotechnical Commission, Geneva.

IEC Market Strategy Board, 2012. Grid Integration of Large-Capacity Renewable Energy Sources and Use of Large-Capacity Electrical Energy Storage. International Electrotechnical Commission, Geneva.

IEEE-PES ESSB, 2016. IEEE Stationary Battery Committee next meeting. [Online] Available from: <http://www.ewh.ieee.org/cmte/PES-SBC/meetings.html> (accessed 05.07.16.).

ISO New England, 2016. 2016 Regional electricity outlook. [Online] Available from: <http://www.iso-ne.com/static-assets/documents/2016/03/2016_reo.pdf> (accessed 18.08.16.).

Lazard, 2015. Levelized cost of storage analysis. [Online] Available from: <https://www.lazard.com/media/2391/lazards-levelized-cost-of-storage-analysis-10.pdf> (accessed 05.07.16.).

Lichtner, S., Brindle, R., Pack, L., 2010. Electric Power Industry Needs for Grid-Scale Storage Applications. Sandia National Laboratories, Sponsored by US Department of Energy.

Luo, X., Wang, J., Dooner, M., Clarke, J., 2015. Overview of current development in electrical energy storage technologies and the application potential in power system operation. Appl. Energy. 137, 511–536.

Massachusetts Office of Energy and Environmental Affairs, 2015. Massachusetts energy storage initiative. [Online] Available from: <http://www.mass.gov/eea/pr-2015/10-million-energy-storage-initiative-announced.html> (accessed 18.08.16.).

Minnesota State House of Representatives, 2015. Minnesota HF 3: Omnibus employment, economic development, jobs, and energy bill. [Online] Available from: <https://www.revisor.mn.gov/bills/bill.php?b=house&f=HF0003&ssn=1&y=2015> (accessed 18.08.16.).

Modern Grid Strategy Team, 2009. Energy Storage–A Key Enabler Of The Smart Grid. National Energy Technology Laboratory.

Mullendore, S., 2016. State energy storage legislation: a look back at 2015. [Online] Available from: <http://www.cleanegroup.org/state-energy-storage-legislation-a-look-back-at-2015/> (accessed 18.08.16.).

New Jersey Board of Public Utilities, 2014. NJ energy resilience bank. [Online] Available from: <http://www.state.nj.us/bpu/newsroom/announcements/pdf/20141020_erb_press.pdf> (accessed 18.08.16.).

New Jersey's Clean Energy Program, 2016. Renewable electric storage program. [Online] Available from: <http://www.njcleanenergy.com/renewable-energy/programs/energy-storage> (Accessed 18.08.16.).

New Mexico Energy, Minerals, and Natural Resources Department, 2015. New Mexico energy policy and implementation plan. [Online] Available from: <http://www.emnrd.state.nm.us/EnergyPolicy/documents/EMNRD_EnergyPolicy.pdf> (accessed 18.08.16.).

New Mexico State Legislature, 2013. New Mexico joint memorial 43. [Online] Available from: <http://www.sos.state.nm.us/uploads/files/SJM43.pdf> (accessed 18.08.16.).

New York Department of Public Service, 2016. Reforming the energy vision. [Online] Available from: <http://www3.dps.ny.gov/W/PSCWeb.nsf/All/CC4F2EFA3A23551585257DEA007DCFE2?OpenDocument> (accessed 18.08.16.).

New York State's Clean Energy Standard, 2010. New York battery and energy storage technology consortium. [Online] Available from: <http://www.nyserda.ny.gov/Cleantech-and-Innovation/Energy-Storage/NY-BEST> (accessed 18.08.16.).

NREL, 2014. Issue brief: a survey of state policies to support utility-scale and distributed-energy storage. [Online] Available from: <http://www.nrel.gov/docs/fy14osti/62726.pdf> (accessed 18.08.16.).

NREL, 2016. Energy storage: possibilities for expanding electric grid flexibility. [Online] Available from: <http://www.nrel.gov/docs/fy16osti/64764.pdf> (accessed 18.08.16.).

Nykvist, B., Nilsson, M., 2015. Rapidly falling costs of battery packs for electric vehicles. [Online] Available from: <http://energy.gov/sites/prod/files/2014/09/f18/Grid%20Energy%20Storage%20December%202013.pdf> (accessed 05.07.15.).

Office of Basic Energy Sciences, 2007. Basic Research Needs for Electrical Energy Storage. Office of Basic Energy Sciences, Germantown.

Oregon State House, 2015. Oregon house bill 2193. [Online] Available from: <http://edocs.puc.state.or.us/efdocs/HAA/haa131044.pdf> (accessed 18.08.16.).

Pasadena Water and Power, 2014. AB 2514 energy storage systems evaluation, Pasadena: s.n.

PR Newswire, 2015. Battery energy storage caucus launches to bipartisan support in congress. [Online] Available from: <http://www.prnewswire.com/news-releases/battery-energy-storage-caucus-launches-to-bipartisan-support-in-congress-300167185.html> (accessed 18.08.16.).

Public Utility Commission of Texas, 2012. Project No. 39917 rulemaking on energy storage issues. [Online] Available from: <http://www.puc.texas.gov/agency/rulesnlaws/subrules/electric/25.501/39917adt.pdf> (accessed 18.08.16.).

Roberts, B.P., Sandberg, C., 2011. The role of energy storage in development of smart grids. IEEE. 99 (6), 1139–1144.

Sandia Corporation, 2016. DOE global energy storage database. [Online] Available from: <http://www.energystorageexchange.org/> (accessed 17.08.16.).

SCOTUSblog, 2016. Federal energy regulatory commission v. electric power supply association. [Online] Available from: <http://www.scotusblog.com/case-files/cases/federal-energy-regulatory-commission-v-electric-power-supply-association/> (accessed 18.08.16.).

Simbolotti, G., Kempener, R., 2012. Electricity Storage Technology Brief, s.l.: Energy Technology Systems Analysis Programme, International Renewable Energy Agency.

Skinner, N., 2010. California assembly bill 2514 energy storage systems. [Online] Available from: <http://leginfo.legislature.ca.gov/faces/billTextClient.xhtml?bill_id = 200920100AB2514> (accessed 18.08.16.).

Stanfield, S., Vanega, A., 2015. Deploying distributed energy storage: near-term regulatory considerations to maximize benefits. [Online] Available from: <http://www.irecusa.org/wp-content/uploads/2015/03/Deploying-Distributed-Energy-Storage-2-27-15-IREC-FINAL-Secure-1-1.pdf> (accessed 05.07.16.).

Stein, A., 2014. Reconsidering regulatory uncertainty: making a case for energy storage. Florida State University Law Review. 41, 697–765.

Texas State Senate, 2011. Texas senate bill 943. [Online] Available from: <http://www.capitol.state.tx.us/billlookup/Text.aspx?LegSess = 82R&Bill = SB943> (accessed 18.08.16.).

The White House, 2016. Fact sheet: Obama administration announces federal and private sector actions on scaling renewable energy and storage with smart markets. [Online] Available from: <https://www.whitehouse.gov/the-press-office/2016/06/16/fact-sheet-obama-administration-announces-federal-and-private-sector> (accessed 18.08.16.).

US Department of Energy Office of Electricity Delivery and Energy Reliability, 2015. Energy storage safety strategic plan. [Online] Available from: <http://energy.gov/sites/prod/files/2014/12/f19/OE%20Safety%20Strategic%20Plan%20December%202014.pdf> (accessed 05.07.16.).

US Department of Energy, 2013. Grid energy storage. [Online] Available from: <http://energy.gov/sites/prod/files/2014/09/f18/Grid%20Energy%20Storage%20December%202013.pdf> (accessed 05.07.16.).

US Energy Information Administration, 2016. Electric power monthly with data for April 2016. [Online] Available from: <https://www.eia.gov/electricity/monthly/epm_table_grapher.cfm?t = epmt_5_6_a> (accessed 05.07.16.).

US House, 2016. US house bill 5350. [Online] Available from: <https://www.congress.gov/bill/114th-congress/house-bill/5350> (accessed 18.08.16.).

US Senate, 2015. US senate bill 1434. [Online] Available from: <https://www.congress.gov/bill/114th-congress/senate-bill/1434> (accessed 18.08.16.).

US Senate, 2016. US senate bill 3159. [Online] Available from: <https://www.congress.gov/bill/114th-congress/senate-bill/3159> (accessed 18.08.16.).

Vermont State House, 2015. Vermont house bill 40. [Online] Available from: <http://legislature.vermont.gov/bill/status/2016/h.40> (accessed 18.08.16.).

Washington State House, 2013a. Washington house bill 1289. [Online] Available from: <http://lawfilesext.leg.wa.gov/biennium/2013-14/Pdf/Bills/House%20Bills/1289.pdf> (accessed 18.08.16.).

Washington State House, 2013b. Washington house bill 1296. [Online] Available from: <http://lawfilesext.leg.wa.gov/biennium/2013-14/Pdf/Bills/House%20Bills/1296.pdf> (accessed 18.08.16.).

Washington State House, 2015. Washington house bill 1115. [Online] Available from: <http://app.leg.wa.gov/billinfo/summary.aspx?bill = 1115&year = 2015> (accessed 18.08.16.).

5

ELECTROMAGNETIC INTERFERENCE CONSIDERATIONS FOR ELECTRICAL POWER SYSTEMS

Benjamin R.T. Cotts[1], Kevin L. Graf[2] and William H. Bailey[1]
[1]Exponent Inc., Bowie, MD, United States [2]Exponent Inc., Menlo Park, CA, United States

CHAPTER OUTLINE

5.1 Introduction

Electromagnetic (EM) fields come from many sources both natural and anthropogenic. In nature, the primary source of EM fields above about 30 megahertz (MHz) is blackbody radiation, while at lower frequencies, the primary source is lightning (ICNIRP, 2009). Anthropogenic EM sources include both the radio and television broadcast signals, which have been around for decades, as well as newer sources such as mobile phones, satellite communications, Wi-Fi, and countless others, which are ubiquitous in modern society. The complex interaction

The Power Grid. DOI: http://dx.doi.org/10.1016/B978-0-12-805321-8.00005-7

among different sources can potentially lead one source to interfere with the operation of another source. In the United States the Federal Communications Commission (Title 47, Part 15) regulates both intentional and unintentional radiators of EM fields at radio frequencies.

Engineers and designers of virtually all electrical and electronic equipment are engaged in ensuring the robustness of design against outside sources of interference, termed electromagnetic susceptibility (EMS), as well as mitigating how much potential interference, termed electromagnetic interference (EMI), their equipment may cause. These terms are often grouped together into a single category of electromagnetic compatibility (EMC) which is defined as "[t]he ability of systems, equipment, and devices that utilize the electromagnetic spectrum to operate in their intended operational environments without suffering unacceptable degradation or causing unintentional degradation because of electromagnetic radiation or response ..." (ANSI, 2009).

The power grid consists of a multitude of components, virtually all of which can be considered from the perspective of either EMI or EMS. In this chapter, the perspective considered is the effect of EMI from components of the power grid to outside systems. EMI from components of the power grid are broken down into three primary sources: (1) radiofrequency noise such as the corona from high-voltage conductors, (2) physical obstruction of other electromagnetic signals, such as the blocking or detuning of other EM signals such as amplitude-modulated (AM) radio signals, and (3) induced or capacitive coupled currents or voltages in other systems such as long parallel pipelines or fences, large conductive objects such as storage containers or semi-trailers, or in medical devices like pacemakers. These EMI sources are discussed in Section 5.2, immediately following this introduction followed by a discussion of some health and safety issues that can arise in the design and permitting process of electrical power system components in Section 5.3. Section 5.4 provides a summary of the chapter including future trends, other references, and conclusions.

5.2 Sources of EMI

EMI from elements of the power grid can be radiated (i.e., propagated through the air as EM waves similar to those from AM or frequency-modulated [FM] radio stations), conducted (i.e., propagated through conductors), capacitively coupled or induced, and occur from a variety of sources. EMI can occur as a result of random noise such as corona or can result from the physical blocking or scattering of existing signals. Effects from quasi-static fields and related induction or capacitive coupling can also be considered sources of EMI. Each of these three sources of EMI—corona, scattering, and induction or capacitive coupling—is discussed below.

5.2.1 Corona

Corona refers to the release of energy at points on the surface of high-voltage conductors in the form of EM fields, audible noise, and light. Corona is a common sources of EMI related to power systems; it is primarily associated with higher-voltage transmission lines, but can also be present on lower-voltage transmission lines and equipment under certain circumstances. Corona is commonly encountered with both Direct Current (DC) and Alternating Current (AC) transmission lines and occurs when the electric field at conductor surfaces exceeds the breakdown field strength of the surrounding air. Another source of EMI associated with transmission lines is gap discharge, which is an electrical discharge typically occurring from defective, damaged or improperly installed insulator hardware or connections. Corona levels may be a design criterion of importance, especially at voltages of 345 kV and higher (IEEE, 1980) although some small level of corona can occur on lines operating above approximately 110 kV (Pakala and Chartier, 1971). For lower voltage transmission lines (and especially distribution lines operating below 70 kV) gap discharges represent the primary source of EMI (Pakala and Chartier, 1971). While gap discharges can cause EMI over a far wider frequency band than corona, when identified, they also are relatively easy to mitigate, while corona on high voltages transmission lines (once constructed) is generally not. Thus, gap discharges are typically a maintenance issue, while corona is a design issue. The primary focus of this section is therefore on corona with brief asides on the characteristics of gap discharges.

The concentrated electric field at points on conductors (often referred to as an electric-field or potential gradient) produces an electrical breakdown of the air resulting in a small spark (e.g., Loeb, 1965). This can occur under certain conditions on high-voltage conductors of transmission lines or substation equipment. The electric field is concentrated by surface irregularities such as water droplets, insects, other debris, or nicks on

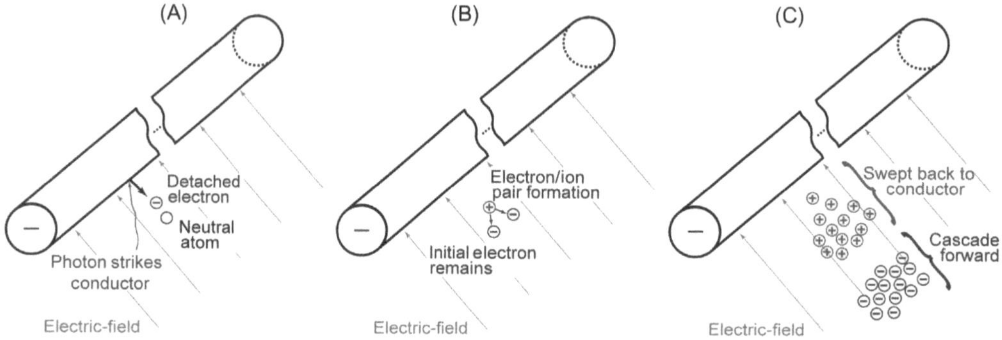

Figure 5.1 Initiation of an electron avalanche from a negative conductor. (A) Initiation, (B) Electron/ion pair formation, and (C) Electron avalanche.

the conductor. While a single spark does not result in appreciable EMI, under certain conditions (such as rain) corona can occur nearly continuously along long stretches of a transmission line conductor and each individual corona discharge serves as a miniature antenna, producing EM energy that can be radiated into space and conducted along the transmission line itself (Newell et al., 1967).

Three basic stages in the process of corona are illustrated in Fig. 5.1. Corona begins with the production of a single electron–ion pair (Fig. 5.1A). Within a few inches of a high-voltage conductor, the local electric field may be strong enough to accelerate the electron to a point that when it impacts a neutral molecule, the collision occurs with sufficient energy to remove an additional electron thus creating an electron–ion pair (Fig. 5.1B). This process continues with the number of electrons increasing exponentially and is thus termed an electron avalanche (Fig. 5.1C).

At sea level and under ideal conditions, corona inception typically occurs at an electric field of approximately 30 kilovolts per centimeter (kV/cm) according to Peek's Law (Peek, 1911) although in practice corona inception can occur at an electric field level of approximately 11–14 kV/cm (EPRI, 1978; Juette and Zaffanella, 1970). At electric field levels below this threshold a free electron will not accelerate quickly enough to dislodge another electron before its next collision. Due to the decreasing air density with increasing altitude, the threshold for corona inception typically decreases with altitude (Chartier et al., 1987).

Under ideal conditions transmission lines will be free of corona because the high electric-fields necessary for corona are eliminated by the smooth conductor surface. The imperfections on the conductor, such as nicks and cuts, which can occur during stringing can cause corona rates to be high initially. However, the ionization resulting from corona itself at these jagged edges tends to remove the protrusions over time through ion bombardment and heating at the site. In addition, the oil coating of a new transmission line conductor can also cause corona rates be higher when a transmission line is first installed during a so-called "burn in" period, but will generally decrease over time so that fair-weather corona on AC lines will be less common after conductor "weathering" (EPRI, 1982). On the other hand, substances including dust, leaf particles, bird droppings, insects, and spider webs can also lead to corona (e.g., Newell et al., 1967, 1968) which, in turn, can lead to a seasonal variation in typical corona levels (Laforest, 1968).

5.2.1.1 Characteristics

Not all corona discharges are the same and in fact the different forms (e.g., from positive or negative corona) have different physical characteristics (Meek and Craggs, 1953; EPRI, 1982). Forms of negative corona include Trichel pulses, pulseless glow, and negative streamers while forms of positive corona include onset pulses, Hermstein's glow, and positive streamers (Hermstein, 1960; Giao and Jordan, 1968; EPRI, 1982).

Both AC and DC transmission lines have positive and negative corona. The difference is that for AC transmission lines, both positive and negative corona will occur on all phase conductors, while for DC transmission lines, positive corona occurs near the positive conductor and negative corona occurs near the negative conductor. This also means that near an AC transmission line the ions created during the positive half-cycle of the AC wave are attracted back to the conductor during the negative half-cycle (and vise-versa) and so there is minimal buildup of space charge (i.e., small air ions and charged aerosols) in air. For a DC transmission line on the other hand, positive ions will build up around the positive conductor and electrons and negative ions will build up around the negative conductor leading to an increase in space charge near DC transmission line conductors of differing polarities (Goldman et al., 1985). This build-up of space charge surrounding a DC transmission line has the further effect of reducing the electric field very near each conductor. Therefore, in contrast to AC transmission lines where corona intensity (and therefore potential EMI) is highest during foul weather, for DC transmission lines, the reduced electric field near the DC conductor produced by increased numbers of

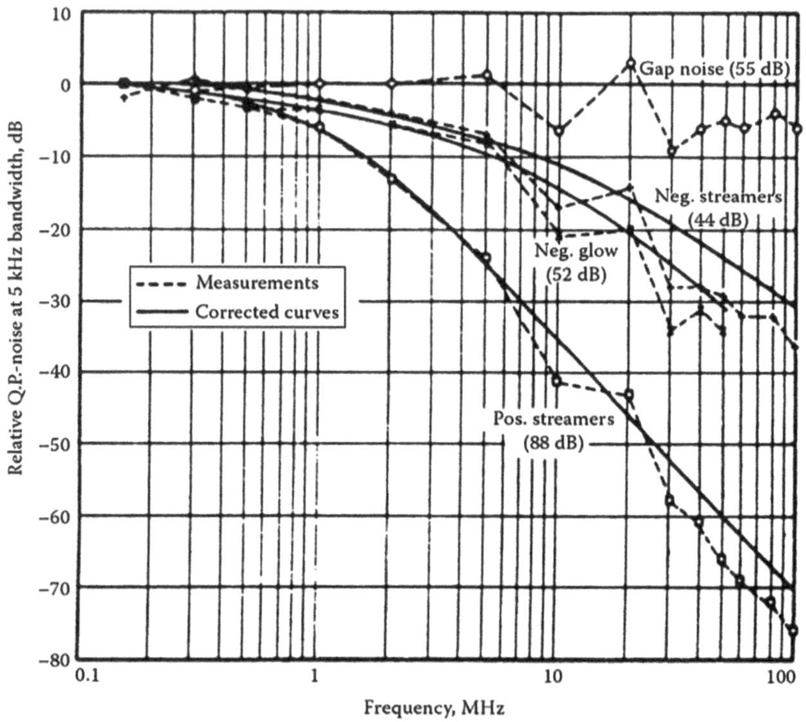

Figure 5.2 Relative frequency spectra for different noise types (Juette, 1972). The numbers in parentheses are the absolute values measured at 150 kHz.

corona sources and thus ions in foul weather, reduces the intensity and extent of the corona discharge in foul weather (and therefore potential EMI). EMI levels are thus generally *lower* during foul weather than fair weather for DC lines.

Given the differences in the physical mechanisms and resulting phenomena from positive and negative corona, it is to be expected that there are some differences in their EMI characteristics. This is demonstrated in both theoretical and laboratory results, which have shown that the EMI from positive corona decreases far more rapidly with frequency than EMI from negative corona (Juette, 1972; Pakala and Chartier, 1971).[1] There are even differences in corona-related EMI among transmission lines of the same voltage class, but constructed in different configurations (Gary et al., 1973). These characteristics can have an effect on the potential susceptibility of other sources to the EMI from transmission lines (Fig. 5.2).

[1]EMI from gap discharges can remain relatively constant with frequency out to a frequency of nearly 100 MHz.

5.2.1.2 Frequency Spectrum of EMI

As discussed above, the frequency spectrum of EMI from discharges is a function of the physical size of the discharge, the frequency at which the discharge is repeated, and the particular characteristics of the underlying physical processes. Laboratory experiments, such as that illustrated in Fig. 6 of Juette (1972), show that there are significant differences in both the intensity and frequency distribution of different types of corona. Similar observations are borne out in some measurements of the EMI of DC lines, which showed that EMI from the positive conductor is significantly greater than EMI from a negative conductor (Pakala and Chartier, 1971). Despite these variations, however, in practical applications there always will be a need to consider both positive and negative forms of corona present from any transmission line and while the intensity of the EMI from transmission lines can increase dramatically with increasing voltage, the spectrum of corona EMI from transmission lines (both AC and DC) tends to be quite similar and follows a decrease of approximately 20 decibels (dB) or more per frequency decade (Pakala et al., 1967; Pakala and Chartier, 1971; EPRI, 1982).

A general spectrum of corona-generated noise from transmission lines is available from a number of sources (e.g., Pakala et al., 1967; Pakala and Chartier, 1971; EPRI, 1982; IEEE Std. 430) and shows that by far the highest levels of EMI are generated at frequencies below approximately 2 MHz and these levels decrease rapidly with frequency (Fig. 5.3A) and distance (Fig. 5.3B).

5.2.1.3 Susceptibility to EMI From Corona

As shown in the discussion above, EMI from transmission lines is a broadband source that can span a range of frequencies from below 100 kilohertz (kHz) to above 1000 MHz, but is predominantly of interest in the 100 kHz to 2 MHz range. Electronic devices that operate at higher frequencies (e.g., tens of MHz or gigahertz [GHz]) and those that use other modulation schemes such as FM radios, cell phones, global positioning systems (GPSs), and satellite receivers, therefore are unlikely to be susceptible to EMI generated by transmission lines (EPRI, 1982; Silva and Olsen, 2002).

In the past, corona-generated EMI from transmission lines could also interfere with the video portion of analog television signals in the form of radio noise (RN) (e.g., Juette, 1972; IEEE, 1980; EPRI, 1982; Claes, 1985). This was a result both of the video portion of analog television signals being broadcast with AM (which is more susceptible to transmission line EMI) as well as the large broadcast area over which some television stations provided

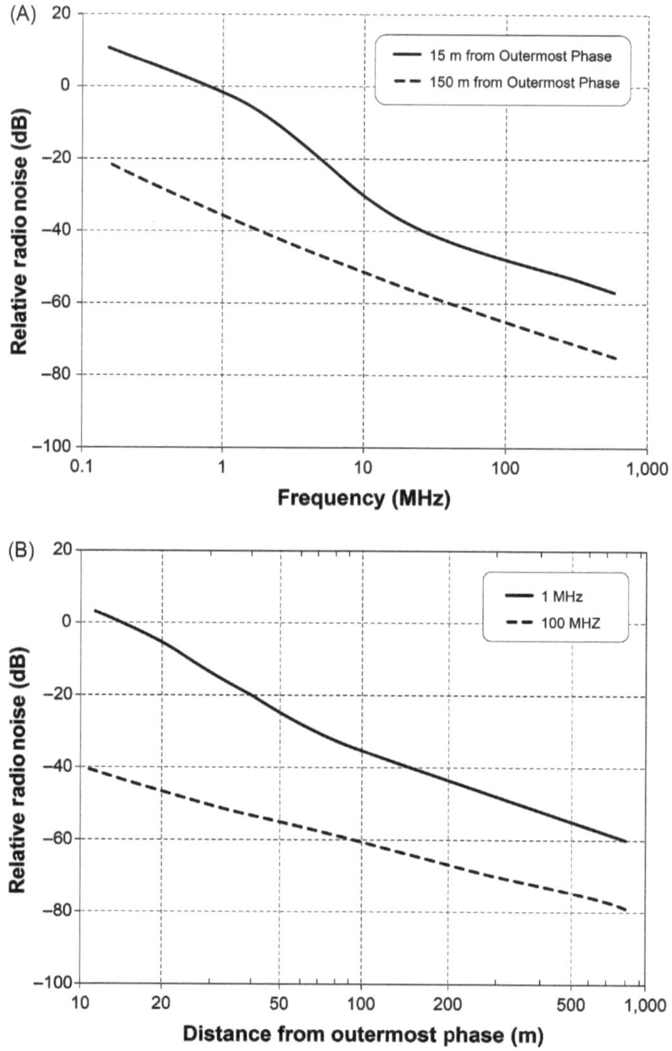

Figure 5.3 (A) Typical frequency spectra of corona-produced radio noise at 15 m and 150 m from outermost phase conductor (modified from Electric Power Research Institute (EPRI), 1982. AC Transmission Line Reference Book—345 kV and Above. second ed. EPRI, Palo Alto, CA). (B) Typical lateral profile of radio noise at 1 and 100 MHz (modified from Electric Power Research Institute (EPRI), 1982. AC Transmission Line Reference Book—345 kV and Above. second ed. EPRI, Palo Alto, CA).

service (and hence had relatively low signal strength in areas distant from the broadcast station). In the United States, however, this is no longer the case for most television stations due to the Digital Television Transition authorized by Congress in 1996, which resulted in digital encoding of both television video and audio signals. This legislation made an additional broadcast frequency available to every full-power television station for the purpose of running digital transmissions. Full-power television stations were required to stop broadcasting analog signals by June 12, 2009, and the transition date for low-power television stations was set for September 1, 2015 (FCC Website, 2016a). These digitally-encoded transmissions are generally far less susceptible to EMI from a transmission line primarily because their digital encoding is inherently more robust to low levels of EMI typical of transmission lines and so EMI from transmission lines is no longer expected to affect television broadcast signals.

The EMI generated by transmission lines may, however, interfere with the reception of lower frequency systems such as AM radio (520–1720 kHz). This is especially the case, e.g., when a car radio is tuned to an AM radio station and the car drives directly adjacent to or beneath a transmission line. In addition, military and other government agencies that operate communications equipment in the high frequency (HF) band also have the potential for interference from transmission line EMI (e.g., Shepherd and Gaddie, 1976).

Other over-the-air transmissions that are occasionally affected by transmission line EMI are ham radios, which can operate in the hundreds of kHz into the MHz bands. Similar to AM radio, at locations very near a transmission line, the noise generated by corona-type EMI may affect the ability of a ham radio operator to receive distant signals. In cases where an issue is identified, a solution such as relocating the existing antenna, using a directional antenna, or installing a repeater nearby can alleviate the issue (USDOE, 1977; USDOE, 1980; Loftness et al., 1981).

While the corona-generated noise from transmission lines may be a source of interference primarily below approximately 2 MHz, defining a particular noise level that is unacceptable is difficult because of wide variations in ambient noise levels, received-signal strengths, and a variety of other factors (EPRI, 1978; IEEE, 1980). Another factor complicating the issue is that increasing EMI levels do not necessarily correspond to increasing interference, at least when people and annoyance are concerned. Previous studies have shown that a quasi-peak measure of the noise provides the best correlation with tolerability of the noise and furthermore that the corona-generated noise from a DC line tends to be considered less objectionable than the same level of noise from an AC line (EPRI, 1978; EPRI, 1982). One study (Gehrig et al., 1967) found that good reception in the presence of corona noise from an AC line required a signal-to-noise ratio (SNR) of as much as 25:1, while near a DC line the SNR could be as low as 9:1. Other studies, illustrated graphically in Fig. 5.4A (for AC transmission lines) and in Fig. 5.4B (for DC transmission

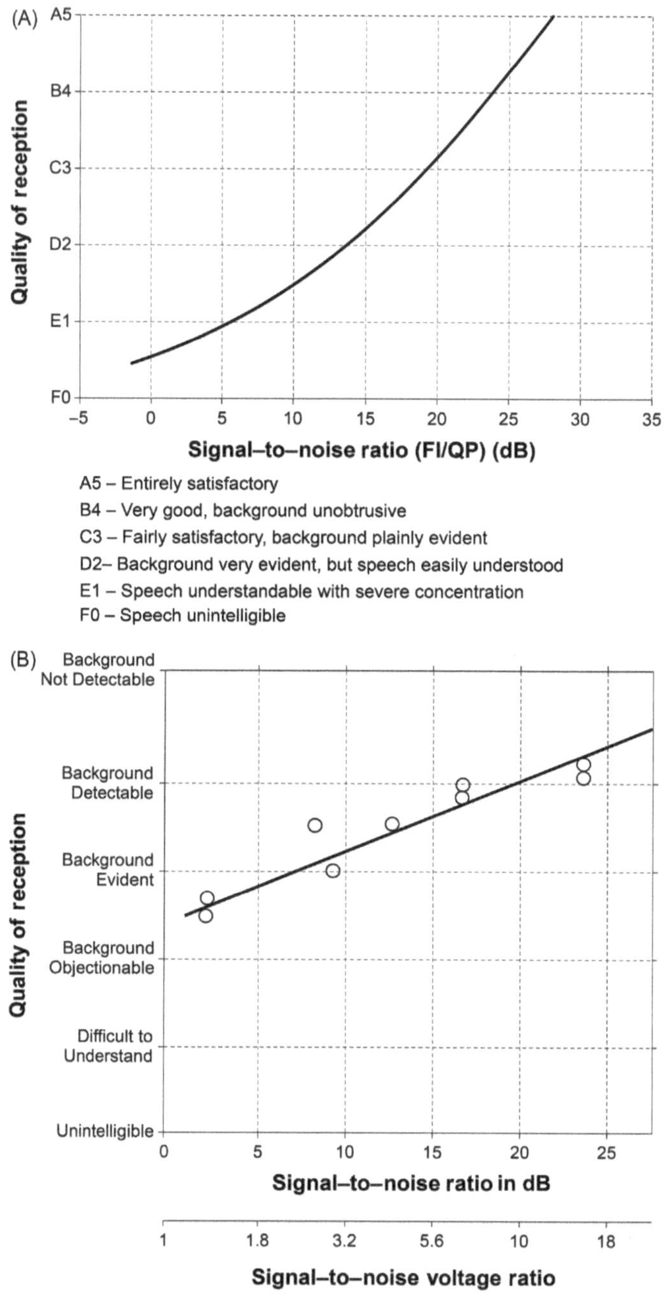

Figure 5.4 (A) AM radio reception quality as a function of signal-to-noise ratio (modified from Electric Power Research Institute (EPRI), 1982. AC Transmission Line Reference Book—345 kV and Above. second ed. EPRI, Palo Alto, CA). (B) Radio interference tolerance level as a function of signal-to-noise ratio (modified from Electric Power Research Institute (EPRI), 1978. Transmission Line Reference Book—HVDC to 600. EPRI, Palo Alto, CA).

lines), demonstrate a similar phenomenon. Additional studies performed by EPRI demonstrated that the perception of annoyance "... depends not only on the peak value, but also on the repetition rate of noise pulses" (EPRI, 1982).

Although some countries including Canada (Industry Canada, 2013), have specific regulations, there are no formal limits defined in the United States. The Institute of Electrical and Electronics Engineers (IEEE) Radio Noise Design Guide (IEEE, 1971) has put forth several metrics found to be acceptable in the past, but the Federal Communication Commission (FCC) Rules and Regulations contain no guideline regarding the RN levels near high-voltage transmission lines. Power transmission lines fall into the FCC category of "incidental radiator," which is defined as "a device that generates radio frequency energy during the course of its operation although the device is not intentionally designed to generate or emit radio frequency energy." Operation of an incidental radiator "is subject to the conditions that no harmful interference is caused and that interference must be accepted that may be caused by the operation of an authorized radio station, by another intentional or unintentional radiator, by industrial, scientific and medical equipment, or by an incidental radiator." Section 15.3(m) of Title 47 (FCC, 2016) defines "harmful interference" as "any emission, radiation or induction that endangers the functioning of a radio navigation service or of other safety services or seriously degrades, obstructs or repeatedly interrupts a radiocommunications service operating in accordance with this Chapter."

Historically, transmission-line operators have not had difficulty operating under the present FCC rules, since most sources of "harmful interference" from power lines in fair weather are due to gap discharges that can be identified and repaired (USDOE, 1980). AM radio reception at residences very near transmission lines, however, may be affected by corona-type RN in foul-weather depending on the strength and quality of the AM radio signal.

5.2.1.4 Mitigation

In many cases, mitigating EMI related to transmission lines is relatively straightforward as "about 90% of all interference complaints are caused by defects in hardware, etc., which can easily be located and rectified" (IEEE, 1980). In these cases the most difficult task is often identifying exactly which element of the transmission system (e.g., loose connection, damaged insulator, etc.) is responsible for the noise. Identification of the offending sources can be performed using either radiofrequency (e.g., Giussani et al., 2012) or through the use of corona cameras that can detect and pinpoint corona in daylight (Phillips et al., 2008), which have made identification of the offending component much easier. Once identified, it is typically a matter of tightening the connection or replacing the damaged or corroded component with a new one (USDOE, 1980).

In contrast to gap discharges, corona-generated EMI can be reduced to a great extent in the design-phase of the transmission line, but once the line is constructed and in service, the noise can be far more difficult to mitigate if the source is due to improper sizing of the conductors). As discussed above, transmission lines are designed to be free of corona under normal operating conditions. This design process involves a trade-off between the constructability and economic viability on one side and the level of RN on the other hand. In the design phase of a transmission line, the expected EMI can be reduced by a number of factors including increasing tower structure height, changing structure configuration, increasing the diameter of phase conductors, using multiple conductors per phase (bundling), optimizing the overall bundle diameter, and lowering the operating voltage (Dallaire et al., 1984; EPRI, 1982).

While the operating voltage is the single most influential factor in determining potential EMI levels, it is typically a design consideration that is determined by other factors prior to analysis of EMI. Of the remaining factors, the two of greatest importance are the size and number of conductors. For example, on a 345-kV transmission line, constructed in an H-frame configuration with 7.5-meter phase spacing and 12.5-meter vertical clearance to the phase conductors, the noise level can decrease by as much as 20 dB or more by increasing the conductor size from approximately one inch to three inches. Similarly, changing from a single conductor to a two-conductor bundle can decrease noise levels by as much as 18 dB. Other factors such as phase spacing, height of conductors above ground, and bundle diameter can all change the EMI level by 1–5 dB (EPRI, 1982).

Once constructed, the corona performance of a transmission line is primarily determined, but will still vary due to environmental factors as discussed above. In rare situations an environmental factor may be dominant enough to semipermanently change the performance of a transmission line. Examples include dust storms, insect migration, or a particularly interesting case that occurred in the northwestern portion of the United States where moss began to grow on a transmission line.

In these extraordinary situations, cleaning the transmission likely will result in some reduction to EMI levels.

Fortunately, reduction of the source of EMI is not the only way to mitigate potential interference and it is often possible to increase the signal reception strength of the desired signal. For example, interference to radio reception can often be mitigated by relocating the existing antenna, using a directional antenna, installing a repeater nearby, or other solutions (USDOE, 1977; USDOE, 1980; Loftness et al., 1981).

5.2.2 Scattering and Reflection

Radio communications such as cell phones, televisions, AM/FM radio, and GPS signals all utilize propagating EM waves (frequently referred to as radio waves) to transmit information wirelessly. While previously discussed forms of EMI focused on sources of RN that could actively interfere with and overwhelm these signals at the receiver, there are also purely passive forms of EMI that can affect the ability of a transmitted signal to distinguishably reach a receiver in the first place.

Transmitted radio waves interact with dielectric and conducting objects in the environment, such as the ground, large buildings, transmission lines, and steel-lattice towers. An incident radio wave can induce electric currents in these objects, leading to scattering of the radio wave. In the case of transmission lines and towers, this can occur regardless of whether the lines are energized and carrying any current. In other words, the mere existence of the conductive lines and towers may result in scattering of certain radio waves. There are several ways that scattering of radio waves may disrupt radio communication:

1. Blocking: The transmitted signal does not reach the receiver due to being scattered by a structure in the path between signal source and receiver.
2. Multipath: Scattering may provide multiple paths by which a radio wave can reach a receiver, leading to multiple copies of the signal received slightly offset in time and overlapping one another.
3. Antenna pattern deformation: Scattering near a transmitting antenna or antenna array may alter the directionality of the antenna radiation pattern, resulting in stronger or weaker signal intensities than intended in certain directions.

The discussion below covers the potential for transmission lines and towers to cause each of these forms of EMI, and covers mitigation techniques, where applicable.

Whether the EMI is through blocking, multipath, or deformation of an antenna radiation pattern, interference will only occur if the transmission lines and towers efficiently interact with the incident radio wave. This is highly frequency-dependent because radio wave scattering tends to be strongest when the scattering object, or a component of the scattering object, is electrically similar in size to the wavelength of the incident radio wave.[2] Scattering may occur from individual transmission line conductors (typically a few centimeters in diameter), conductor bundles (tens of centimeters spacing between conductors), or tower structures (tens of meters in height, with subcomponents of varying size). Scattering may also occur from currents flowing in the closed loop formed by two adjacent towers, the interconnecting shieldwire, and the ground. This loop is typically several hundred meters in circumference.

At frequencies greater than approximately 10 MHz, blocking and multipath EMI due to scattering from transmission lines and towers is a potential concern. If blocking were experienced, loss of signal may occur. If multipath were experienced, signal degradation may take on varying forms, including ghost images or screen loss in the case of older amplitude modulated television broadcasts (Toyada and Hashimoto, 1979), or variation in range measurement in the case of GPS (Gerdan et al., 1995). In practice, however, these EMI effects are rarely experienced in relation to transmission lines and towers. Scattering of 1 GHz signals by individual transmission line conductors is unlikely to cause significant signal degradation (Silva and Olsen, 2002). Scattering by conductor bundles can be significant when resonant conditions occur (Takeshita et al., 1979), but these resonant conditions require specific combinations of radio wave frequency, sub-conductor spacing, and incidence angle, as well as appropriate geometry between the transmitter, scattering transmission lines, and receiver.

Scattering from transmission line towers can be significant, but typically only when very close to the tower. Toyada and Hashimoto (1979) showed in one study that very high frequency (VHF) scattering from a typical steel transmission line tower at

[2]If a conducting object is much larger than the wavelength of the incident radio wave, then it can reflect the incident wave, which can also result in EMI. If a single conducting object is much smaller than the wavelength of the incident radio wave, then its effect on the radio wave will be comparatively small. Scattering from many small objects can add up to an appreciable effect, however, such as in Rayleigh scattering. Rayleigh scattering effects are typically negligible for transmission lines and towers.

an approximate 100-meter distance from a receiver produced a received scattered field approximately 16 dB less than the field incident on the tower. In the case of GPS signals, studies have shown little to no impeding effects of transmission lines or towers on GPS code measurements (e.g., Bancroft et al., 2012; Silva and Olsen, 2002). Only nonimpeding effects on GPS, in the form of cycle slips, were detected by Bancroft et al. (2012), and those occurred near the transmission line towers. If line-of-sight signal blocking does occur close to a transmission line tower, one option for mitigation is to relocate the receiving antenna to outside the shadow of the blocking tower. These potential blocking and multipath effects at frequencies greater than approximately 10 MHz are usually not significant and usually do not factor in the design of transmission lines, but may occasionally arise in the form of complaints after line construction and are handled on a case-by-case basis (EPRI, 1982; Loftness, 2003).

At frequencies below approximately 10 MHz, for AM radio broadcasts in particular, scattering from transmission lines and towers may interfere with the antenna radiation pattern of a broadcast tower. To avoid interfering with other radio stations in overlapping frequency bands, or to avoid wasting power on radio transmissions into less populated areas, medium frequency (300 kHz to 3 MHz) broadcast stations often require specific antenna radiation patterns that focus their transmissions into specific directions. Scattering due to re-radiation from nearby transmission lines can alter this intended radiation pattern. In situations where proposed construction of a tower may affect the directionality of a nearby AM broadcast station, FCC Rules 47 CFR Part 1, Subpart BB of FCC Title 47 (FCC, 2016), mandate that the party proposing these additions investigate any potential effects on the AM broadcast station and take measures to mitigate those effects. In the case of transmission lines, the re-radiation of AM radio waves may be significant when the electrical length of the closed loop formed by adjacent transmission line towers, the shieldwire, and the ground is near a multiple of the wavelength of the radio station. In one scenario, re-radiation from a transmission line at a distance of 450 m from an omnidirectional AM broadcast antenna introduced peaks as high as 2 dB and minima as deep as −8 dB into the nominal radiation pattern of the antenna (Trueman and Kubina, 1981). To mitigate such effects, the electrical length of the closed loops in the transmission lines can be altered by installing de-tuning skirts or stub circuits on the transmission line towers (see, e.g., Trueman et al., 1989; Madge and Jones,

1986), or by isolating select towers from the interconnecting shieldwire (Trueman and Kubina, 1986).

5.2.3 Field Effects and Induction

In addition to EMI effects from corona, quasi-static electric and magnetic fields (EMFs) produced by transmission lines can also affect nearby facilities and devices.

The quasi-static fields associated with transmission lines are referred to in some literature as power frequency EMF or sometimes extremely low frequency (ELF) EMF. ELF EMF can couple to objects both within and near the right of way and can take the form of induced currents or induced voltages (the exposure of persons to EMF is discussed in Section 5.3). The two primary interactions are called electrostatic coupling and magnetic induction, although both are sometimes referred to simply (and erroneously in the case of electrostatic coupling) as induction effects. While electrostatic coupling can occur with DC transmission lines, magnetic induction cannot. Both electrostatic coupling and magnetic induction are of interest with respect to AC transmission lines, and so the discussion below focusses on AC transmission lines.

5.2.3.1 Electrostatic Coupling

Electrostatic coupling is a capacitive charging process where the electric field from transmission lines couple to nearby conductive objects thereby raising their electric potential relative to ground by causing the electric charges within the conductive object to redistribute. The degree of charge buildup depends on the size and electrical isolation of the conductive object with respect to ground as well as the intensity of the field. For the case of good isolation, a relatively large field, and a sufficiently large object, an electrical discharge may occur if a person touches both the object and ground whereby an electric current may flow through the person to ground (EPRI, 1982). A prime example of this is a large vehicle parked under a transmission line. Two forms of discharge may occur in this situation. The first is a spark discharge (also sometimes called a nuisance shock), which may occur just prior to or just after the moment of direct contact with the object. The second is often called a steady state current and is calculated assuming a short-circuit path to ground (a conservative assumption). Spark discharges, while annoying (as their pseudonym suggests), are not generally considered dangerous, while steady state currents can, in particular instances, lead to conditions where muscle contraction and potentially pain and are therefore limited by standards.

5.2.3.2 Magnetic Induction

Similar to the coupling of the electric field to nearby conductive objects, magnetic fields from AC transmission and distribution lines can also induce voltages and currents in nearby objects (magnetic induction does not occur for DC transmission lines). Magnetic induction typically occurs when nearby objects parallel the transmission line for a long distance and can form a conducting loop allowing current to flow. Magnetic induction can occur, e.g., on wire fences, irrigation pipes, telephone lines, or pipelines (Jaffa, 1981; Jaffa and Stewart, 1981). For induction to occur, the object must form a closed loop or be grounded at both ends (to allow a current to flow). It is also possible for an induced voltage to develop if only one end of the object is grounded; in this case, current can flow only when the loop is completed, e.g., by a person touching the object and ground at the same time. The magnitude of the induced voltage and current depends on several factors including the length of the parallel object, the total resistance of the closed loop, the orientation relative to the transmission line, and the magnitude of the field. Perpendicular objects, short objects, or high-resistance loops will result in no or small induced voltages/currents, while long parallel objects with low resistance will result in higher induced voltages/currents.

5.2.3.3 Potential Field Effect Issues

5.2.3.3.1 Pipelines

There is an increasing trend to locate underground oil and gas pipelines on the same right-of-way as electric transmission lines. Pipelines in the vicinity of a transmission line can experience induced voltages both from electrostatic coupling and magnetic induction, typically during pipeline construction or if the pipeline is constructed above ground for a significant distance (Kirkpatrick, 1995). Often of more interest, however, are the pipelines that, although buried underground and therefore shielded from electrostatic effects, can still be affected due to magnetically induced currents from power lines (Nelson, 1986). This typically occurs where the pipeline runs parallel to the transmission line for a long distance, as well as where the pipeline may enter or exit the shared right-of-way (Dabkowski and Taflove, 1979). In addition to the distance, the pipeline and transmission line parallel one another: the magnitude of the induced currents is a complex function of transmission line configuration, distance from transmission line to pipeline, and the potential imbalance of transmission line phase currents (Nelson, 1986).

One of the primary concerns with induced currents on underground pipelines is the possibility of enhanced corrosion of the pipeline by an electrochemical effect (Bruckner, 1964) or degradation of the cathodic protection systems (Taflove, 1979). In an effort to mitigate these mechanisms the National Association of Corrosion Engineers (NACE) has published standards to control corrosion and safety issues (NACE, 2014). Standards covering similar topics exist in Canada (CSA, 2003) and Europe (CENELEC, 2012).

A series of studies, funded jointly by the Electric Power Research Institute (EPRI) and the American Gas Association (AGA), resulted in comprehensive prediction methods and mitigation techniques for induced voltages and currents on pipelines (Dabkowski and Taflove, 1979; Taflove, 1979; Taflove and Dabkowski, 1979). Some of the effective techniques that can be applied to mitigating these induction issues include design of the joint corridor, pipeline grounding methods, insulating devices, and pipeline extensions (Taflove, 1979). During the design phase, an effort is made to maintain the configuration of the transmission line and pipeline relative to one another to avoid electrical discontinuities. To accomplish this, as described by Taflove (1979),

> *[a] joint-use right-of-way should be designed to: avoid any changes of separation between individual utilities; avoid pipeline insulating joints; avoid changes in pipeline coating quality; avoid the use of power line phase transpositions; and avoid any other geometric or electrical discontinuities of utilities sharing the right-of-way.*

Although developed specifically for pipelines, similar computations and analyses can be applied to problems of a similar nature, including irrigation pipes and fences (e.g., Jaffa and Stewart, 1981).

5.2.3.3.2 Medical Devices

At sufficiently high EMF levels, medical devices such as pacemakers or implantable cardioverter defibrillators (ICDs) can experience EMI from both AC and DC sources. There are generally two categories defined by medical device standards and medical device manufacturers. The first is a threshold below that the device will operate without change in functional; above this level (but below the second threshold) the device will cease normal operation and will enter a default mode (such as automatic pacing at a predetermined rate), but will return to normal operation when the incident field is removed. Finally, above the second threshold, the device may be damaged to such an extent that it cannot return to normal operation even after removal of the source.

The variability in susceptibility among different medical device manufacturers, and in particular, among older versus newer devices can be very wide. Older models are often more susceptible to interference; however, the immunity of these devices to AC EMF is continually improving. Each manufacturer can give specific information regarding the limits of interference; however, in the absence of information regarding the susceptibility limits of a particular device, other standards may be consulted. For DC magnetic fields, Standard ANSI/AAMI/ISO 14117:2012 14117:2012 adopted by the American National Standards Institute (ANSI) specifies that devices tested to this standard should not be affected by static magnetic fields of flux density up to 1 millitesla (mT) and no changes in the function of the pacemaker or the ICD should occur. Up to a static magnetic flux density of 50 mT an ICD should not remain functionally affected after discontinuation of the exposure.

For AC fields between 16.6 Hz and 1 kHz, ANSI/AAMI/ISO 14117:2012 specifies that a 2 mV peak-to-peak voltage at the device terminals shall be unlikely to change the therapeutic behavior of the device. While this internal device voltage threshold may be difficult to relate directly to an external EMI signal, it can be done with information regarding the magnitude of the EMI signal, the relative orientation of the patient with respect to the EM field, and the design and location of the specific device itself. To address this aspect of exposure, ANSI/AAMI/ISO 14117:2012 presents a "correlation between the voltage test levels [defined by peak-to-peak voltage] and EM field strength levels (amps/meter, peak). This correlation uses Faraday's law and reflects an effective induction loop area of 225 cm^2, considered worst case for left pectoral implants." ANSI/AAMI/ISO 14117:2012 presents an annex that details magnetic-field levels from 10 Hertz (Hz) to 10 MHz using this correlation, which if not exceeded, would comply with test limits.

European Standard EN 45502:2004 also specifies an immunity level, between 16.6 Hz and 1 kHz, of 2 mV peak-to-peak at device terminals, although it does not have an informative annex similar to ANSI/AAMI/ISO 14117:2012 explicitly calculating equivalent magnetic-field levels. Other European standards, such as the European Committee for Electrotechnical Standardization (CENELEC), also give conservative general limits, stating that implanted medical devices should function without interference below levels specified in EN 519/1999, which at a frequency of 60 Hz correspond to of 4.2 kilovolts per meter (kV/m) and 833 milligauss (mG) for ELF electric fields and magnetic fields, respectively, although the standard goes on to state that "[f]or higher fields the voltage can cause EMI effects but often this is not clinically significant ... and transient exposure [to higher fields] can be permitted" (EN 50527-1, 2010).

The manufacturer of any particular medical device is the ultimate authority on the susceptibility of their device and can give precise information regarding the limits of interference to that device. If specific situations arise it is always prudent to confer with the device manufacturer and one's physician for specific guidance on medical questions.

5.3 Worker and Public Safety

The previous sections focused on EMI to electrical devices and equipment from corona on transmission lines, the scattering of radiofrequency communication signals by transmission lines and structures, and coupling of power frequency EMF. While preventing interference from energy facilities to devices is a key design goal for engineers, it is also important in the design and operation of energy facilities to prevent overexposure of workers and the public (IEEE, 1978).

Exposure to EM fields from transmission line conductors at frequencies above a few MHz is so extraordinarily low from a safety perspective it can be regarded as *de minimis* (Pakala et al., 1967; Chartier et al., 1986).[3,4] A potential exception might be where a utility line worker enters a high field zone around a radiofrequency communications antenna mounted on a utility structure where exposure (or coupling of fields to very nearby objects) may exceed radiofrequency limits within a few meters of an antenna (see Mann et al., 2000).

5.3.1 Coupling of EMF to Persons

While the exposures to power frequency EMF from energy facilities also are low, the potential responses of workers and the public to capacitive-coupled electric fields and contact with materials and equipment exposed to inductively coupled magnetic fields can lead to perception of the electric field and annoying or potentially hazardous shocks from either electric or

[3]Evaluation of human exposure at frequencies above a few 10s of MHz are made in terms of body heating through energy absorption, called specific absorption rate or SAR in health-based literature.

[4]Pakala et al. (1967) does not assess exposure to human health. Measurements from this study between 100 kHz and 1000 MHz were used to calculate a conservative upper bound EM field levels close to the conductor which were far below human health guidelines (ICNIRP 1998; ICES, 2005).

magnetic fields. For utility workers, the risks of shocks related to power frequency EMF are managed by adherence to U.S. Occupational Health and Safety standards, transmission line designs that limit induced currents in accordance with the National Electric Safety Code (NESC) (IEEE, 2012), and standard utility safety practices.

For a DC transmission line, the perception of the electric field arises because of the ability of a static electric field to move charges on the surface of the body (possibly enhanced by space charge in the form of air ions). Because static fields are nonoscillating they do not cause any significant currents or fields within the body (Polk and Postow, 1986). In contrast, if a person is extremely well insulated from ground, then accumulated static charge can discharge when contact is made with a grounded object, which is a potential form of a nuisance shock. While nuisance shocks can occur for either AC or DC transmission lines, the threshold for perception of DC shocks is higher and the physiological reactions are less than for AC shocks (Dalziel and Mansfield, 1950; Dalziel and Massoglia, 1956). The primary focus below is therefore on AC fields.

The likelihood and magnitude of AC and DC shocks is often limited by regulations. For example, in the United States and Canada, the NESC (NESC, 2012) and Canadian Standards Association (CSA, 2015), respectively, provide guidelines to limit approach to energized conductors and specifically limit the induced current under overhead transmission lines to a maximum of 5 milliamperes (mA). This current level was selected as a limit because it is below the threshold at which most people can still let go of a charged object (Dalziel and Massoglia, 1956). For current levels above 5 mA, it is possible that the current could cause "grip tetanus" or muscle contractions that leave a person unable to let go of the charged object. In practical cases, however, induced currents are typically lower than these short-circuited values because the average individual wearing shoes does not provide a short-circuited path to ground (e.g., IEC, 2005).

5.3.2 Standards and Guidelines

A number of states have regulations on the levels of EMF permitted on the rights-of-way of new transmission lines, as summarized in Table 5.1.

The principal rationale for the limits on electric fields listed in Table 5.1 is to minimize perception of the field and electric shocks. Two states, Florida and New York, also have enacted

Table 5.1 State Standards and Guidelines for EMF for New Transmission Lines and Substations

State	Transmission Line Rating	Property Boundary at Substation		At Edge of Transmission Line ROW		On Transmission Line ROW	
		Electric Field (kV/m)	Magnetic Field (mG)	Electric Field (kV/m)	Magnetic Field (mG)	Electric Field (kV/m)	Magnetic Field (mG)
New York	All	–	–	1.6	200[d]	11.8 11.0[b] 7.0[c]	–
	$\leq$230 kV	2.00	150	2.00	150	8	–
Florida	$>$230 kV – $\leq$500	2.00	200	2.00	200	10	–
	$>$500	5.5	250	5.5	250	15	
Minnesota	All					8	
Montana	All			1[a]		7[b]	
New Jersey	All					3	
Oregon	All					9	

[a]Can be waived by landowner.
[b]Maximum for private road crossings.
[c]Maximum for highway crossings.
[d]Evaluated at winter normal conductor rating.

standards to limit magnetic fields at the edge of transmission line rights-of-way (150 and 200 mG, respectively) (NYPSC, 1978, 1990; FDER, 1989; FDEP, 1996). The basis for limiting magnetic fields from transmission lines was not based upon a finding that EMF cause health effects, but rather was to maintain the status quo so that fields from new transmission lines would be no higher than those produced by existing transmission lines.

Apart from shock, the possibility of stimulation of neural and muscle tissues by noncontact exposure to high-strength EMF also is a theoretical possibility. Two international scientific organizations, the International Commission on NonIonizing Radiation Protection (ICNIRP) and the International Committee on Electromagnetic Safety (ICES), have published guidelines for limiting public exposure to ELF EMF to protect against these effects (ICNIRP, 1998, 2010; ICES, 2002). ICNIRP is an independent organization of scientists from various disciplines with expertise in the field of nonionizing radiation assembled from around the world. It is the formally recognized, nongovernmental organization that develops safety guidance for nonionizing radiation for the World Health Organization, the International Labor Organization, and the European Union.

The ICES is sponsored by the ANSI and the IEEE. The mandate for ICES is the "[d]evelopment of standards for the safe use of electromagnetic energy in the range of 0 Hz to 300 GHz relative to the hazards of exposure to man ... to such energy."[5] The ICES encourages a balanced international volunteer participation from several sectors: the interested general public; the scientific, health and engineering communities; agencies of governments; energy producers; and energy users.

The limits on exposure to static and ELF fields by these organizations are designed to provide protection from known adverse effects of ELF fields. The practical relevance of these limits, however, is quite low as the exposures of workers and the public from transmission and substation facilities to EMF at power frequencies are well below levels that would cause internal electric fields to exceed limits set by these organizations. Both ICNIRP and ICES have specified limits on exposure to EMF by placing a limit on the maximum electric field induced inside the body from an external source of EMF. This internal exposure limit is called the Basic Restriction and is important because sufficiently high electric fields in tissue can potentially result in a physiological effect such as nerve or muscle stimulation. Electric field measurements inside the body, however, are difficult to perform and so both ICNIRP and ICES have provided screening values of external EMF levels for comparison to measurements and calculations that are universally guaranteed to not exceed the Basic Restrictions. ICNIRP calls the screening levels reference levels (RLs), while ICES uses the term maximum permissible exposure. In cases where the exposure exceeds the RL, evaluation of compliance then needs to be made by measurements in human phantom or via mathematical modeling such as that described by Kavet et al. (2012). Table 5.2 summarizes the exposure levels that would produce electric fields in the brain and central nervous system equal to these Basic Restrictions.

[5]The ICES is a 50-year-old internationally recognized, EMF standard-setting organization, which is sponsored by the IEEE that, itself, was established in 1884. The ICES should not be confused with a group of scientists who have acted together as an advocacy group and banded together in 2003 under the similar name of the International Commission for Electromagnetic Safety.

Table 5.2 Reference Levels (ICNIRP), Maximum Permissible Exposures (ICES), and Exposure Levels Estimated to Produce Electric Fields in Tissue Equal to Basic Restrictions for 60 Hz. All Values Are rms

Agency	Magnetic Field (mG)	Electric Fields (kV/m)
ICNIRP (2010)		
Reference Level		
Public	2000	4.2
Occupational	10,000	8.3
Basic Restriction		
Public[a]	12,422[b]	36[b]
ICES (2002)		
Maximum Permissible Exposure		
Public	9,040	5 or 10[c]
Occupational	27,100	20
Basic Restriction		
Public[d]	9,146[b]	27[b]

[a]Basic Restriction is 24 mV/m in CNS tissue of the head.
[b]Field levels calculated from Kavet et al. (2012).
[c]ICES determined an exception of 10 kV/m within transmission line rights-of-way because persons do not spend any significant amount of time here and very specific conditions are needed for a response to occur (ICES, 2002, p. 27).
[d]Basic Restriction is 17.7 mV/m in the brain.

5.3.3 Public Health Issues

The siting and construction of transmission lines, substations, and in some instances low-voltage distribution lines has prompted public concern that exposure to EMF might adversely affect health. Concern about health effects first emerged in the 1960s with reports, later unconfirmed, that workers in 400-kV and 500-kV high-voltage switchyards in the former Soviet Union experienced a variety of nonspecific health complaints. Years later, the publication of community epidemiology studies in Denver, Colorado, sparked concerns about magnetic fields because of statistical associations between adult and childhood cancers and proximity of homes to high-current carrying distribution and service lines (Wertheimer and Leeper, 1979; 1982). Over the next three decades many other human epidemiology studies as well as laboratory studies of animals and isolated tissues and cells were conducted. Because of public concern, the widespread exposure of populations to EMF from the transport and use of electricity, and the amount and complexity of the scientific studies in this area, comprehensive evaluations of the available scientific evidence have been performed for health and scientific agencies by panels composed of scientists with expertise in relevant scientific disciplines. Reviews published just since 2006 include those of the World Health Organization (WHO, 2007); the Scientific Committee on Emerging and Newly Identified Health Risks (SCENIHR, 2007, 2009, 2015), ICNIRP (2010); the European Health Risk Assessment Network on Electromagnetic Fields Exposure (EFHRAN, 2010, 2012); the Swedish Radiation Protection/Safety Authority (SSM, 2015 and preceding 6 years); and the Ministry of Health of New Zealand (NZMH, 2015). In 2015, SCENIHR published the most recent and comprehensive review of EMF research and concluded that the research did not confirm any adverse health effects of EMF exposure (SCENIHR, 2015), a conclusion shared with previous reviews. With respect to cancers among children and adults, these reviews noted that the statistical associations with EMF reported in some of the epidemiologic studies could not be distinguished from effects of bias, confounding, and chance; no mechanism had been confirmed to explain adverse effects at low exposure levels; and the animal studies did not confirm adverse effects, including cancer. Altogether, these considerations persuaded the reviewing bodies that the associations reported in some epidemiologic studies did not reflect causal relationships.

Thus, like any environmental exposure, very high exposures to EMF have been found to produce adverse effects (nerve and muscle stimulation), but compliance with exposure limits set by national and international scientific agencies prevent such acute effects and are protective of public health. Reviews by health and scientific agencies have not found that exposures at levels below these standards are the cause of any effects on health.

5.4 Summary

5.4.1 Future Trends and Other References

As the demand for electricity continues to increase, so too will the necessity for new power generation and transmission.

While the nature of the sources of EMI described in this chapter are not likely to change significantly in the future, corona in particular is likely to become an increasingly-important consideration during the design of transmission lines, as utilities move toward increasing transmission line voltages such as 800-kV DC transmission lines (CAE, 2009). Greater detail regarding the repetition rate, physical characteristics, rise time, current amplitudes, and voltage dependence of each type of corona are available in EPRI's AC Transmission Line Reference Book (EPRI, 2006), while additional information regarding the potential scattering and detuning of transmission line towers can be found in IEEE, (1996, under revision). Finally, as discussed in Section 5.3.2, numerous and detailed reviews of the potential effects on worker and public safety were detailed by the World Health Organization (2007), the Scientific Committee on Emerging and Newly Identified Health Risks (2009, 2015), ICNIRP (2010), the European Health Risk Assessment Network on Electromagnetic Fields Exposure (2010, 2012), the Swedish Radiation Protection/Safety Authority (2010), and the Ministry of Health of New Zealand (NZMH, 2015).

The continued development of the Smart Grid, in which users, appliances, and devices connect and interact with the electric grid, will result in more and more types of devices connected to the electric grid, and additional evaluations of EMC will become increasingly necessary and are being addressed in publications such as Smolenski (2012). Organizations such as the IEEE, IEC, and the International Special Committee on Radio Interference (CISPR) are all immersed in evaluating potential for EMC and standardizing requirements. For example, the IEEE has an active project (IEEE, n.d.) to establish testing requirements for intelligent electronic devices to be connected with the grid. Likewise the IEC has established what it describes as a "Road Map" which is designed to provide an overview of all IEC standards relevant for the Smart Grid, as well as to identify potential gaps in EMC standards that will be relevant to the implementation of the Smart Grid (IEC, 2010). CISPR has also developed a guidance document designed to ensure that proper EMC performance is considered in the Smart Grid (CISPR, 2014).

In addition to the typical EMI scenarios discussed here, a number of reviews, studies, and reports have been published in recent years regarding the susceptibility of the electric power grid to natural or intentional EMI including catastrophic threats from various sources such as electromagnetic pulses, high-impact low-frequency events, and geomagnetically induced currents from solar storms (e.g., Radasky and Savage, 2010; SGIP, 2012; EPRI, 2013; Zanetti, 2013). Other less dramatic but no less important issues of the interoperability and EMC of the electric power grid are discussed in multiple standards by the IEC and IEEE (e.g., Ramie, 2013; Smart Grid Strategic Group, 2010; IEEE, n.d.).

5.4.2 Conclusion

This chapter has discussed some of the common sources of potential EMI from electrical power systems including both radiated EM sources, physical sources and inductively or capacitively coupled sources.

Section 5.2.1 discussed potential EMI from corona, a common source of EMI related to power systems, which is primarily associated with higher-voltage AC and DC transmission lines. Corona is a broadband source that can span a range of frequencies from below 100 kHz to above 1000 MHz, but is predominantly of interest in the 100 kHz to 2 MHz range. Degradation of AM radio may be encountered beneath or in very close proximity to a transmission line although the effect decreases rapidly with distance so is not typically an issue in homes or businesses. Some military and other government agencies also operate communications equipment in this frequency band and so have the potential for interference from transmission line EMI. Devices and systems that operate at higher frequencies and those that use other modulation schemes such as FM radios, cell phones, GPS, and satellite receivers, are unlikely to be susceptible to EMI generated by transmission lines. Fortunately in many cases, interference from these types of signals can be mitigated either through various methods of reducing the magnitude of the offending source or by altering the characteristics of the receiver.

Section 5.2.2 discussed that while radiated EM fields can actively interfere with and overwhelm transmitted radio waves from devices such as such as cell phones, televisions, AM/FM radio, and GPS, various forms of blocking and scattering can also be considered forms of EMI due to their ability to impede a transmitted signal from reaching a receiver in the first place. Whether the EMI is through blocking, multipath, or deformation of an antenna radiation pattern, interference will only occur if the transmission lines and towers efficiently interact with the incident radio wave; a scenario which is both highly dependent on frequency and relative location. Above approximately 10 MHz, blocking and multipath EMI due to scattering from transmission lines and towers is rare and while scattering from transmission line towers can be significant, this typically occurs only when very close to the tower. The most commonly affected transmitters are those operating below approximately 10 MHz and using AM broadcast schemes. For this reason FCC Rules mandate the construction of a tower near an AM broadcast station requires investigation of any potential effects on the AM broadcast station signal pattern and to take measures to mitigate those effects. Such measures may include installing detuning skirts or stub circuits on the transmission line towers or by isolating select towers from the interconnecting shieldwires.

Section 5.2.3 discussed potential EMI from quasi-static EMF produced by transmission lines, which can affect nearby facilities and devices through either electrostatic coupling or magnetic induction. In particular, this section covered potential interaction of EMF with pipelines and medical devices. The interaction between electric transmission lines and oil or gas pipelines, particularly safety and corrosion concerns, have received special attention due to the increasing trend to locate the two on the same right-of-way. A series of studies, funded jointly by EPRI and the AGA, resulted in comprehensive prediction methods and mitigation techniques for induced voltages and currents on pipelines and standard-setting organizations such as NACE have addressed ways to mitigate corrosion and safety issues. Some medical devices, such as implantable pacemakers or ICDs can experience interference from sufficiently high EMF levels (either AC or DC). For this reason, medical device manufacturers comply with standards such as ANSI/AAMI/ISO 14117:2012 or European Standard EN 45502:2004, which specify specific immunity levels to EMI that device manufacturers must meet for compliance.

While Sections 5.2 focuses primarily on EMI to electrical devices and equipment, Section 5.3 discusses EMF interactions with people. Section 5.3.1 discusses the forms in which EMF can couple to persons including nuisance shocks and steady-state currents as well as the regulations surrounding these potential effects, primarily the NESC and CSA, which provide guidelines to limit approach to energized conductors and specifically limit the induced current under overhead transmission lines to a maximum of 5 mA. This current level was selected as a limit because it is below the threshold at which most people can still let go of a charged object. Section 5.3.2 discusses the standards and guidelines for EMF adopted by various U.S. states, developed primarily to minimize perception of the field and shocks or, in the case of Florida and New York, to maintain the status quo compared to existing fields. The section also discusses health-based guidelines of ICNIRP and ICES for public exposure, which are designed to provide protection from known adverse effects of EMF. The Section (5.3.3) concludes with a discussion of the public health issues associated with EMF and the reviews of the body of scientific literature performed by various health and scientific agencies. Based on thorough reviews of the body of scientific literature these agencies have not found that exposures at levels below these standards are the cause of any effects on health.

References

American National Standard Dictionary of Electromagnetic Compatibility (EMC) including Electromagnetic Environmental Effects (E3), 2009. ANSI C63.14-2009 c1–46. <http://dx.doi.org/10.1109/IEEESTD.2009.5297139>.

American National Standards Institute (ANSI), 2012. ANSI/AAMI/ISO 14117/Ed. 1, Active implantable medical devices—Electromagnetic compatibility—EMC test protocols for implantable cardiac pacemakers, implantable cardioverter defibrillators and cardiac resynchronization devices.

Bancroft, J.B., Morrison, A., Lachapelle, G., 2012. Validation of GNSS under 500,000 V direct current (DC) transmission lines. Computers and Electronics in Agriculture. 83, 58–67.

Bruckner, W.H., 1964. The effects of 60 Hz alternating currents on the corrosion of steels and other metals buried in soil. Engineering Experiment Station Bulletin. 470, University of Illinois, Urbana, Illinois.

Canadian Academy of Engineering (CAE), 2009. Report of the Canada Power Grid Task Force—Volume II Background and Assessment. Canadian Academy of Engineering.

Canadian Standards Association (CSA), 2003. CAN/CSA-C22.3 NO. 6-13: Principles and Practices of Electrical Coordination Between Pipelines and Electric Supply Lines. Canadian Standards Association, Etobicoke, ON.

Canadian Standards Association (CSA), 2015. C22.3 NO. 1-15: Overhead Systems. Canadian Standards Association, Toronto, ON.

Chartier, V.L., Lee, L.Y., Dickson, L.D., Martin, K.E., 1987. Effect of high altitude on high voltage AC transmission line corona phenomena. IEEE Trans. Power Delivery. 2, 225–237. Available from: http://dx.doi.org/10.1109/TPWRD.1987.4308094.

Chartier, V.L., Sheridan, R., DiPlacido, J.N., Loftness, M.O., 1986. Electromagnetic interference measurements at 900 MHZ on 230-KV and 500-KV transmission lines. IEEE Trans. Power Delivery. 1, 140–149. Available from: http://dx.doi.org/10.1109/TPWRD.1986.4307944.

Claes, D., 1985. Radio Influence (RI) and Television Influence (TVI) Voltage contribution of a 345kV transmission line. IEEE Trans. Broadcast. BC-31, 66–77. Available from: http://dx.doi.org/10.1109/TBC.1985.266557.

Dabkowski, J., Taflove, A., 1979. Prediction method for buried pipeline voltages due to 60 Hz AC inductive coupling Part II–Field test verification. IEEE Trans. Power Appar. Syst. PAS-98, 788–794. Available from: http://dx.doi.org/10.1109/TPAS.1979.319291.

Dallaire, R.D., Maruvada, P.S., Rivest, N., 1984. HVDC monopolar and bipolar cage studies on the corona performance of conductor bundles. IEEE Trans. Power Appar. Syst. PAS-103, 84–91. Available from: http://dx.doi.org/10.1109/TPAS.1984.318581.

Dalziel, C.F., Mansfield, T.H., 1950. Effect of frequency on perception currents. Trans. Am. Inst. Electr. Eng. 69, 1162–1168. Available from: http://dx.doi.org/10.1109/T-AIEE.1950.5060274.

Dalziel, C.F., Massoglia, F.P., 1956. Let-go currents and voltages. Trans. Am. Inst. Electr. Eng., Part II: Appl. Ind. 75, 49–56. Available from: http://dx.doi.org/10.1109/TAI.1956.6367148.

Electric Power Research Institute (EPRI), 1978. Transmission Line Reference Book—HVDC to ±. EPRI, Palo Alto, CA, p. 600.

Electric Power Research Institute (EPRI), 1982. AC Transmission Line Reference Book—345 kV and Above. second ed. EPRI, Palo Alto, CA.

Electric Power Research Institute (EPRI), 2006. AC Transmission Line Reference Book—200 kV and Above. third ed. EPRI, Palo Alto, CA.

Electric Power Research Institute (EPRI), 2013. *Electromagnetic Pulse and Intentional Electromagnetic Interference (EMI) Threats to the Power Grid: Characterization of the Threat, Available Countermeasures, and Opportunities for Technology Research* (No. 3002000796), Power Delivery & Utilization—Transmission. Palo Alto, CA: EPRI.

European Committee for Electrotechnical Standardization (CENELEC), 2010. Procedure for the Assessment of the Exposure to Electromagnetic Fields of Workers Bearing Active Implantable Medical Devices—Part 1: General (EN 50527-1). CENELEC, Brussels, Belgium.

European Committee for Standardization (CEN), 2012. CEN EN 50443:2012. Effects of Electromagnetic Interference on Pipelines Caused by High Voltage AC Electric Traction Systems and /or High Voltage AC Power Supply Systems.

European Health Risk Assessment Network on Electromagnetic Fields Exposure (EFHRAN), 2010. Report on the Analysis of Risks Associated to Exposure to EMF: *In Vitro* and *In Vivo* (Animals) Studies. EHFRAN, Milan, Italy.

European Health Risk Assessment Network on Electromagnetic Fields Exposure (EFHRAN), 2012. Risk Analysis of Human Exposure to Electromagnetic Fields (Revised). Report D2 of the EFHRAN Project. EFHRAN, Milan, Italy.

Federal Communications Commission (FCC). Title 47: Telecommunication, Electronic Code of Federal Regulations, 2016. <http://www.ecfr.gov/> (accessed 28.12.16.).

Federal Communications Commission (FCC), 2016a. *Digital television*. <https://www.fcc.gov/general/digital-television> (accessed 29.08.16.).

Florida Department of Environmental Regulation (FDER). 1989. *Electric and Magnetic Fields*. Chapter 17–274. FDER.

Florida Department of Environmental Protection (FDEP), 1996. *Electric and Magnetic Fields*. Chapter 62–814. FDEP.

Gary, C., Kolcio, N., Moreau, M., Reichman, J., Taylor, E.R., 1973. CIGRE/IEEE survey on extra high voltage transmission line radio noise. IEEE Trans. Power Appar. Syst. PAS-92, 1019–1028. Available from: http://dx.doi.org/10.1109/TPAS.1973.293668.

Gehrig, E.H., Peterson, A.C., Clark, C.F., Rednour, T.C., 1967. Bonneville power administration's 1100-kV direct current test project: II—radio interference and corona loss. IEEE Trans. Power Appar. Syst. PAS-86, 278–290. Available from: http://dx.doi.org/10.1109/TPAS.1967.291954.

Gerdan, G.P., Coombe, L.J., Takac F., 1995. The effects of RF interference, multipath and signal obstruction on the GPS observables. Prepared for the State Data Centre (Number SDC95/1). Melbourne, Australia: Royal Melbourne Institute of Technology.

Giao, T.N., Jordan, J.B., 1968. Modes of corona discharges in Air. IEEE Trans. Power Appar. Syst. PAS-87, 1207–1215. Available from: http://dx.doi.org/10.1109/TPAS.1968.292211.

Giussani, R., Cotton, I., Sloan, R., 2012. Detection of corona with RF methods and spectra analysis. In: *Conference Record of the 2012 IEEE International Symposium on Electrical Insulation* (ISEI). Presented at the Conference Record of the 2012 IEEE International Symposium on Electrical Insulation (ISEI), pp. 132–136. Available from: http://dx.doi.org/10.1109/ELINSL.2012.6251442.

Goldman, M., Goldman, A., Sigmond, R.S., 1985. The corona discharge, its properties and specific uses. Pure Appl. Chem. 57. Available from: http://dx.doi.org/10.1351/pac198557091353.

Hermstein, D.-I.W., 1960. Die Stromfaden-Entladung und ihr Übergang in das Glimmen. Archiv f. Elektrotechnik. 45, 209–224. Available from: http://dx.doi.org/10.1007/BF01580597.

Industry Canada, 2013. *ICES-004—alternating current high voltage power systems.* <http://www.ic.gc.ca/eic/site/smt-gst.nsf/eng/sf00021.html> (accessed 29.08.16.).

Institute of Electrical and Electronics Engineers (IEEE), 1978. Electric and magnetic field coupling from high voltage AC power transmission lines—classification of short-term effects on people. IEEE Trans. Power Appar. Syst. PAS-97, 2243–2252. Available from: http://dx.doi.org/10.1109/TPAS.1978.354728.

Institute of Electrical and Electronics Engineers (IEEE), 2012. *National Electrical Safety Code (NESC)*. CS-2012. IEEE, New York, p. 2012.

Institute of Electrical and Electronics Engineers (IEEE), IEEE Radio Noise Subcommittee, 1971. Radio noise design guide for high-voltage transmission lines. IEEE Trans. Power Appar. Syst. PAS-90, 833–842. Available from: http://dx.doi.org/10.1109/TPAS.1971.293115.

Institute of Electrical and Electronics Engineers (IEEE), IEEE Radio Noise and Corona Committee, 1980. Review of technical considerations on limits to interference from power lines and stations. IEEE Trans. Power Appar. Syst. PAS-99, 365–388. Available from: http://dx.doi.org/10.1109/TPAS.1980.319648.

Institute of Electrical and Electronics Engineers (IEEE), 1996. Guide on the Prediction, Measurement, and Analysis of AM Broadcast Re-radiation by Power Lines. IEEE, New York, NY.

Institute of Electrical and Electronics Engineers (IEEE), International Committee on Electromagnetic Safety (ICES), 2002. Standard for Safety Levels with Respect to Human Exposure to Electromagnetic Fields 0 to 3 kHz. IEEE, Piscataway, NJ.

Institute of Electrical and Electronics Engineers (IEEE). International Committee on Electromagnetic Safety (ICES), IEEE Recommended Practice for Radio Frequency Safety Programs, 3 kHz to 300 GHz, 2005, IEEE; Piscataway, NJ.

Institute of Electrical and Electronics Engineers (IEEE), n.d. P1613.1-Standard Environmental and Electromagnetic Compatibility (EMC) Testing Requirements for Intelligent Electronic Devices (IEDs) Installed in Transmission and Distribution Facilities. Accessed 29 August 2016. <https://standards.ieee.org/develop/project/1613.1.html> (accessed 02.07.16.).

International Commission on Non-Ionizing Radiation Protection (ICNIRP). Exposure to high frequency electromagnetic fields, biological effects and health consequences (100 kHz-300 GHz), 2009. <http://www.icnirp.de/documents/RFReview.pdf> (accessed 06.01.16.).

International Commission on Non-Ionizing Radiation Protection (ICNIRP), 1998. Guidelines for limiting exposure to time-varying electric, magnetic, and electromagnetic fields (up to 300 GHz). Health Phys. 74, 494–522.

International Commission on Non-Ionizing Radiation Protection (ICNIRP), 2010. Guidelines for limiting exposure to time-varying electric and magnetic fields (1 Hz to 100 kHz). Health Phys. 99, 818–836.

International Electrotechnical Commission (IEC), 2005. *Effects of current on human beings and livestock—Part 1: General aspects.* IEC 60479-1:2005-07.

International Electrotechnical Commission (IEC), Prepared by SMB Smart Grid Strategic Group (SG3), 2010. *IEC smart grid standardization roadmap* (No. 2009–12). <http://www.iec.ch/smartgrid/downloads/sg3_roadmap.pdf> (Accessed August 2016.).

International Special Committee on Radio Interference (CISPR), 2014. CISPR/1270/INF, CISPR Guidance document on EMC of equipment connected to the SmartGrid (No. 2014-02-21).

Jaffa, K.C., 1981. Magnetic field induction from overhead transmission and distribution power lines on parallel fences. IEEE Trans. Power Appar. Syst. PAS-100, 1624–1636. Available from: http://dx.doi.org/10.1109/TPAS.1981.316557.

Jaffa, K.C., Stewart, J.B., 1981. Magnetic field induction from overhead transmission and distribution power lines on buried irrigation pipelines. IEEE Trans. Power Appar. Syst. PAS-100, 990–1000. Available from: http://dx.doi.org/10.1109/TPAS.1981.316634.

Juette, G.W., 1972. Evaluation of television interference from high-voltage transmission lines. IEEE Trans. Power Appar. Syst. PAS-91, 865–873. Available from: http://dx.doi.org/10.1109/TPAS.1972.293419.

Juette, G.W., Zaffanella, L.E., 1970. Radio noise currents and audible noise on short sections of UHV bundle conductors. IEEE Trans. Power Appar. Syst. PAS-89, 902–913. Available from: http://dx.doi.org/10.1109/TPAS.1970.292653.

Kavet, R., Dovan, T., Reilly, J.P., 2012. The relationship between anatomically correct electric and magnetic field dosimetry and published electric and magnetic field exposure limits. Radiat. Prot. Dosim. 152, 279–295.

Kirkpatrick, E., 1995. Basic concepts of induced AC voltages on pipelines. Mater. Perform. 34, 14–18.

Laforest, J.J., 1968. Seasonal variation of fair-weather radio noise. IEEE Trans. Power Appar. Syst. PAS-87, 928–931. Available from: http://dx.doi.org/10.1109/TPAS.1968.292066.

Loeb, L.B., 1965. Electrical Coronas, Their Basic Physical Mechanisms. University of California Press, Berkeley, CA.

Loftness, M.O., 2003. AC Power Interference Handbook. Percival Technology, Tumwater, WA.

Loftness M.O., Chartier V., Reiner G.L., 1981. EMI Correction techniques for transmission line corona. *Proceedings of the 1981 IEEE International Symposium on Electromagnetic Compatibility*, Boulder, CO.

Madge, R.C., Jones, D.E., 1986. Effect of power lines on AM radio broadcast radiation patterns. IEEE Power Eng. Rev. PER-6, 44–144. Available from: http://dx.doi.org/10.1109/MPER.1986.5527725.

Mann, S., Cooper, T., Allen, S., Blackwell, R., Lowe, A., 2000. Exposure to Radio Waves near to Mobile Phone Base Stations. National Radiological Protection Board, Chilton, Didcot, UK.

Meek, J.M., Craggs, J.D., 1953. Electrical Breakdown of Gasses. Oxford University Press, Clarendon Press.

NACE International, 2014. SP0177-2014-Mitigation of Alternating Current and Lightning Effects on Metallic Structures and Corrosion Control Systems. NACE, Houston, TX.

Nelson, J.P., 1986. Power systems in close proximity to pipelines. IEEE Trans. Ind. Appl. IA-22, 435–441. Available from: http://dx.doi.org/10.1109/TIA.1986.4504739.

Newell, H.H., Liao, T.-W., Warburton, F.W., 1967. Corona and RI caused by particles on or near EHV conductors: I-fair weather. IEEE Trans. Power Appar. Syst. PAS-86, 1375–1383. Available from: http://dx.doi.org/10.1109/TPAS.1967.291815.

Newell, H.H., Liao, T.-W., Warburton, F.W., 1968. Corona and RI caused by particles on or near EHV conductors: II-foul weather. IEEE Trans. Power Appar. Syst. PAS-87, 911–927. Available from: http://dx.doi.org/10.1109/TPAS.1968.292065.

New York Public Service Commission (NYPSC), 1978. Opinion No. 78-13. Opinion and Order Determining Health and Safety Issues, Imposing Operating Conditions, and Authorizing, in Case 26529, Operation Pursuant to those Conditions: NYPSC.

New York Public Service Commission (NYPSC), 1990. Statement of Interim Policy on Magnetic Fields of Major Transmission Facilities. Cases 26529 and 26559 Proceeding on Motion of the Commission: NYPSC.

New Zealand Ministry of Health (NZMH), 2015. Interagency Committee on the Health Effects of Non-ionising Fields: Report to Ministers. Ministry of Health, Wellington, New Zealand.

Pakala, W.E., Chartier, V.L., 1971. Radio noise measurements on overhead power lines from 2.4 to 800 KV. IEEE Trans. Power Appar. Syst. PAS-90, 1155–1165. Available from: http://dx.doi.org/10.1109/TPAS.1971.292880.

Pakala, W.E., Taylor, J., Harrold, R.T., 1967. *High voltage power line siting criteria. Volume I.* Technical Report No. RADC-TR-68-316.

Peek Jr., F.W., 1911. Law of corona and the dielectric strength of air. AIEE Trans. Power Appar. Syst. 30 (pt. III), 1889–1965, 1911.

Phillips, A.J., Kuffel, J., Baker, A., Burnham, J., Carreira, A., Cherney, E., et al., 2008. Electric fields on AC composite transmission line insulators. IEEE Trans. Power Delivery. 23, 823–830. Available from: http://dx.doi.org/10.1109/TPWRD.2007.911127.

Polk, C., Postow, E., 1986. CRC Handbook of Biological Effects of Electromagnetic Fields. CRC Press, Boca Raton, FL.

Radasky, W., Savage, E., 2010. *Intentional Electromagnetic Interference (IEMI) and Its Impact on the U.S. Power Grid* (No. Meta-R-323). Prepared for the Oak Ridge National Laboratory. Metatech Corporation, Goleta, CA.

Ramie, J., 2013. Smart grid EMC standards harmonization. IEEE Electromagn. Compat. 2, 79–82. Available from: http://dx.doi.org/10.1109/MEMC.2013.6714704.

Scientific Committee of Emerging and Newly Identified Health Risks (SCENIHR) for the Directorate-General for Health & Consumers of the European Commission, 2007. Possible Effects of Electromagnetic Fields (EMF) on Human Health. European Commission, Brussels, Belgium.

Scientific Committee on Emerging and Newly Identified Health Risks (SCENIHR), 2009. Health Effects of Exposure to EMF. European Commission, Brussels, Belgium.

Scientific Committee on Emerging and Newly Identified Health Risks (SCENIHR), 2015. SCENIHR opinion on Potential Health Effects of Exposure to Electromagnetic Fields (EMF). European Commission, Brussels, Belgium.

Shepherd, R.A., Gaddie, J.C., 1976. Measurement of the APD and the Degradation Caused by Power Line Noise at HF. (accessed 20.08. 16.).

Silva, J.M., Olsen, R.G., 2002. Use of Global Positioning System (GPS) receivers under power-line conductors. IEEE Trans. Power Delivery. 17, 938–944. Available from: http://dx.doi.org/10.1109/TPWRD.2002.803791.

Smart Grid Interoperability Panel (SGIP) Electromagnetic Interoperability Working Group, 2012. *Electromagnetic Compatibility and Smart Grid Interoperability Issues* (No. 2012-005). Smart Grid Interoperability Panel.

Smart Grid Strategic Group, 2010. *IEC smart grid standardization roadmap* (No. 2009-12). International Electrotechnical Commission.

Smolenski, R., 2012. Conducted Electromagnetic Interference (EMI) in Smart Grids, Power Systems. Springer, London.

Swedish Radiation Safety Authority (SSM), 2010. Recent Research on EMF and Health Risks: Seventh Annual Report From SSM's Independent Expert Group on Electromagnetic Fields. Report Number: 2010:44. Swedish Radiation Safety Authority; Stockholm, Sweden.

Swedish Radiation Safety Authority (SSM), 2015. Research 2015:19-Recent Research on EMF and Health Risk-Tenth report from SSM's Scientific Council on Electromagnetic Fields. Swedish Radiation Safety Authority, Stockholm, Sweden.

Taflove, A., 1979. Mitigation of buried pipeline voltages due to 60 Hz AC inductive coupling Part I-Design of joint rights-of-way. IEEE Trans. Power Appar. Syst. PAS-98, 1806–1813. Available from: http://dx.doi.org/10.1109/TPAS.1979.319498.

Taflove, A., Dabkowski, J., 1979. Prediction method for buried pipeline voltages due to 60 Hz AC-inductive coupling Part I-Analysis. IEEE Trans. Power Appar. Syst. PAS-98, 780–787. Available from: http://dx.doi.org/10.1109/TPAS.1979.319290.

Takeshita, K., Takeshita, S., Hashimoto, H., 1979. Scattering characteristics of VHF/UHF television broadcasting waves by overhead power transmission conductors. IEEE Trans. Electromagn. Compat. EMC-21, 33–40. Available from: http://dx.doi.org/10.1109/TEMC.1979.303794.

Toyada, S., Hashimoto, H., 1979. Scattering characteristics of VHF television broadcasting waves by steel towers of overhead power transmission lines. IEEE Trans. Electromagn. Compat. EMC-21, 62–65. Available from: http://dx.doi.org/10.1109/TEMC.1979.303799.

Trueman, C.W., Kubina, S.J., 1981. Numerical computation of the reradiation from power lines at MF frequencies. IEEE Trans. on Broadcasting BC. 27, 39–45. Available from: http://dx.doi.org/10.1109/TBC.1981.266402.

Trueman, C.W., Kubina, S.J., 1986. Detuning power lines by isolating towers for the suppression of resonances. IEEE Trans. Broadcast. BC-32, 44–55. Available from: http://dx.doi.org/10.1109/TBC.1986.266583.

Trueman, C.W., Roobroeck, T.M., Kubina, S.J., 1989. Stub detuners for free-standing towers. IEEE Trans. Broadcast. 35, 325–338. Available from: http://dx.doi.org/10.1109/11.40833.

US Department of Energy (USDOE) and Bonneville Power Administration (BPA), 1977. A Practical Handbook for the Location, Prevention, and Correction of Television Interference from Overhead Power Lines. USDOE/BPA, Portland, OR.

US Department of Energy (USDOE) and Bonneville Power Administration (BPA), 1980. A Practical Handbook for the Correction of Radio Interference from Overhead Powerlines and Facilities. USDOE/EPA, Portland, OR.

Wertheimer, N., Leeper, E., 1979. Electrical wiring configurations and childhood cancer. Am. J. Epidemiol. 109, 273–284.

Wertheimer, N., Leeper, E., 1982. Adult cancer related to electrical wires near the home. Int. J. Epidemiol. 11, 345–355.

World Health Organization (WHO), 2007. Environmental Health Criteria 238: Extremely Low Frequency (ELF) Fields. World Health Organization, Geneva, Switzerland.

Zanetti, L.J., 2013. Review of North American Electric Reliability Corporation (NERC) interim report: effects of geomagnetic disturbances on the bulk power system—February 2012. Space Weather. 11, 335–336. Available from: http://dx.doi.org/10.1002/swe.20060.

6

HVDC TRANSMISSION FOR RENEWABLE ENERGY INTEGRATION

Benjamin R.T. Cotts[1], Jay R. Prigmore II[2] and Kevin L. Graf[3]

[1]Exponent Inc., Bowie, MD, United States [2]Exponent Inc., Warrenville, IL, United States [3]Exponent Inc., Menlo Park, CA, United States

CHAPTER OUTLINE

6.1 Introduction

High Voltage Direct Current (HVDC) technology is commonly used as a method to isolate two independent Alternating Current (AC) power grids while permitting the exchange of

The Power Grid. DOI: http://dx.doi.org/10.1016/B978-0-12-805321-8.00006-9

power between both AC grids. HVDC technology can disconnect and isolate each AC grid from the other in the event of a blackout of one of the AC grids to avoid losing both AC grids. This isolation technique is commonly used at the interconnection points of AC grids for two countries. HVDC can also prevent unwanted transients and other system anomalies from affecting the "healthy" AC grid. HVDC acts similarly to a filter and can "filter" out many of the unwanted and imposed anomalies on the grid to prevent any operational difficulties. The general applications for an HVDC transmission link are listed below (Molburg et al., 2007; Biggs & Jewell, 1984):

- Transmitting large amounts of power over long distances.
- Transmitting power by cable over moderate to long distances.
- Ability to transmit a greater amount of power versus AC-based transmission systems with same conductor size.
- Transmitting power between asynchronous systems (e.g., isolation technique).
- Forcing power into an area (e.g., avoid loop flow, stability issues).
- Providing additional capacity to congested transmission corridors.
- Improving system stability and reliability (provide voltage support).
- Enabling stronger, smarter, and more flexible grids by controlling power flow.

In this chapter, we review the key HVDC technologies that enable AC/DC (Alternative Current/Direct Current) conversion in power distribution systems, overview the common HVDC transmission configurations, and discuss the advantages and disadvantages in using DC versus AC for power distribution. First, we review the history of HVDC and some key technical advancements.

6.2 History of HVDC

The beginning stages of DC power transmission first came into existence by Marcel Deprez in Paris in 1881 (Tiku, 2014). A year later, Deprez helped to design a 15-kW, 2 kV DC transmission line for 35 miles (56.3 km) in Germany (Tiku, 2014). DC power continued to be developed by a select few but DC had lost the "war of currents" to AC current transmission partially due to the ability to change voltages in an AC system rather easily. It was not until the mercury-arc valve was used in HVDC applications that major progress was made. The mercury-arc valves allowed for increased power transmission and power-handling capability. Experimental installations in Europe in the 1930s between Switzerland and Germany were not economical compared to the improvements made in AC transmission. In 1941, the first commercial HVDC transmission system was ordered in Germany and would connect the Vockerode generating station to Berlin over a distance of 71.5 miles (115 km) and would transmit 60 MW at ± 200 kV (Tiku, 2014). In 1945, the Swedish State Power Board and ASEA company installed a 50-km test line in Sweden (Long & Nilsson, 2007). The development of HVDC technology became very active in 1929 and continued for 25 years of extensive development and pioneering research work before the first HVDC link was commissioned between the Swedish mainland and the Island of Gotland in the Baltic Sea in 1954 and was rated 20 MW at 110 kV (Hingorani, 1996; Long & Nilsson, 2007). The first HVDC installations that occurred in the 1950s utilized mercury-arc valves for the converter stations (Rudervall et al., 2000). Mercury-arc valve technology consists of a cold-cathode gas filled tube that employs a pool of liquid mercury as the cathode material. The mercury-arc valves are automatically reset after each operation due to the inherent feature of the design. The use of mercury-arc valves continued until the 1970s, when semiconductor switching devices began to reach the same power levels as the mercury-arc valves (Kirkham et al., 2014). The next installation for HVDC was the "Cross Channel" project commissioned in 1961 between England and France which transmitted 160 MW at ± 100 kV (Long & Nilsson, 2007). A few more projects began to initiate such as the Benmore hydropower station in New Zealand (600 MW, ± 250 kV), a 250-MW, ± 275 kV link between Sweden and Denmark and the Sakuma project in Japan (300 MW, ± 125 kV) (Long & Nilsson, 2007).

The next major advancement in HVDC incorporated digital process control computers into the control system of the converter stations. The first installation to use digital process control computers was the Pacific HVDC Intertie between the Columbus River in Oregon and Los Angeles, California in the United States. The Pacific HVDC Intertie project was an ASEA-General Electric joint venture for an 856-mile, bipolar overhead line operating at ± 400 kV and transmitting 1.44 GW. The Pacific Intertie also included the adaptation of new surge arrestor technologies specifically for DC use (Long & Nilsson, 2007). The last mercury-arc value based HVDC installation was built by English Electric for the Nelson River and Kingsnorth system (Long & Nilsson, 2007). Mercury-arc valves were then replaced by thyristor-based designs. The trend for HVDC has continued to grow in both voltage levels and power transmission capabilities through the capabilities of thyristor technologies. The next advancement in HVDC technology is expected to occur through the use of different semiconductor technologies with more advantageous turn-on and turn-off capabilities that will be described in more detail later in this chapter. Some of the most notable installations, longest installations, and highest voltage and power levels are listed in Table 6.1.

6.3 High Voltage Direct Current Converter Station Technologies

HVDC is a well-proven technology used to transmit electrical power long distances worldwide. The basic HVDC setup consists of two AC transmission grids interconnected and isolated by the HVDC link. At the interconnection between the AC grid and the HVDC transmission line is a converter station, also called a rectifier that converts (rectifies) AC to DC. After the converted DC current is transmitted along the HVDC transmission line, it reaches another converter station, called an inverter station, where the DC power is converted (inverted) back to AC power. Fig. 6.1 shows the basic topology of an HVDC transmission link.

The HVDC installations that occurred in the 1950s utilized mercury-arc valves for the converter stations (Rudervall et al., 2000). Mercury-arc valve technology consists of a cold-cathode gas filled tube that employs a self-restoring pool of liquid mercury as the cathode material. The use of mercury-arc valves continued until the 1970s when semiconductor switching devices began to reach the same power levels as the mercury-arc valves (Kirkham et al., 2014).

Semiconductor technology provided a significant improvement for HVDC technology, granting improved control of the HVDC system, design flexibility, and installation flexibility. The semiconductor device of choice was the thyristor, also known as a silicon-controlled rectifier (SCR). The thyristor is a three-terminal device consisting of an anode, cathode, and a gate terminal. Electrical current flows from the anode to the cathode once a gate signal is present (Tolbert et al., 2005). The thyristor turns on when the anode–cathode voltage is positive and a firing pulse is supplied to the gate terminal. However, it cannot be turned off by removing the firing pulse; the thyristor remains on until the anode current falls below the

Table 6.1 Notable HVDC Installations

Project Name	Location	Length (km)	Year	Voltage (kV)	Power (MW)	Type	Remarks
Pacific DC Intertie	North America	1362	1970	500	3100	Merc	Upgraded to 500 kV, replaced mercury-arc valves with thyristors
Nelson River Bipole 1	North America	1835	1971	463.5	1854	Merc	Used largest mercury-arc valves ever built. Thyristor conversion in 1993
Trans Bay Cable	North America	85	2010	200	400	IGBT	First HVDC using modular multi-level converters
Plains & Eastern Clean Line	North America	1207	2018[a]	600	4000	Thyr	Approved in 2016
Miesbach–Munich	Europe	58	1882	2	0.0025	Single Machine	First HVDC installation in the world. It was a demonstration facility
Gotland	Europe	98	1954	200	20	Merc	World's first commercial HVDC link
Western HVDC Link	Europe	422	2017[a]	600	2200	Thyr	First Subsea 600 kV link
NSN Link	Europe	730	2020[a]	515	1400	IGBT	Submarine cable
Rio Madeira	South America	2375	2013	600	7100	Thyr	Longest HVDC transmission link in the world at the time of commissioning
Hokkaido–Honshu	Asia	193	1979	250	300	Thyr	One of the oldest HVDC installations in Asia
Three Gorges	Asia	860	2003	500	3000	Thyr	Benchmark installation when installed
Jinping-Sunan	Asia	2100	2012	800	7200	Thyr	Highest power HVDC installation in the world
HVDC Inter-Island 1,2,3	Oceania	611 (each)	1965 1992 2013	270 350 350	600 700 735	Merc Thyr Thyr	New Zealand
Basslink	Oceania	370	2006	400	500	Thyr	One of the longest HVDC cables in operation at the time
Cahora Bassa	Africa	1420	1979	533	1920	Thyr	First HVDC above 500 kV
Caprivi Link	Africa	950	2010	350	300	IGBT	First IGBT installation in Africa

[a]Expected completion date.

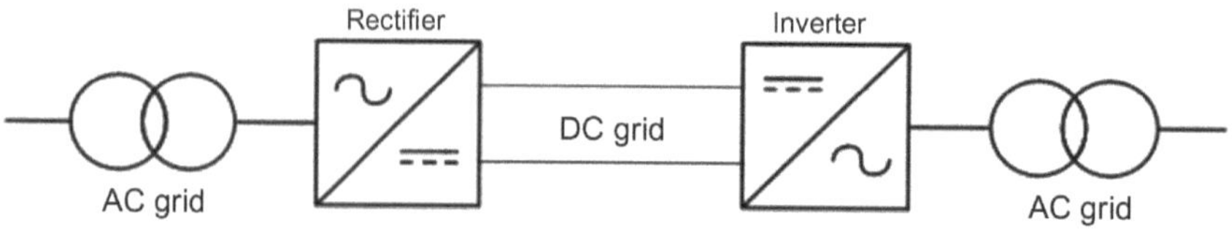

Figure 6.1 Basic topology of an HVDC transmission link.

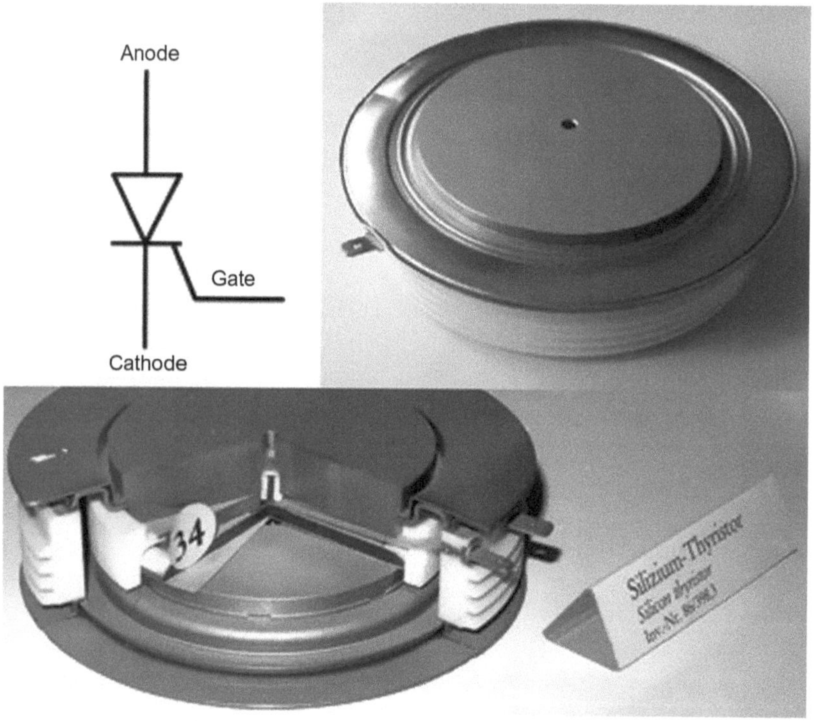

Figure 6.2 Thyristor and its electrical symbol (Anon, 2012, 2015).

thyristor holding current. The thyristor may turn on at any electrical angle of the AC sinusoidal wave as long as the anode–cathode voltage is positive flowing through the thyristor and the gate terminal has been triggered. To commutate the entire sinusoidal wave (both positive and negative portions), two thyristors are required, typically mounted in antiparallel orientations. Compared to mercury-arc valves, thyristor-based valves are physically lighter and are more cost effective, efficient, and reliable. A thyristor is shown below in the Fig. 6.2.

The thyristor serves as the fundamental switch element in line commutated converter (LCC) systems while newer semiconductor switch types are used in voltage sourced converter

(VSC) systems. Each of these system types are discussed below, along with a brief review of state-of-the-art developments in HVDC technologies.

6.3.1 LCC Based HVDC Applications

A LCC is a converter station that utilizes thyristor valves that are commutated by the line voltage itself. The thyristors are enabled at the desired electrical angle to convert the power from AC to DC at the appropriate power level. LCCs require a synchronous voltage source to properly function (Bahrman & Johnson, 2007). A six-pulse bridge LCC consists of a three-phase, full-wave bridge rectifier topology for each converter station. Therefore, there will be six thyristors (two per phase) to convert the AC signal to DC. Another full-wave bridge rectifier (six thyristors) is required to invert the DC back into an AC signal. A typical LCC topology is shown below in Fig. 6.3 where P is the real power, Q is the reactive power, and I_d is the DC current flowing in the transmission line.

Although the LCC based HVDC installations have a successful installation history, there are some disadvantages that ultimately led to the investigation of different technologies. One main disadvantage is the fundamental design characteristic of the line voltage forcing commutation of the thyristors, which requires adequate strength of both AC grids. An LCC installation may not function properly with a "weak" AC grid such as a small microgrid. Determining the strength of the connected AC grid, and ultimately the feasibility of an LCC HVDC installation, is dependent on the short circuit ratio (SC Ratio). The SC Ratio is the ratio of the short-circuit power of the AC grid (P_{SC-AC}) at the converter station terminals over the rated power of the HVDC link (P_{DC}). The SC Ratio is shown in the equation below.

$$\text{SC Ratio} = \frac{P_{SC-AC}}{P_{DC}}$$

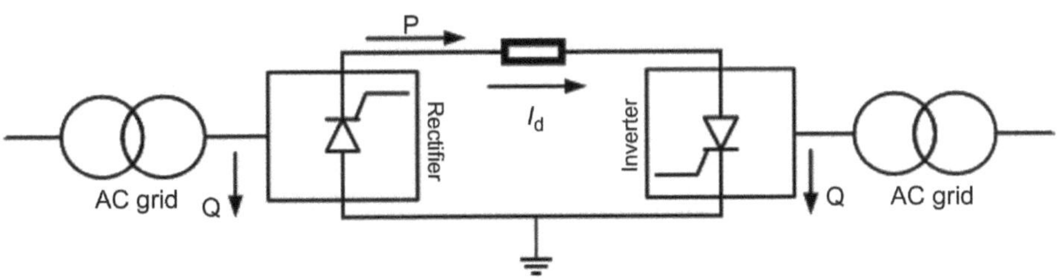

Figure 6.3 Schematic diagram of LCC HVDC.

A robust AC grid is given by an SC Ratio greater than four while an intermediate AC grid is determined by a SC Ratio between two and four. Furthermore, a weak AC grid is indicated by a SC Ratio less than two, which increases the risk for commutation failure of the thyristors (Woodford, 1998; Eremia, 1987). Another disadvantage of LCC systems is that power can only flow in a single direction unless a terminal polarity change is performed to reverse the DC power flow. Terminal polarity changes are not instantaneous, as they require maintenance and the physical reversal of connections. An LCC system also requires reactive power compensation at the converter terminals due to the reactive power draw from the connected AC grids at the converter stations. The thyristors have a delay in their gate signals, which yields a delay in phase current from the corresponding phase voltage. Typically, the reactive power draw is approximately 60% of the DC power flow (Siemens AG, 2011). The amount of reactive power compensation is a function of the voltage drop due to the reactive power flow.

6.3.2 Voltage Sourced Converter (VSC)

A VSC HVDC system is composed of a semiconductor switch that can turn on and turn off at any electrical angle of the AC sinusoidal wave (Sea Breeze Olympic Converter, 2006). While the switches employed in LCC installations can only be turned on by a control signal, the newer semiconductor switches employed in VSC installations can both be turned on and turned off by control signals. Due to this added turn-off feature, newer semiconductor devices are the most prominent technology used today in HVDC applications. The semiconductor device of choice has changed as technology has improved over time. There are presently three main semiconductors switches that are being used in HVDC applications: the Gate Turn-Off Thyristor (GTO), the Integrated Gate Commutated Thyristor (IGCT), and the Insulated Gate Bipolar Transistor (IGBT).

The GTO is still used in the highest power VSC installations due to its excellent turn- off characteristics, its power-handling capabilities, and its relatively low cost. An example of a GTO is shown below in Fig. 6.4. The GTO can be turned off by applying a current at the gate in the reverse direction.

The IGCTs, which improved on the technology of the GTO, are starting to reach the same power levels as the GTOs but they have a simpler gate drive requirement of only about 20% of the current flowing through the anode to the cathode to turn-off the device. The main advantage of the IGCT versus the

Figure 6.4 Gate Turn-Off Thyristor (GTO) (Anon, 2013).

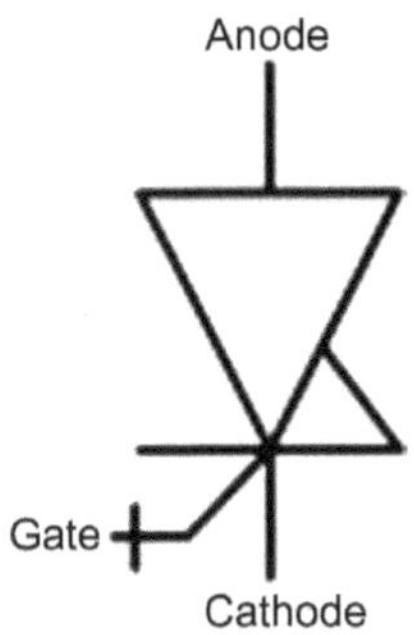

Figure 6.5 Integrated Gate Commutated Thyristor (IGCT) electrical schematic.

GTO is the ability for the IGCT to switch at a higher frequency than the GTO. The electrical schematic diagram of an IGCT is shown below in Fig. 6.5.

The third device, which should eventually replace both the GTO and IGCT if it can reach the highest power levels, is the IGBT. The IGBT electrical symbol and a photograph are shown below in Fig. 6.6.

The IGBT has the intrinsic properties of a transistor but can conduct and switch large amounts of power. The IGBT's main advantages over the GTO and IGCT are the capability of higher frequency switching and the much simpler gate drive requirements. The GTO and IGCT both require a current applied at the gate to switch off the devices while the IGBT requires a voltage at the terminals to initiate the switching function. A voltage signal is much simpler to implement when designing gate drives and requires physically less space than a gate current design.

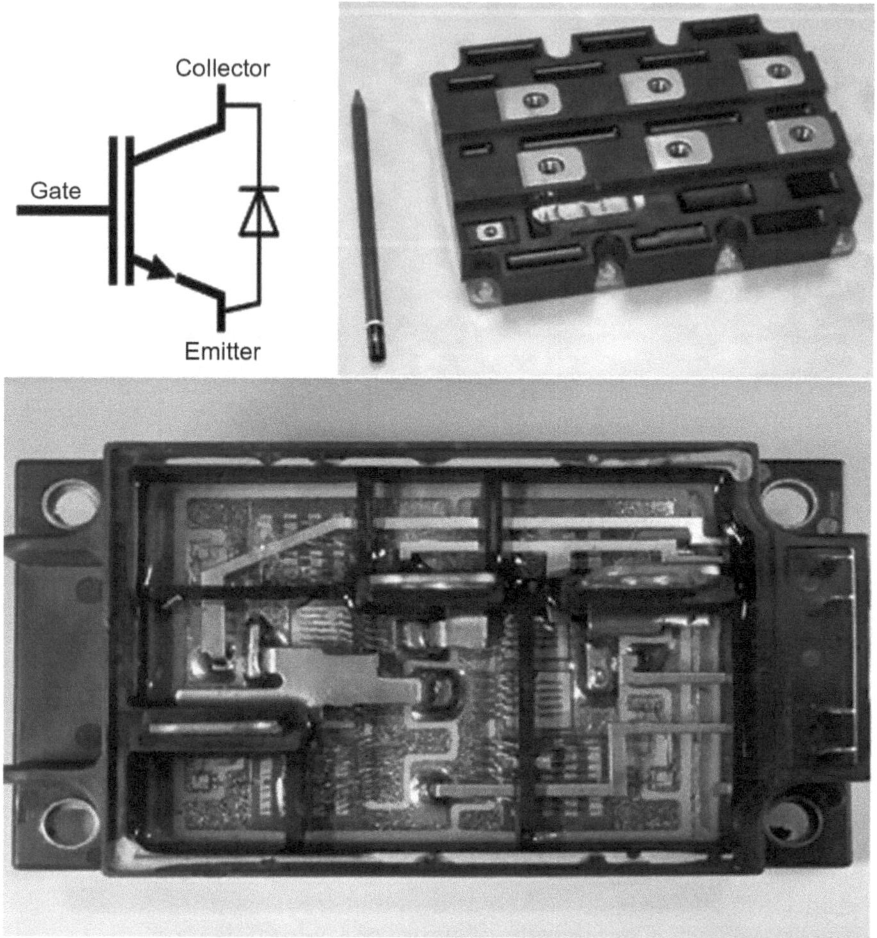

Figure 6.6 Integrated Gate Bipolar Transistor (IGBT) and its electrical symbol with antiparallel diode (Anon, 2007, 2008).

The advantage of using semiconductor switches with a forced turn-off capability in HVDC applications is the ability to force commutation of the converter stations. Forced commutation eliminates the SC Ratio rule used for LCC HVDC installations that allows VSC HVDC applications to connect and operate properly even when connected to a "weak" AC grid. In fact, VSC HVDC installations can still operate with an SC Ratio of zero—a passive grid that has no self-voltage support. The switching capability of the semiconductors can provide independent control of real and reactive power in both DC power flow directions by controlling the gate signals to each of the semiconductors in the converter stations. VSC HVDC installations eliminate the terminal reversal requirement of LCC HVDC installations and provide improved operational flexibility. While their maximum achievable power ratings are currently lower, VSC HVDC overcomes several limitations of LCC HVDC:

- Can connect to weak AC grids.
- Provides independent control of real and reactive power.
- Provides voltage support to a passive grid (black-start capability).
- Can support multi-terminal HVDC grids and provides control flexibility.
- Integration of renewable energy sources, and capable of handling their intermittent nature.
- Physically smaller footprint than LCC HVDC.

Due to the advantages of the VSC HVDC, a shift in semiconductor of choice is beginning to occur between the thyristor and newer power electronic switch technologies, most notably the IGBT. Lower power HVDC installations are trending toward the use of IGBTs due to their excellent control capabilities and operational benefits. Higher power HVDC installations may adopt IGBTs as well if the technology can advance to reach the power levels of the thyristor. At the time of this book, the most practical and economical choice for high power HVDC transmission is the thyristor. However, many more installations are utilizing the IGBT to meet these higher power levels.

6.3.3 HVDC Technologies: State-of-the-Development

Improved power-handling capabilities of semiconductor technologies may allow VSC HVDC to reach the power levels of LCC HVDC installations. Presently, LCC HVDC installations have power ratings up to approximately 8 GW at $\pm$800 kV while VSC HVDC installations are approximately at 1 GW. The aim of VSC HVDC development is to achieve the power levels of LCC HVDC while still reducing switching and conduction losses of the converter stations. The benefits of VSC HVDC installations and the trend in the industry indicates that VSC HVDC is the preferred technology for the future once the semiconductor devices that have turn-on and turn-off capabilities reach the corresponding ratings of thyristors.

A new semiconductor switch that may end up challenging the GTO, IGCT, and IGBT for future dominance of the HVDC market is the Emitter Turn-Off Thyristor (ETO), which has the improved characteristics of high frequency operations and high

Figure 6.7 Emitter Turn-Off Thyristor (ETO) and its electrical schematic symbol.

power capabilities. The ETO is essentially a GTO with additional Metal-Oxide Field Effect Transistors to assist with the switching process. An example of an ETO is shown below in Fig. 6.7 along with its electrical schematic symbol.

The ETO combines the switching capabilities of the IGBT (gate voltage and higher frequencies) with the power-handling capabilities of the GTO and IGCT.

Research and development efforts are ongoing to improve power-handling capabilities of a single semiconductor-switching device. The present material of choice for high power installations is Silicon Carbide (SiC), which is poised to replace Silicon (Si) as the semiconductor material of choice due to its improved power capabilities and higher maximum thermal temperature. SiC devices are considered wide-band gap semiconductors and as manufacturing methods continue to improve, SiC devices should allow for a smaller number of semiconductor switches to be required to meet the same power requirements of Si-based installations. Converter losses should further reduce and the overall efficiency of HVDC installations should improve. The SiC-based HVDC installations also should require a much smaller footprint than the Si-based installations.

6.4 HVDC Transmission Configurations

HVDC systems can be used to connect between the converter stations of two or more power networks. In point-to-point HVDC systems, an HVDC transmission line connects one converter station to another. The most common configurations for these point-to-point HVDC systems are the monopolar and bipolar configurations. As an alternative to the point-to-point HVDC system, a back-to-back HVDC link can be established between two converters located at the same site. In back-to-back HVDC links, no HVDC transmission line is required. If more than two converter stations are to be connected within a single HVDC system, a multi-terminal configuration can be employed (Eremia, 2006). In this section, we will discuss each of these configuration options for HVDC systems.

6.4.1 Monopolar

A monopolar HVDC transmission line requires only a single high voltage conductor. Current flows from one converter, along the high voltage line, to another converter. There are two options for the current return path: either the current return path is through the earth or seawater, or it is through a metallic return. A monopolar HVDC transmission configuration with an earth return is illustrated in Fig. 6.8, where an electrode line and ground electrode connects the return terminal to earth at each end of the line. This electrode line is typically medium voltage and several kilometers long, allowing the earth connection to be located some distance from the converter station. The earth return current may interfere with other structures or systems, so this must be considered in the design of the HVDC system (Eremia, 2006). In ground, the return current may cause corrosion of buried metal objects. In seawater, an exposed return electrode may locally develop chlorine and affect water chemistry (Ardelean and Minnebo, 2015). Another aspect of HVDC systems, which is often required in environmental impact statements, is the magnetic fields emanating from the transmission line. In transmission configurations with multiple parallel conductors very close to one another, such as three-phase AC systems or monopolar HVDC with metallic return, electric currents flowing in opposite directions can induce magnetic fields that partially cancel one another, resulting in lower magnetic field levels. Along most of the transmission line

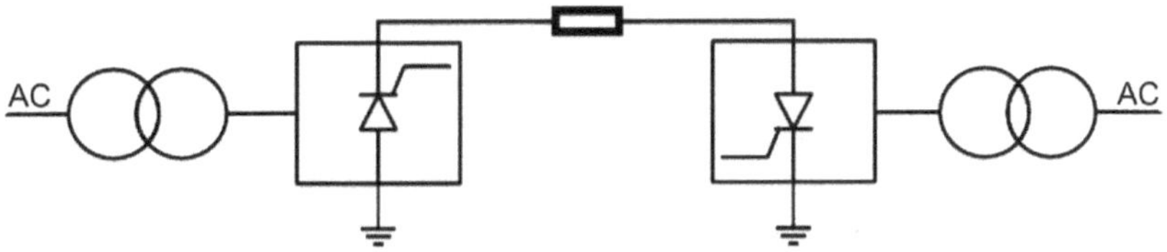

Figure 6.8 Monopolar HVDC transmission configuration with earth return.

route in a monopole configuration with earth return, however, there is effectively only a single conductor, and thus there is no mutual cancellation of the induced magnetic field. This behavior can result in relatively larger magnetic field levels, depending on the design.

In a monopolar system with metallic return path, a second conductor runs alongside the high voltage conductor. This second conductor can be low voltage and provides a metallic return path for the current. The second conductor eliminates the concerns associated with an earth return current, and the magnetic field induced by the electric current flowing in this second conductor can partially cancel the magnetic fields induced by the transmission line, thus reducing magnetic field levels near the line. If the metallic return path conductor is installed with proper insulation for high voltage use, then the pair of conductors could later be converted into a bipolar system.

6.4.2 Bipolar

A bipolar HVDC transmission line consists of two high voltage conductors, one of positive polarity with respect to ground, and the other of negative polarity. A ground connection at each converter station provides an earth return path for any current imbalance, or a third conductor can be installed with the line to provide a metallic return path. Under normal load with both poles in operation, the unbalanced current in the return path should be very low. Additionally, if a fault occurs on one line of the high voltage conductors, the system can be configured to continue to operate in a monopolar mode. A bipolar HVDC transmission configuration with a ground connection at each converter station is illustrated in Fig. 6.9.

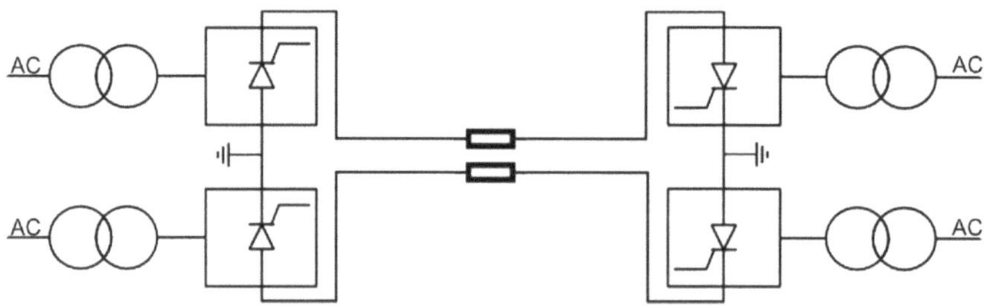

Figure 6.9 Bipolar HVDC transmission configuration.

6.4.3 Back-to-Back

A back-to-back HVDC link connects two converters located at the same site without the need for a long distance transmission line. Because of the short conductor length, resistive losses (I^2R) in the DC system are less of a concern and lower DC voltages can be used. Since the cost of valves for converters depends more on voltage rating than on current rating back-to-back HVDC systems typically use lower DC voltages (Eremia, 2006). Lower DC voltage on the converter transformer also facilitates smaller reactive power switching steps. One main reason to use a back-to-back configuration is to provide isolation between two AC grids in the event one of the AC grids experiences a problem that could affect the other AC grid. This technique is commonly used in interconnection points between different countries while still allowing power transfer to occur between the countries. This configuration also provides security and a quick means of disconnecting from the abnormal conditions imposed by the faulty AC grid. Once both AC grids are healthy again, the converters are enabled and the interconnection is re-established.

6.4.4 Multi-Terminal

Multi-terminal HVDC systems connect more than two converter stations in a series, radial, or mesh configuration, as illustrated in Figs. 6.10–6.12, respectively. In a series-connected

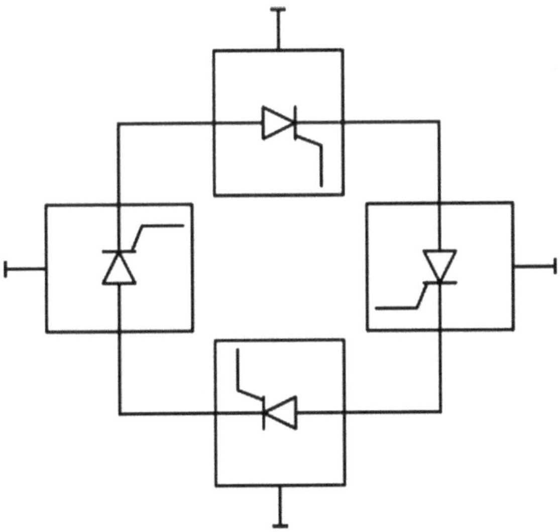

Figure 6.10 Series HVDC transmission configuration.

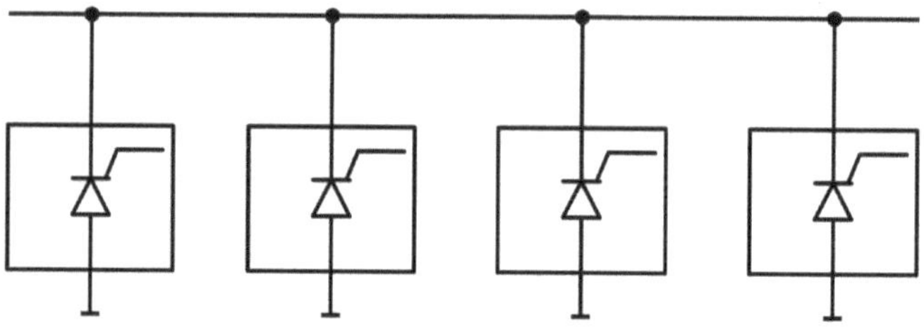

Figure 6.11 Radial HVDC transmission configuration.

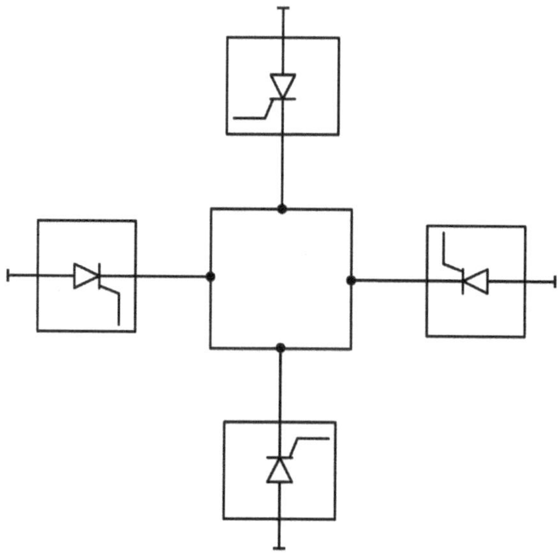

Figure 6.12 Mesh HVDC transmission configuration.

system, all converters are connected to the HVDC system along a ring-shaped DC line. The series connection requires that the DC current remain equal throughout the system. If any part of the DC line is disconnected, then the energy transfer is interrupted (Eremia, 2006). In a radial-connected system, each converter is connected to a single DC line, which requires that the DC voltage remain equal throughout the system. A disconnection of any part from the radial-connected system can result in increased energy flow between the parts that remain connected. A meshed-connected system connects each converter station to one or more DC lines and can facilitate the removal of parts without disrupting or inducing energy flow between other parts of the system.

6.5 AC Versus DC

6.5.1 Benefits of DC

As discussed above, there are significant advantages in using HVDC over HVAC in certain scenarios. While the choice of HVDC or HVAC is often dictated by economic considerations, there can be additional factors to consider. For example, two asynchronous systems may not accommodate interconnection by an HVAC line or a congested right-of-way (ROW) may preclude the use of an HVAC line of necessary capacity. However, the smaller required clearances of an HVDC line may make the HVDC link feasible for the asynchronous grid interconnection. In addition to these potential benefits of HVDC (each of which is discussed below), an HVDC interconnection between power systems can provide additional controllability, power delivery rate, or system capacity (e.g., Andersen and Barker, 2000).

Presently, the biggest advantage of HVDC over HVAC is in the bulk transmission of electricity over long distances such as from remote hydroelectric generation or remote large-scale wind farms. While the power capacity of long HVAC lines is limited without widespread use of series capacitors, there is no such limitation when using HVDC lines (Weimers, 2011). In addition, HVAC lines require three conductors while bipolar HVDC lines require only two, and while the transmission capacity of HVAC lines is often limited by the reactive power consumption (charging current), the capacity of HVDC has no such charging current limitation (Bahrman, 2006). Bipolar HVDC (discussed in greater detail above) has an additional benefit that during scheduled maintenance or outages of only a single pole the transmission line (with overload capacity) can still be used with more than 50% of the full transmission capacity (Bahrman, 2008).

An additional benefit of HVDC transmission is that the current in an HVDC conductor flows in the entire conductor while at a frequency of 50–60 Hz typical of AC, the current in a large cable flows only in the outer shell of the conductor, approximately 9 mm thick. This phenomenon, known as the skin effect, is related to the driving electric and magnetic fields within and outside the conductor of an AC system. The limitation of current to flow only in the outer portion of the conductor leads to its being termed as a current flow in the "skin" of the conductor or the skin depth (Inan and Inan, 1999) and is often assigned a symbol of δ. The skin depth or skin effect is especially important for larger conductors common to HVAC transmission lines because limitation of current to this 9 mm shell can result in significant portions of the center of the

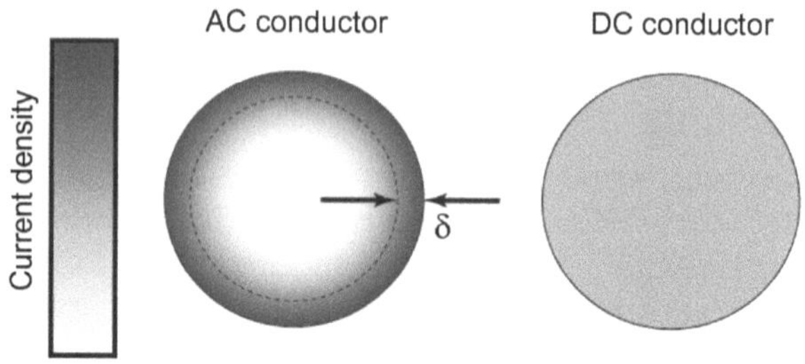

Figure 6.13 Current density in an AC conductor compared to a DC conductor.

conductor being unused for power transfer and therefore to higher power loss. In contrast, since HVDC operates at an effective frequency of 0 Hz there is no skin effect and the entire cross sectional area of a conductor is used for power transfer. An illustrative example of the practical meaning of skin depth is shown in Fig. 6.13.

The resulting greater capacity of HVDC lines also means that a single HVDC transmission line can provide as much power as an HVAC line of much higher voltage or even of multiple HVAC lines. For instance, a total transmission system capacity of 2000 MW can be carried by a single ±500 kV DC line situated on a 50-meter ROW. To transfer the same 2000 MW of power for an HVAC line would require either an 800 kV line on an 85-meter ROW or a pair of 500 kV AC lines on a 100 m ROW (Arrillaga, 1998). A recent example is the first stage (6000 MW) of generation at the Three Gorges Project in China, which used two ±500 kV DC lines compared to five 500 kV AC lines (Bahrman, 2006). This significant savings in required ROW width can be especially important in locations where only a limited ROW is available such as when land is unavailable or expensive to acquire.

An additional use of HVDC systems is in connecting two asynchronous systems. An AC transmission system operates at a single voltage and frequency and locally at a particular phase angle and power factor. Two systems are said to be asynchronous if, e.g., one operates at a frequency of 60 Hz and the second operates at a frequency of 50 Hz. Two systems of the same frequency may also be asynchronous if the two operate at different relative phase angles from one another. In this situation, an HVDC link is the only practical way to connect the two asynchronous systems together (Wang and Redfern, 2010). In North America, for instance, there are several HVDC asynchronous links including those between the Electric Reliability Council of Texas and Mexico, between Quebec and New England,

and between the eastern and western interconnected systems (Bahrman, 2006). An additional benefit of using an HVDC link between asynchronous systems is that it can provide protection from a cascading power outage. For example, the August 14, 2003 blackout in the northeast portion of the United States was stopped at the asynchronous interface with Quebec. Since Quebec was unaffected, the HVDC link from Quebec was able to continue delivering power to the New England area (Bahrman and Johnson, 2007).

6.5.2 Cost Structure: DC Versus AC Transmission Lines

In the construction of any transmission line project, many factors are considered in making decisions of design and implementation. Not least among these is the cost of the various options. The interconnection of an HVDC line with the predominantly AC power grid requires that the HVDC line have converter stations (often at each end of the line). These converter stations represent a large initial capital investment that is largely independent of the transmission line length (Eremia, 2006). The economic benefit from HVDC comes when considering the costs of construction for the transmission lines themselves, which are significantly lower for HVDC lines than for HVAC lines.

In addition to the greater transfer capacity and smaller/fewer conductors discussed in the above section, the cost of an HVDC line is less than that of an HVAC line because for the same rated voltage the insulation level is lower for HVDC than for HVAC and the losses with distance are also lower. The higher insulation level of an AC line is due to the changing voltage of the AC sine wave that will increase by a factor of $\sqrt{2}$ (approximately 1.41) at the peak of the sine wave compared to the root-mean-square voltage level. The lower losses per unit distance of an HVDC line are due both to the elimination of reactive power (discussed above) and to the skin effect (discussed above) which limits the current in a large AC cable to flow primarily in the outer-shell of the conductor (Inan and Inan, 1999). Because the current flowing within a large AC conductor is restricted primarily to the outer $\sim$1 cm shell, the current density (and hence losses) are higher. In contrast, there is no skin effect at the 0 Hz frequency of an HVDC line and so the flowing current can take up the whole cross section leading to lower current density and lower losses. The combination of these factors leads to a lower per-length cost of HVDC lines compared to HVAC.

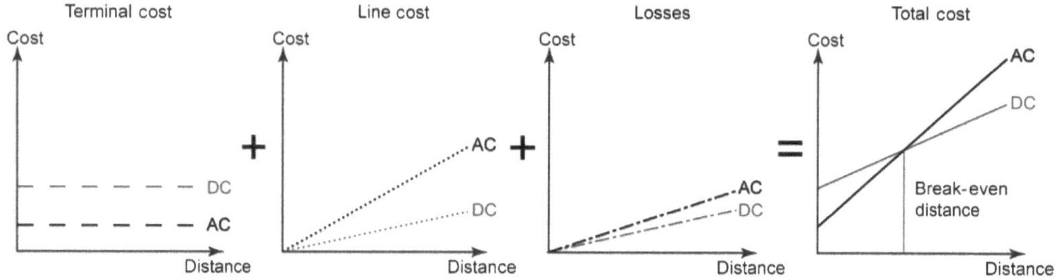

Figure 6.14 Cost structure comparison of AC and DC transmission lines.

The higher initial costs, but lower per-length cost of HVDC means there is typically a break-even distance beyond which the HVDC line becomes a more economical option than the HVAC. Though additional factors such as region-specific costs, interest rates, cost of right of way, and others can influence individual projects, the break-even distance is typically near 500 km (320 miles) (Bahrman, 2006). This cost structure is shown qualitatively in Fig. 6.14 below. In all panels the blue line represents the qualitative cost of the DC line while the green line shows the cost of a similar AC line. The dashed lines in the first panel show the initial (terminal) cost associated with the respective HVAC or HVDC lines, the cost of the HVAC and HVDC line itself is shown by the dashed lines in the second panel, and the cost associated with the losses of the HVAC and HVDC lines are shown by the dash-dot lines in the third panel. Summing all three costs together gives the total cost of the HVAC and HVDC systems, shown by the solid lines in the last panel. The point of intersection is the break-even distance beyond which the HVDC option is more economical.

6.5.3 Limitations of DC

While the potential benefits of HVDC are numerous, there are also a number of potential disadvantages that still limit its viability for widespread use. As discussed in the previous section, the initial cost of constructing an HVDC line is far higher than for HVAC (e.g., Wang and Redfern, 2010). Much of the cost of an HVDC line is spent on the converter transformers, power electronics, and AC filters, none of which is typically needed for an HVAC line. A second limitation of HVDC transmission lines is that harmonics are generated during the conversion process that can potentially cause interference with nearby electronic devices and in some circumstances can affect the power quality of the HVAC system, even leading to system oscillations (e.g., Amin et al., 2015).

An additional limitation of HVDC is that the vast majority of HVDC lines are point-to-point systems and not a "network" as has been developed for HVAC systems. An HVDC network, commonly referred to as multi-terminal HVDC, would provide the additional advantages of HVDC discussed above, but would also reduce the number of required terminals. Multi-terminal HVDC would also provide the redundancy enjoyed by HVAC whereby an outage of a single line would not necessarily result in loss of power at any terminal (Franck, 2011).

The limiting factor in enabling the future of multi-terminal HVDC networks is widely considered to be HVDC circuit breakers (Bucher and Franck, 2016). The biggest challenges in the implementation of an HVDC circuit breaker are that the current must be forced to a zero crossing point, and the magnetic energy stored in the transmission line in the form of system inductance needs to be dissipated. In addition, the breaker must have sufficient dielectric strength to survive the resultant large voltage that appears across the terminals of the circuit breaker, known as the Transient Recovery Voltage (Bucher and Franck, 2016). Given the many potential benefits of multi-terminal HVDC network there is currently an enormous amount of research, development, testing, and implementation of HVDC circuit breakers (e.g., Prigmore et al., 2010; Smeets et al., 2015; Magnusson et al., 2015; Hassanpoor et al., 2015; Gaxiola and Vogelsang, 2015; Berggren et al., 2015).

6.6 Summary

6.6.1 Future Trends

HVDC may permit the integration of renewable energy sources such as wind and solar in remote locations. Presently, many offshore wind applications are transmitting power back to shore before being transformed back to AC and connected to the power system. These offshore applications utilize a submarine cable that may either lay on the sea floor or be buried some small distance beneath.

Future trends for HVDC technology and installations are shifting toward VSC HVDC designs due to the improved operational flexibility inherent in this type of design. The key enabling technology has been the controllability of the power electronic switching devices. While the thyristor's power level ratings at its given economic costs have yet to be matched, the advent of SiC in place of Si may permit the IGBT or other devices to overtake the thyristor as the high power HVDC semiconductor device of the future. The industry trend is to use the IGBT devices when power ratings and economics are attractive and use the thyristor when it becomes economically more attractive at higher power levels. As the ever-increasing demand for cheap reliable energy grows, the demand on the power industry is to increase both voltage levels and power levels. Future installations may reach and exceed ratings of 1000 kV and 10 GW.

6.6.2 Additional References

Some additional references on HVDC may be found on the Institute of Electrical and Electronic Engineers' (IEEE) website, conference, and journals at www.ieee.org including an HVDC Design Guide, which, at the time of this publishing, is in preparation and nearing initial publication by the Overhead Transmission Lines Subcommittee of the Power Engineering Society (PES). Another worldwide-recognized entity is the International Council on Large Electric Systems (CIGRE) and more information can be found at www.cigre.org. Technical Committee SC4 specializes in HVDC and Power Electronics (http://b4.cigre.org/What-is-SC-B4). At the time of this book writing, CIGRE is in the process of publishing a "Green Book" under the topic of "HVDC and Power Electronics." Additionally, the Electric Power Research Institute (EPRI) has published an HVDC Reference Book: Overhead Lines for HVDC Transmission. Furthermore, there are numerous universities, government agencies, and laboratories that publish innovative research and industry trends in HVDC technologies. Attending transmission-based conferences put forth by the above-mentioned organizations is an excellent way to learn more about HVDC. Many manufacturers of components used in HVDC applications publish "best practices" or "application guides" on using their equipment in an HVDC application.

6.7 Conclusion

While DC lost the original "War of the Currents" between Edison Electric Company and Westinghouse Electric Company in the early part of the 19th century, there are many potential advantages, both technical and economical, to transmit electricity using DC instead of AC. With the advent of HVDC and its increasing popularity worldwide for the highest power transmission applications, DC transmission is beginning to make a comeback in the "War of the Currents." HVDC lines are currently constructed primarily as point-to-point transmissions between two or more power networks in either monopolar or bipolar configurations but are also used in back-to-back HVDC configurations to link two converters located at the same site. If more than two converter stations are to be connected within a single HVDC system, a multi-terminal configuration can be employed. Meshed multi-terminal HVDC configurations can add redundancy and improve reliability of the installation.

The primary benefits of HVDC include transmitting large amounts of power over long distances, and transmitting power underground and underwater over even moderate distances. HVDC also has a greater capacity than AC for systems operating at the same voltage or with the same conductor size and can be used to connect asynchronous systems. The greater capacity of HVDC lines also means that a single HVDC transmission line can provide as much power as an HVAC line of much higher voltage or even of multiple HVAC lines. This significant savings in required ROW width can be especially important in locations where only a limited ROW is available such as when land is unavailable or expensive to acquire.

While early HVDC installations utilized mercury-arc valves for the converter stations, they were eventually replaced when semiconductor-switching devices began to reach the same power levels as the mercury-arc valves and granted improved control of the HVDC system as well as design and installation flexibility. The first semiconductor device of choice was the thyristor, also known as a SCR, which served as the fundamental switch element in LCC systems.

LCC systems also had some shortcomings that ultimately led to the investigation of different technologies such as VSC in which the semiconductor switches can turn on and turn off at any electrical angle of the AC sinusoidal wave. The three main semiconductor switches that are currently being used in HVDC applications are the GTO, IGCT, and IGBT. The GTO is still used in the highest power VSC installations due to its excellent turn-off characteristics, its power-handling capabilities, and its relatively low cost, but should also eventually be replaced by the IGBT provided the IGBT can reach sufficiently high power levels. The main advantages of the IGBT over the GTO and IGCT are the capability of higher frequency switching (reduced losses) and the fact that it can be operated using a much simpler gate voltage rather than a gate current to initiate switching.

A new semiconductor switch that may end up challenging the GTO, IGCT, and IGBT for future dominance of the HVDC market is the ETO, which has the improved characteristics of high frequency operations and high power capabilities.

The present limiting factor inhibiting the widespread use of multi-terminal HVDC networks is widely considered to be HVDC circuit breakers, which require a significant engineering challenge of forcing the current to zero and dissipating the massive magnetic energy stored in the transmission line. The HVDC breaker must also have sufficient dielectric strength to survive the resultant large voltage that appears across the terminals during and after interruption. However, given the many potential benefits of multi-terminal HVDC networks, there is currently an enormous amount of research, development, testing, and implementation of HVDC circuit breakers that may bring the worldwide power grid into a new era of power distribution as an end result of these efforts.

References

Amin, M., Molinas, M., Lyu, J., 2015. Oscillatory phenomena between wind farms and HVDC systems: The impact of control. In: 2015 IEEE 16th Workshop on Control and Modeling for Power Electronics (COMPEL)., pp. 1–8. Available from: http://dx.doi.org/10.1109/COMPEL.2015.7236513.

Andersen, B., Barker, C., 2000. A new era in HVDC? IEE Rev. 46, 33–39. Available from: http://dx.doi.org/10.1049/ir:20000208.

Anon, 2007. IGBT 2441. *Wikimedia commons*. Available from: <https://commons.wikimedia.org/wiki/File:IGBT_2441.JPG>.

Anon, 2008. IGBT 3300V 1200A Mitsubishi. *Wikimedia commons*. Available from: <https://commons.wikimedia.org/wiki/File:IGBT_3300V_1200A_Mitsubishi.jpg>.

Anon, 2012. Thyristor Westcode. *Wikimedia commons*. Available from: <https://commons.wikimedia.org/wiki/File:Thyristor_Westcode_N3476TC420.jpg>.

Anon, 2013. GTO Press-pack. *Wikimedia commons*. Available from: <https://commons.wikimedia.org/wiki/File:GTO_press-pack.jpg>.

Anon, 2015. Thyristor Schnitt. *Wikimedia commons*. Available from: <https://commons.wikimedia.org/wiki/File:Thyristor_Schnitt.jpg>.

Ardelean, M., Minnebo, P., European Commission, Joint Research Centre, Institute for Energy and Transport, 2015. HVDC Submarine Power Cables in the World State-of-the-Art Knowledge. Publications Office, Luxembourg.

Arrillaga, J., 1998. High Voltage Direct Current Transmission. IET.

Bahrman, M.P., 2006. Overview of HVDC transmission. In: Power Systems Conference and Exposition, 2006. IEEE PES, pp. 18–23. Available from: http://dx.doi.org/10.1109/PSCE.2006.296221.

Bahrman, M.P., 2008. HVDC transmission overview. In: Transmission and Distribution Conference and Exposition, 2008. IEEE/PES. Presented at the Transmission and Distribution Conference and Exposition, 2008. IEEE/PES, pp. 1–7. Available from: http://dx.doi.org/10.1109/TDC.2008.4517304.

Bahrman, M.P., Johnson, B.K., 2007. The ABCs of HVDC transmission technologies. IEEE Power and Energy Magazine. 5, 32–44. Available from: http://dx.doi.org/10.1109/MPAE.2007.329194.

Berggren, B., Wang, J., Pan, J., Linden, K., Nuqui, R., 2015. Breaker failure protection of HVDC circuit breakers. US9042071 B2.

Biggs, R.B., Jewell, W.T., 1984. Summary of Multiterminal High-Voltage Direct Current Transmission Technology.

Bucher, M.K., Franck, C.M., 2016. Fault current interruption in multiterminal HVDC networks. IEEE Trans. Power Delivery. 31, 87–95. Available from: http://dx.doi.org/10.1109/TPWRD.2015.2448761.

Eremia, M., 1987. In: Marian, M., Monica, S., Chiriac, C. (Eds.), Electric Power Systems, first ed. Publishing House of the Romanian Academy, Bucharest.

Eremia, M., 2006. In: Marian, M., Monica, S., Chiriac, C. (Eds.), Electric Power Systems. Volume 1. Electric Networks. Publishing House of the Romanian Academy, Bucharest.

Franck, C.M., 2011. HVDC circuit breakers: a review identifying future research needs. IEEE Trans. Power Delivery. 26, 998–1007. Available from: http://dx.doi.org/10.1109/TPWRD.2010.2095889.

Gaxiola, E., Vogelsang, J., 2015. DC voltage line circuit breaker. US9178348 B2.

Hassanpoor, A., Häfner, J., Jacobson, B., 2015. Technical assessment of load commutation switch in hybrid HVDC breaker. IEEE Trans. Power Electronics. 30, 5393–5400. Available from: http://dx.doi.org/10.1109/TPEL.2014.2372815.

Hingorani, N.G., 1996. High-voltage DC transmission: a power electronics workhorse. IEEE Spectrum. 33 (4), 63–72.

Inan, U.S., Inan, A., 1999. Electromagnetic Waves. first ed. Prentice Hall, Upper Saddle River, NJ.

Kirkham, H., Elizondo, M., Dagle, J., 2014. An Introduction to High Voltage DC Networks.

Long, W., Nilsson, S., 2007. HVDC transmission: yesterday and today. IEEE Power and Energy Magazine. 5 (2), 22–31.

Magnusson, J., Bissai, A., Engdahl, G., Martinez, J.A., Liljestrand, L., 2015. Experimental study of the current commutation in hybrid DC-breakers. In: 2015 Third International Conference on Electric Power Equipment Switching Technology (ICEPE-ST). pp. 506–511. Available from: http://dx.doi.org/10.1109/ICEPE-ST.2015.7368342.

Molburg, J.C., Kavicky, J.A., Picel, K.C., 2007. The design, construction, and operation of long-distance high-voltage electricity transmission technologies. Available from: <http://solareis.anl.gov/documents/docs/APT_61117_EVS_TM_08_4.pdf> (Accessed 29.07.16.).

Prigmore, J., Tcheslavski, G., Bahrim, C., 2010. An IGCT-based electronic circuit breaker design for a 12.47kv distribution system. Power and Energy Society General Meeting, 2010. IEEE, pp. 1–5.

Rudervall, R., Charpentier, J., Sharma, R., 2000. High Voltage Direct Current (HVDC) Transmission Systems Technology Review Paper. In: Energy Week, Washington, DC. pp. 1–17.

Sea Breeze Olympic Converter, 2006. Appendix A: HVDC light technical background.

Smeets, R.P.P., Yanushkevich, A., Belda, N.A., Scharrenberg, R., 2015. Design of test-circuits for HVDC circuit breakers, in: 2015 Third International Conference on Electric Power Equipment Switching Technology (ICEPE-ST). pp. 229–234. Available from: http://dx.doi.org/10.1109/ICEPE-ST.2015.7368338.

Siemens AG, 2011. High Voltage Direct Current Transmission–Proven Technology for Power Exchange, Available from: <http://www.energy.siemens.com/us/pool/hq/power-transmission/HVDC/HVDC_Proven_Technology.pdf>.

Tiku, D., 2014. DC power transmission: mercury-arc to thyristor HVDC valves [History]. IEEE Power and Energy Magazine. 12 (2), 76–96.
Tolbert, L.M. et al., 2005. Power Electronics for distributed energy systems and transmission and distribution applications.
Wang, H., Redfern, M.A., 2010. The advantages and disadvantages of using HVDC to interconnect AC networks. In: Universities Power Engineering Conference (UPEC), 2010. 45th International, pp. 1–5.
Weimers, L., 2011. Bulk power transmission at extra high voltages, a comparison between transmission lines for HVDC at voltages above 600 kV DC and 800 kV AC. Indian Journal of Power and River Valley Development. 61, 107.
Woodford, D.A, 1998. HVDC Transmission, Winnipeg.

7

ELECTRIC AND PLUG-IN HYBRID ELECTRIC VEHICLES AND SMART GRIDS

Noshirwan K. Medora

Exponent Inc., Phoenix, AZ, United States

CHAPTER OUTLINE

The Power Grid. DOI: http://dx.doi.org/10.1016/B978-0-12-805321-8.00007-0

7.1 Introduction

The typical utility-transmission grid is a highly complex electrical network that routinely incorporates technologies using electromechanical components and with an aging infrastructure. The modern grid is undergoing a significant change driven by a need for improved operational efficiency, increased reliability, modernization initiatives, and reliable customer service. Manual operations using traditional centralized control and using the technology of the past can no longer cope with the onset of new and rapid demanding responses and accommodate the environmental, reliability, consistency, and safety goals of the future. It is estimated that 70% of large power transformers and transmission lines are over 25 years old and 60% of circuit breakers are over 30 years old (OECD/IEA, 2011).

Fig. 7.1 shows a conceptual view of the present US electric grid system.

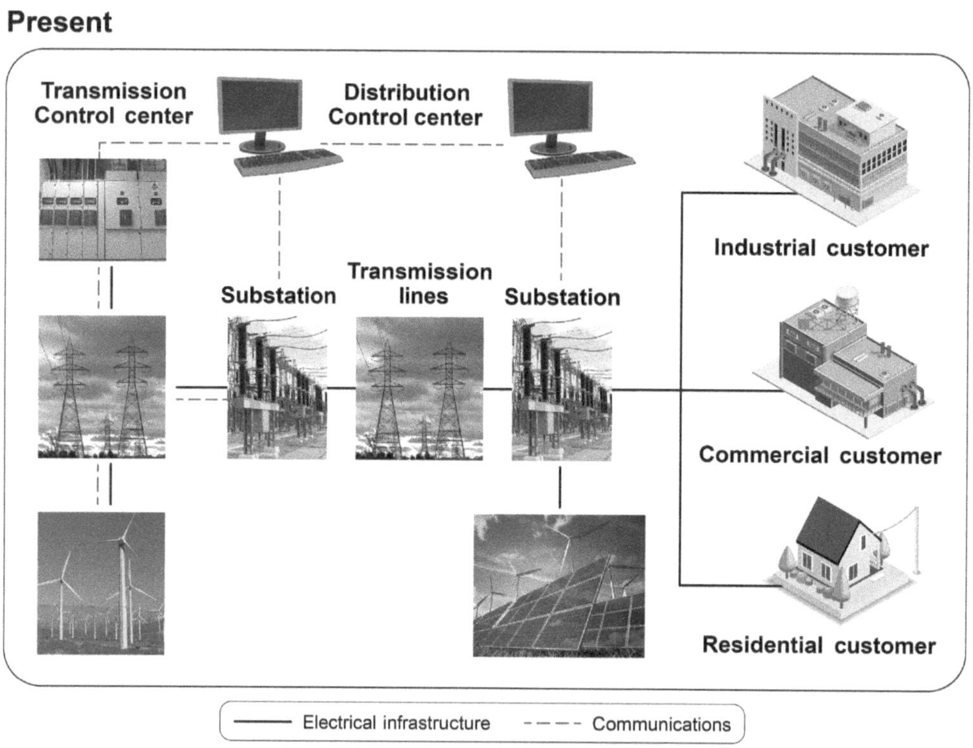

Figure 7.1 The present US electrical grid system. Adapted from OECD/IEA, 2011. Technology Roadmap: Smart Grids. IEA Publishing, p. 8.

7.2 System Architecture

The transmission grid in the United States is a complex electrical network and is the leading contributor in interfacing various technologies. The existing transmission grid currently uses a variety of new and emerging energy sources to reduce energy cost, and increase reliability, security and efficiency. Some of these sources include:

- Nuclear
- Solar
- Wind
- Hydroelectric
- Geothermal
- Natural gas
- Coal
- Oil
- Biomass

The electric power grid is undergoing modernization of its infrastructure using upcoming information technologies and with major advancements in solid-state devices in new and existing fields. The road map goal of the smart grid plan is to enhance and expand the aging, inefficient and congested system to improve resilience and reliability and to permit accommodating the anticipated additional loading as residential and industrial consumer demand increases.

As of 2015, the US electrical grid has approximately 19,000 generators (OECD/IEA, 2011). The electrical grid includes the electrical distribution system which is difficult to upgrade. Furthermore, it is estimated that 90% of outage minutes originate on the distribution system due to the 145 million customers and 6.3 million miles of distribution lines. In the United States, in 2014, the total installed electricity generation summer capacity was in excess of 1068 GW (Wikipedia).

The advent of new state-of-the-art technologies in the industry has resulted in the emergence of larger and more efficient power electronic devices with superior characteristics such as higher switching frequency, lower switching losses, lower on-state losses, and more rapid turn-on and turn-off control. The proliferation of these devices in various converter topologies has permitted a surge of promising sustainable and renewable energy sources which are being implemented and integrated into existing power systems.

Consequently, new industry trends are being driven by both, the integration of power generation and by the vast number of newly developed renewal energy technologies, while attempting to compensate for their limitation of uncontrolled, unscheduled and random power production. One prominent method to offset this limitation is to utilize a renewal energy storage source, with adequate storage capacity and which is distributed all across the country. One emerging energy storage source that has adequate storage capacity and is dispersed all over the country is the electric vehicle (EV) and the plug-in hybrid electric vehicle (PHEV).

The US utility power grid has a multitude of generators for generating electric power. However, the current power grid has almost no storage capacity, other than a few percent capacity in pumped storage. Due to this limitation, the generation and transmission of power must be continuously monitored and endlessly controlled to match the time-varying loads. This is accomplished by turning large generators on and off, or ramping them up and down, sometimes on a minute-by-minute basis.

On the other hand, vehicles have to be designed to have a large and frequent power fluctuation since that is the nature of this mode of transportation. Furthermore, personal vehicles are utilized only about 4% of the time for transportation and thus they are available for the remaining 96% of the time for other functions. These features open the electric-drive vehicles such as the PHEVs and EVs for performing a unique function—providing quick response utility power with large, inexpensive, and dispersed storage capacity, thus enhancing the stability and reliability of the utility grid (Kempton and Tomic, 2005a,b).

For a safe and feasible connection to the utility grid, an electric-drive vehicle must fulfill at least three requirements:

1. A standardized connection to the utility grid with power and control circuitry to allow bi-directional power flow
2. Standardized communication protocol to permit bi-directional communication between the control system of the vehicle and the utility grid
3. Monitoring and control system in the vehicle to permit user friendly interaction

7.2.1 Alternate Power Sources

Many alternate power sources are intermittent in nature and their output is highly dependent on several natural parameters, none of which are very consistent or predictable. Wind generation is an example of such an alternate power source with total US-installed wind capacity through 4Q 2015 of approximately 74 GW, and the total number of operating utility-scale wind turbines exceeding 48,800—with 39 states in the United States (plus Puerto

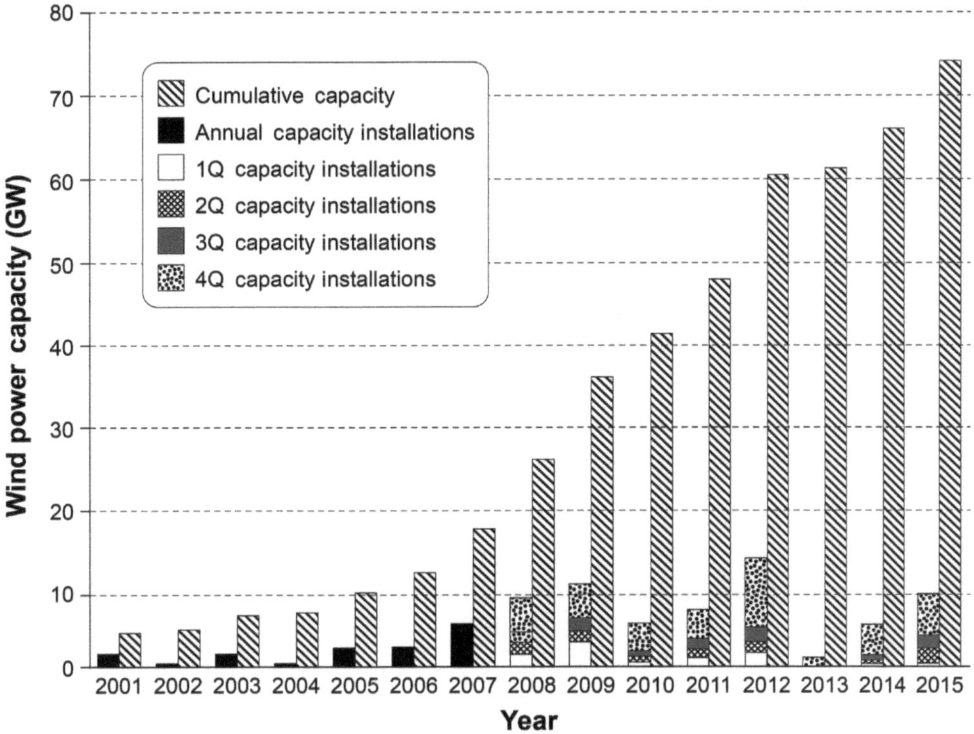

Figure 7.2 Total installed wind generation in the United States through 2015 (American Wind Energy Association, 1996–2016).

Rico) operating utility-scale wind energy projects (American Wind Energy Association, 1996–2016). Fig. 7.2 shows the total installed wind generation in the United States from 2001 to 2015.

Consequently, although wind farms show great promise, this power generation is intermittent in nature and highly dependent on wind speeds, wind gusts, daily thermal cycles, and movement of hot and cold front weather cycles. Because of the fluctuating nature of the high level of thermal and weather cycles, wind generation typically operates either at low output or high output with relatively fast rate-of-rise and rate-of-fall in wind power. Such a rapid ramping up and down can result in severe voltage fluctuations, and can also cause rigorous power quality problems. This rapid ramping can also result in severe stability issues in the electrical grid. However, it is persuasively beneficial to the industry to actively incorporate wind generation power, as this is a low-cost power generation commodity with low pollution effects, and when coupled with an energy storage source, can effectively be used as a peak shaver.

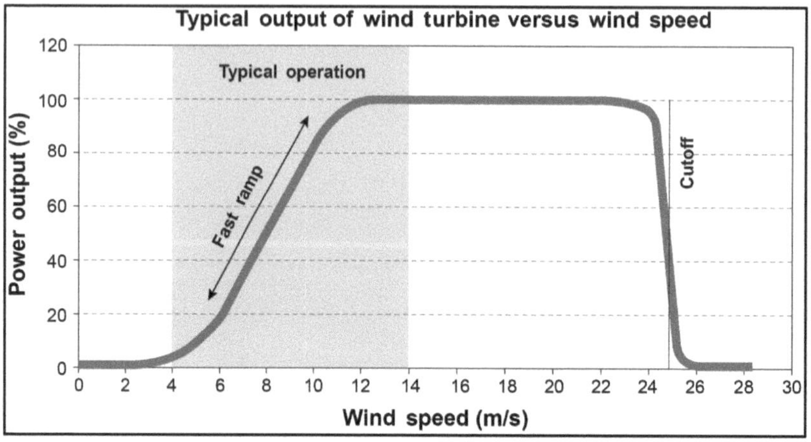

Figure 7.3 Power generation curve for a typical wind generator (Ipakchi and Albuyeh, 2009).

Wind generation also has other issues. At high wind speeds, the wind turbine protection-control system activates and cuts off power generation to prevent damage to the blades and the turbine generator, due to overspeed and possible tensional oscillation. This sudden power cutoff creates additional challenges due to the sudden step changes in reduced power generation. Consequently, although wind generation has a number of benefits, there are some drawbacks that need to be addressed.

Fig. 7.3 shows a typical wind turbine power curve showing the fast on and off power generation ramps.

7.3 Energy Load to the US Grid System

The US transmission line voltages vary and include lines with transmission voltages of 115, 138, 161, 230, 345, 500, and 765 kV (Sliverstein, 2011). It also includes high-voltage direct-current transmission lines. The key intention is to have safe, but rapid control of the volts and vars at multiple hundreds of nodes in the grid system. Fig. 7.4 shows a typical feeder with time-varying loads with changing power factor (Divan and Moghe, 2014).

The line voltage drops and the distribution transformer voltage drops are dependent on the line current, line impedances, and the transformer impedances. The secondary voltage drop may be as high as 10–15 V, causing a high load current dependent voltage variation. Furthermore, other factors causing a voltage variation include alternate power generation

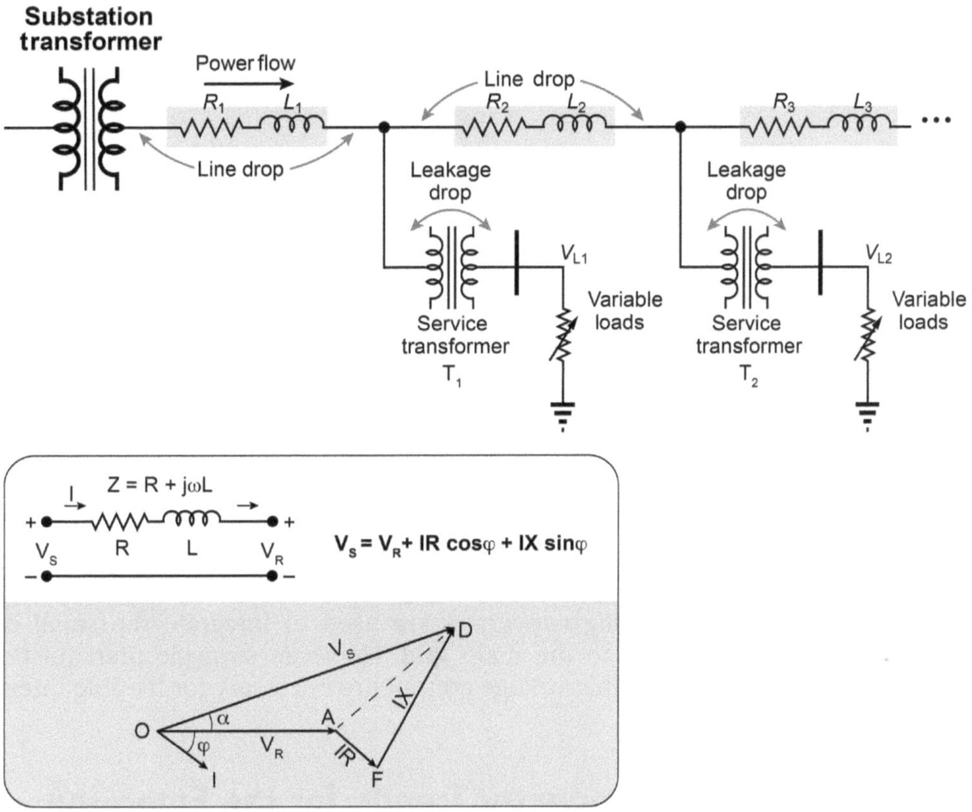

Figure 7.4 Typical feeder with time-varying loads with changing power factor. Load parameters determine the current and voltage drop along lines and distribution transformer impedance. Adapted from Divan, D., Moghe, R., 2014. Power electronics at the grid edge. IEEE Power Electronics Magazine, December 2014.

sources connected to the grid, including solar power, wind power, and others. Fig. 7.5 shows a typical line and distribution voltage drop on a hypothetical feeder system.

Power electronics has been extensively used in the transmission system for efficiently controlling large amounts of power flow. In 2009, there were 15,700 transmission substations in the United States and over 450,000 circuit-miles of transmission in North America (US, Canada, and Mexico). (Circuit-mile = miles of line x circuits/line) (Sliverstein, 2011). There is now an increasingly large number of a variety of small-capacity electrical sources such as EVs and PHEVs that have the potential to be integrated into microgrids and to the main grid. However, unlike large power sources with generating capacities of tens to hundreds of mega-watts, which are integrated at the

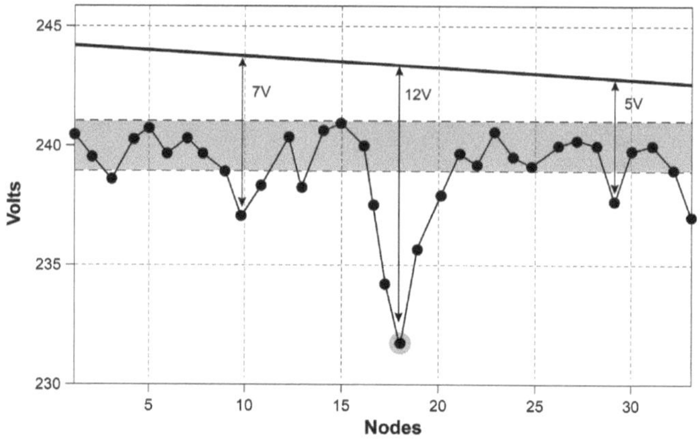

Figure 7.5 Line and distribution voltage drop on a hypothetical feeder system. Adapted from Divan, D., Moghe, R., 2014. Power electronics at the grid edge. IEEE Power Electronics Magazine, December 2014.

generating and transmission levels, these small-capacity sources of power are essentially at the distribution level. Typically grid-interfacing converters are used to integrate the small capacity sources to the main grid. However, strategic prerequisites such as dynamic voltage control are necessary for feasible integration.

7.4 Industry Trends for the Emerging Smart Grid

The industry trends for the embryonic smart grid are many and varying. The current industry focus is on the use of EVs and PHEVs as a dynamic on-line, real-time, distributed active energy storage system, permitting the bi-directional flow of power to reduce the system voltage fluctuation, and permit the control system to average out the peaks and valleys of power flow. Also there is the industry expectation of dependability, reliability and consistency, determined by conducting case studies using a sizeable dissemination of EVs and PHEVs with investigations and analyses to determine the effect of the power flow on the utility grid, and how it may impact the grid of the future.

This emerging endeavor will also focus on the effect of variable but low-cost renewal energy sources such as wind and solar power and the electrical stresses they will impose on the power grid, while emphasizing the security, efficiency, and reliability, as well as ensuring the power quality of the grid system. Studies are also being conducted to determine the effect of circuit loading if the charging and discharging cycles of the EVs and PHEVs are not monitored continuously, comprehensively controlled, and governed accordingly. There may be grid management issues that may need to be addressed and which may require the exploration and implementation of new technologies incorporating novel major advancements in the power semiconductor industry. This exploration would be conducted to ensure that the modern, rapid control systems using new and evolving control algorithms affirm that the overall smart grid is inherently stable, and yet allows a rapid bidirectional power flow and further validates that the grid can safely handle the increased capacity. New and existing standardization will be required to address the challenges to wide-area monitoring and allow a variety of users to judiciously employ the new and existing facilities to enhance the power grid infrastructure.

7.5 Unequal Utilization of the Distribution Transformer Thermal Life With Vehicle Charging

The PHEVs and EVs use stored electrical energy that needs to be periodically replenished by an external power source, which is typically the utility. High-usage PHEVs and EVs need to be charged on a regular basis with the charging current being supplied by the utility distribution transformers. It is expected that in most residential areas, during the charging time-interval, these vehicles will probably be asymmetrically grouped resulting in a diverse and heterogeneous distribution of vehicles plugging into the power grid. This would result in some utility distribution transformers seeing a large number of PHEVs and EVs in one location and others seeing few or none in an adjacent location.

This unequal distribution can cause a severe overloading of some distribution transformers resulting in a major reduction in operating service life. The reduction in service life is due to the generation of hotspots in the transformer. The hotspot temperature is typically the hottest winding temperature in the device. The higher hotspot temperature can result in increased core and winding temperatures with thermal degradation of the insulating paper and insulating oil. This would result in reduced operating life of the transformer winding insulation.

7.6 Hotspot Temperature in a Distribution Transformer With Vehicle Charging

Several studies using computer simulations and analyses have been conducted to ascertain the effect of PHEVs and EVs being plugged into distribution transformers and the resulting increased load and hotspot temperature.

One study involved four PHEVs charging at 3.3 kW with the assumption that each vehicle travels 40 miles each day. The vehicles have 11.6 kWh of useable energy storage with a charging efficiency of 81%. The vehicle's propulsion system (electric motor) provides 100% of the traction power when in charge-depleting mode. When the battery state-of-charge drops to 35%, the gasoline engine powers a generator that maintains the battery's SOC for the rest of the trip. The study was performed over a 1-year simulation period and used standard residential load profiles. A random Monte Carlo simulation was also plotted to show the temporal diversity in charging events (Kuss et al., 2010).

The results of the study of the transformer load and hotspot temperature is shown in Figs. 7.6 and 7.7. It is observed that the

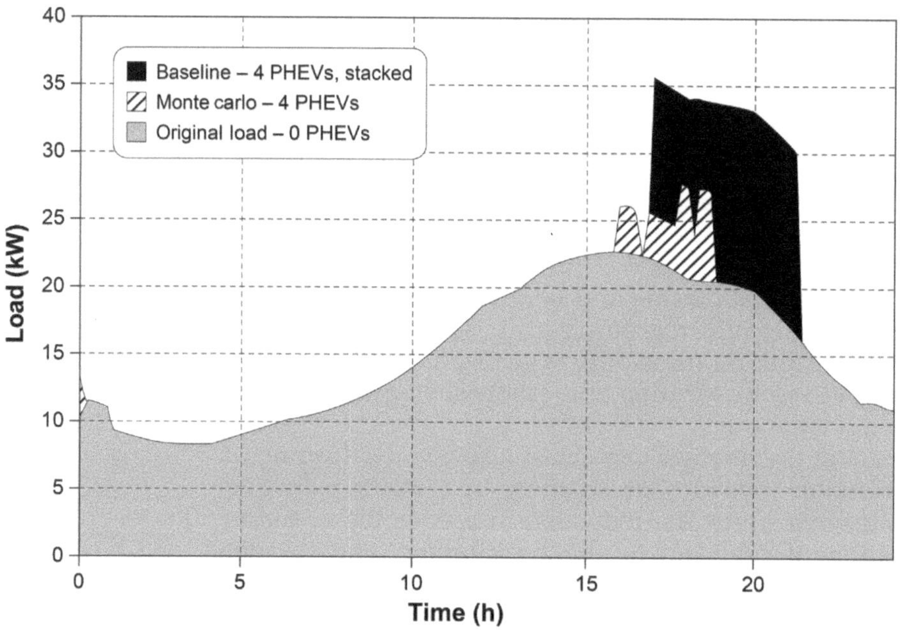

Figure 7.6 Transformer load as a function of time, before and after PHEV charging load contribution. The distribution transformer serves four PHEVs charging at 3.3 kW. The baseline assumes each vehicle travels 40 miles per day. The random Monte Carlo simulation illustrates the temporal diversity in the charging events (Kuss et al., 2010).

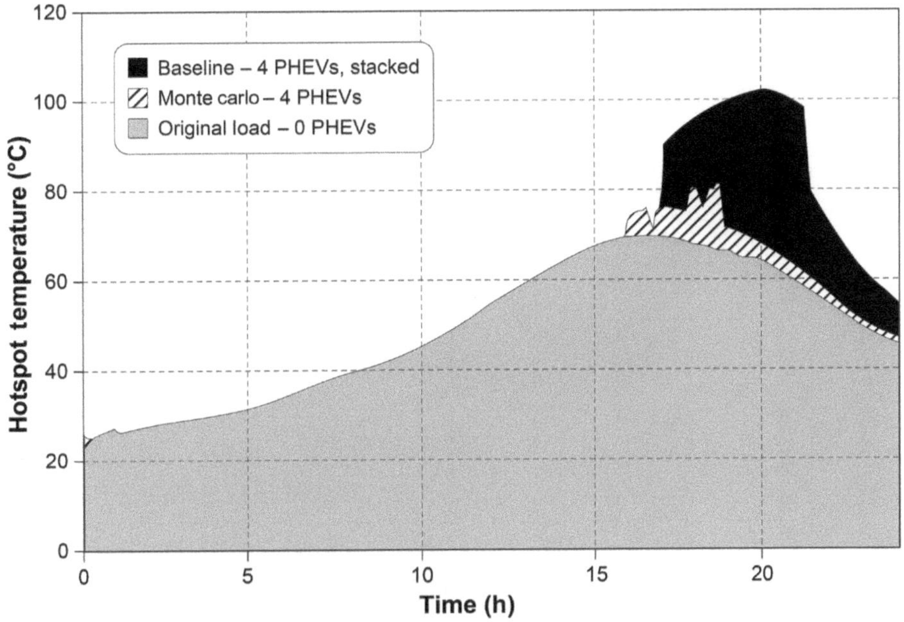

Figure 7.7 Transformer hotspot temperature as a function of time before and after PHEV charging load contribution. As expected, the maximum hotspot temperature appears sometime after the peak on the load curve due to the thermal inertial of the system and other factors (Kuss et al., 2010).

transformer peak load increases from approximately 22 kW to approximately 35 kW and the hotspot temperature increases from approximately 68°C to over 100°C. As expected, the hotspot temperature is not superimposed on the peak load curve due to the thermal inertia of the system and other factors. Furthermore, even after the charging load had dropped to zero, the transformer hotspot temperature continues to remain higher than normal for a period of time.

7.7 Overview of Plug-in Hybrid Vehicles

The gas-fueled internal combustion engine has long held a monopoly in road transportation in western nations. However, environmental conditions, global warming, air pollution, low efficiencies, and imported oil costs have resulted in a wave of research conducted to investigate alternate modes of transportation, with low emissions, higher efficiency, and less impact on climate change and less dependence on the economics of unstable and varying prices of oil. Several different types of

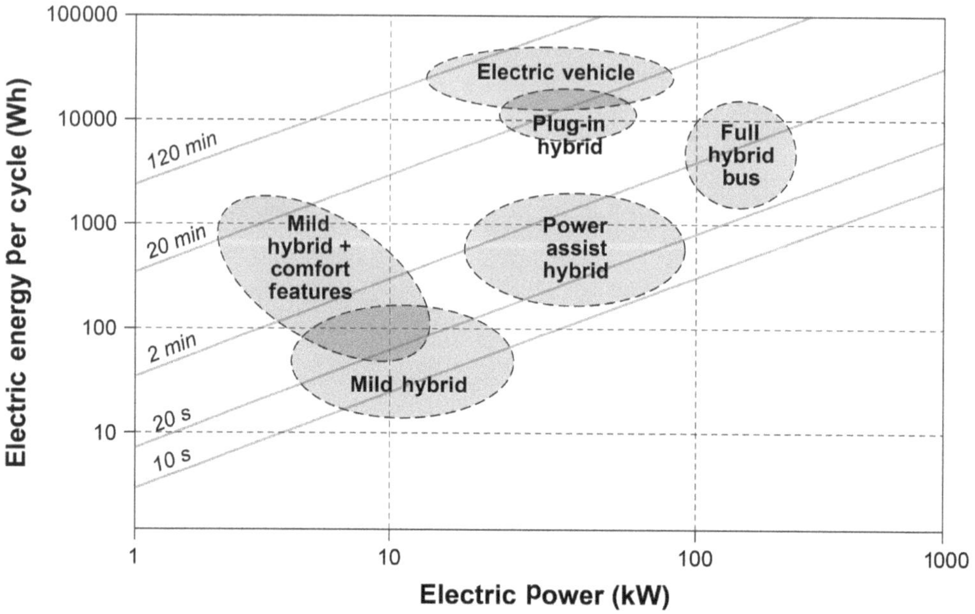

Figure 7.8 Description of the power and energy characteristics required from the battery by different types of electric vehicles (Broussely, 2010).

electric propulsion vehicles are currently available, including mild hybrid, power assisted hybrid, full hybrid, and full electric. One of the most popular is the PHEV. Fig. 7.8 presents the general power and energy ranges required from the battery by different types of EVs.

PHEVs exploit dual sources of energy for their drive train, including electric power from the utility grid and combustion energy from the different hydrocarbon fuels (e.g., gasoline or diesel) in the vehicle fuel tank. In the PHEV, the electrical drivetrain and internal combustion drivetrain are mechanically interconnected and real-time controlled by the engine control module such that the highest efficiency is obtained with low emissions within the constraints imposed by the vehicle parameters. The propulsion energy from the two sources is continuously monitored and is under software control and is supplied to the wheels as either in parallel, or in series or in a combination of the two.

For EVs, the total emission involves emission at the generating station (grid level). For PHEVs, the total emission involves emission at the generating station (grid level), and the emission during the use of the internal combustion engine and thus emission at the vehicle level.

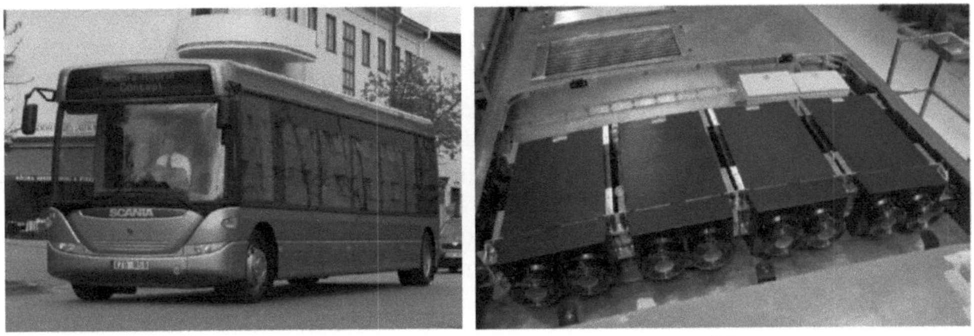

(A) (B)

Figure 7.9 Hybrid bus with supercapacitors on the roof of the bus: (A) Hybrid bus with supercapacitors on the roof (B) Supercapacitors on the roof of the bus (Medora and Kusko, 2012).

Supercapacitors, also known as ultra capacitors, are also used in hybrid vehicles to enhance the performance of the batteries and extend battery life. Supercapacitors have a low internal resistance and a high cycle rate of up to 1,000,000 cycles of operation and can readily handle the charge/discharge cycles in a hybrid vehicle. Thus, a supercapacitor can increase the life of the battery by handling the tens of thousands of high peak charge and discharge currents while the battery is used to supply the substantially lower average load current.

However, when the EV or PHEV is connected to the smart grid during the charge/discharge energy storage activity, supercapacitors would result in a higher fault current resulting in requiring higher interrupting capacity fuses and circuit breakers to handle a low-resistance fault condition. Thus, increased fault current protection requirement would need to be taken into consideration when a supercapacitor-based hybrid vehicle is connected to the power grid. Fig. 7.9 shows supercapacitors installed on the roof of a hybrid bus (Medora and Kusko, 2012).

7.8 Distributed Power Generation With Super Capacitor Storage

Renewal power sources such as wind generation and solar are, at best, intermittent and unpredictable in nature. Consequently, if these power sources are used as the predominant energy source for the unpredictable, but high-power demand of PHEVs and EVs, it can result in undesirable consequences to the smart grid. Power quality issues may arise with voltage surges, voltage sags, brownouts, and voltage flicker.

To reduce the effect of the interaction between the renewal energy sources and the PHEVs and EVs, it is proposed that a short-term storage system such as supercapacitors may be used. Use of this proposed storage system would help to reduce power quality problems and further enhance the reliability of the smart grid during the momentary transient state (around 1 minute) (Cheng, 2011). This proposed methodology would provide a measure of additional stability to the smart grid during the short-term transient conditions.

7.9 Load Flow and Power Flow Demand

New trends, new standards and restrictions envisioned for power flow are now in the pipeline and are also being enforced by regulatory organizations. These are becoming apparent in the utility grid system at the distribution level from the demand side and from the supply side. The utility requirement is for an efficient, reliable, and stable electric system. On the other hand, the supply side requirement is predominately to assist the base load generation, and to integrate peak shaving of power during high-power demand periods.

Currently, the electric power utility is predominately subdivided for meeting four power market demands (Kempton and Tomic, 2005a,b):

1. Base load power

   Utility generators such as hydroelectric power stations which are running almost all the time.
2. Peak power demand

   Historically predictable peak power demand such as power for air-conditioning systems on hot summer afternoons.
3. Spinning reserve

   These are generating systems set up for rapid response in case of equipment failure or to meet contractual requirements. This may be up to 20 times a year with a typical duration of 10 minutes to 1 hour.
4. Regulation

   Required to keep constant voltage and frequency and are required for a few minutes, but up to 400 times a day.

Spinning reserve and regulation have opposing requirements. Maximum spinning reserve is obtained with the EV and PHEV vehicle battery fully charged; however, this would negate any regulation down. Consequently, for up and down regulation to be effective, the vehicle battery should be partially charged. One strategy would be to initially charge the vehicle to about 50% state-of-charge, and after peak hours, shift to full charge.

The major automotive manufacturers are actively investigating electric-propulsion vehicles using a multiplicity of electrical power sources such as a large battery, or a fuel cell, or a hybrid vehicle using a gasoline engine integrated to an electric motor/generator system. Irrespective of the electric power source, the current accepted trend is to use state-of-the-art battery methodologies and diversity of configurations in these relatively low-power–capacity propulsion systems. The current EV and PHEV propulsion system investigation also includes an analytical review of the existing topologies for the design and implementation of an efficient, bi-directional grid-interfacing power electronic system.

Currently, for the EV and PHEV, the typical vehicle-to-grid (V2G) interface is only at the distribution level. However, the existing trend in the industry is that the grid-to-vehicle (G2V) interface would use sophisticated and redundant software controlled voltage and current control strategies. This system would include layers of active safeguards and protection with a prime focus on active power flow adaptation and control, sophisticated energy controlling techniques, power quality management and control, G2V interface security, and integrated climate control and greenhouse gas emissions.

7.10 Synergistic Operation of Hybrid Vehicles and the Smart Grid

As mentioned earlier, in 2014, the total installed electricity generation summer capacity in the United States was in excess of 1068 GW (Wikipedia). This energy is obtained from fossil fuel, hydroelectric, nuclear, and alternate sources such as wind power and solar.

The US cumulative sales of PHEVs are rapidly increasing, and it is predicted that this increase will continue in the future. The cumulative sales of PHEVs increased from about 25,000 vehicles in 2012 to 250,000 in 2014 and to 400,000 in 2015. This rapid increase in EVs (including cars and trucks) on the road will result in a well-distributed, efficient, large energy storage system in the United States and in the world, which is currently relatively untapped (Fig. 7.10).

As of 2014, the number of vehicles on the road in the United States is over 250 million (Hirsch, 2014). An EPRI prediction of the total PHEV penetration by 2050 is 62% of the US fleet (Kulshrestha et al., 2009).

If in the future just 25% of these vehicles use electric propulsion, and if each vehicle is assumed to be only about

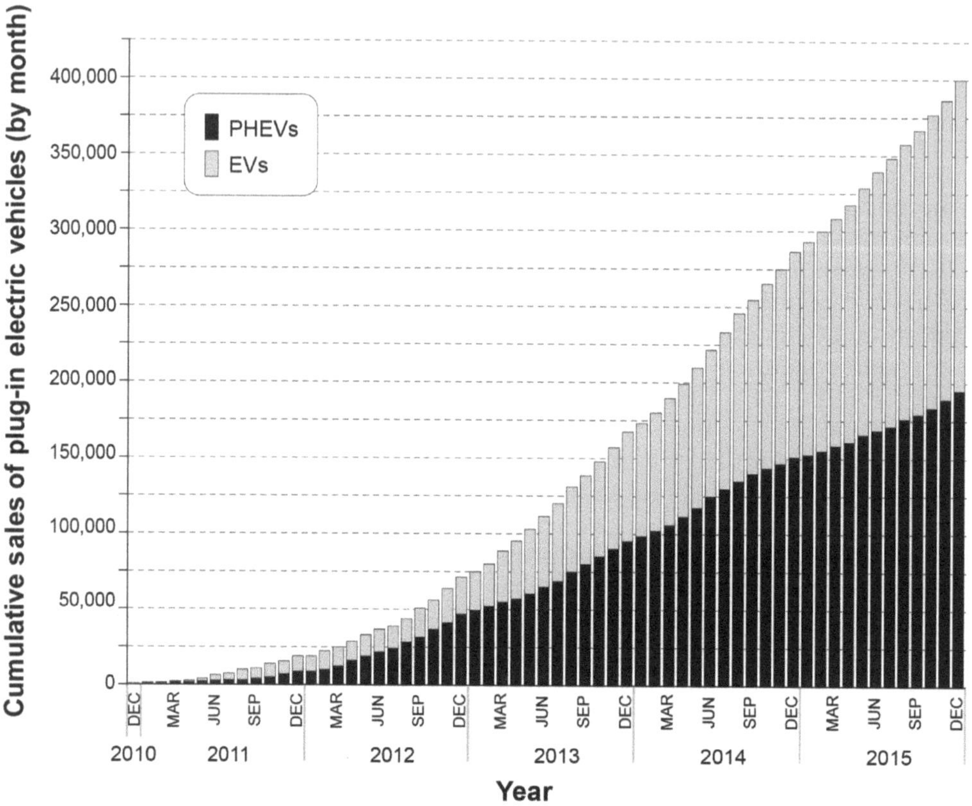

Figure 7.10 United States cumulative sales of plug-in electric vehicles (US PEV Sales).

50 kW, which is about 50% of the representative vehicle rated capacity (typical EV capacity is of the order of 100 kW (Fig. 7.8), the total vehicle generating capacity will be approximately 3125 GW. This is almost three times the total installed electric generating capacity of 1063 GW in the United States in 2014 (Wikipedia).

Furthermore, even if it is assumed that only 15% of the vehicles use electric propulsion and each vehicle is assumed to only supply 25 kW, which is about 25% of the vehicle rated capacity, the vehicle generating capacity is still about 938 GW, which is quite comparable to the total installed electric generating capacity in the United States.

It is expected that this unique, distributed renewable energy source can be tapped to synergistically use these strengths to provide peak shaving power and thus enhance and augment the existing electric utility system without a major change in the electric utility generation network.

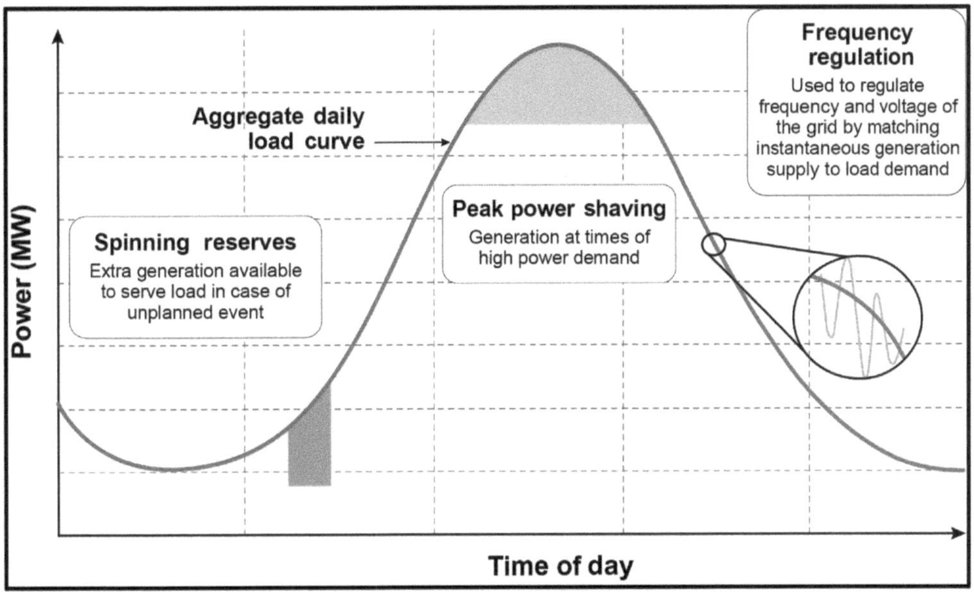

Figure 7.11 Illustrative diagram of V2G power and the use of spinning reserves and peak power shaving (Gorguinpour, 2013).

In summary, it is envisioned that when the EV or the PHEV is not on the road, it would be connected to the utility grid. In this connected mode, the EV's battery would be charged during nondemand hours and would provide rapid response to peak power demands made by other connected loads to the electrical grid. Thus, this V2G power, would act as an energy storage device and provide peak shaving power by storing the electric power when the utility demand is low and providing a rapid response to the utility demand for additional power during peak demand loads. The vehicle battery would provide storage of electric power from renewal energy sources such as solar and wind in the vehicle battery in a chemical form thus permitting a rapid response (fraction of a second to a few seconds) to utility power demands.

Fig. 7.11 shows the illustration of V2G power and the use of spinning reserves and peak power shaving.

Thus for a hybrid vehicle at least two primary power flows are possible between the vehicle and the grid:

- G2V when charging the vehicle batteries
- V2G when discharging the vehicle batteries

This unique system infrastructure permitting the ability to interconnect multiple vehicles and provide bi-directional power flow allows each vehicle to be used as a miniature power system. The feasibility of this high-power storage facility depends on the smart grid's demand for reliability, dependability, and consistency and for the vehicle owner's demand for a strong financial return-on investment.

7.11 Availability of the Hybrid Vehicle for V2G Ancillary Services

A key factor in the feasibility of the V2G dual power flow system is the availability of the hybrid or EV as a function of the time of day. To quantify the availability of the V2G ancillary services, calculations have been performed to determine the Availability Factor (AF). "AF is a North American Electric Reliability Corporation (NERC) metric of the ability of an individual generating resource to enter into a contract with the grid system operator" (Quinn et al., 2009).

Using AF as a qualifying factor, a comparison can now be made between the AF of a comparable generating competitor to the AF from the V2G ancillary services. Using gas turbine power plants as a benchmark for ancillary power sources, comparisons have been made to determine the feasibility of V2G ancillary services. For gas turbine power plants in operation from 2003 to 2007, the NERC reports an AF of 92.91%.

The National Household Transportation Survey data was used to determine the availability of the hybrid or EV as a function of the time of day for two different V2G scenarios:

- Charging from home only
- Charging from home and work

Fig. 7.12 indicates that during a 24-hour period, for both scenarios, the availability is high for the early portion of the day and the later portion of the day. The figure shows that for charging from home only, the availability of the vehicles decreases as they drive to work. From 10:45 a.m. to mid-afternoon, the minimum vehicle availability is only 62.7% and the long-term average AF is 83.6%. However, for home and work charging scenarios, the availability of the V2G vehicles is substantially increased with a minimum vehicle availability of 82.0% and a long-term AF of about 91.7%. This AF value is comparable to the gas turbine power plants AF of 92.91%.

These computations clearly indicate that for V2G power to be feasible, the charging stations need to be available at home and also at work and also at other locations.

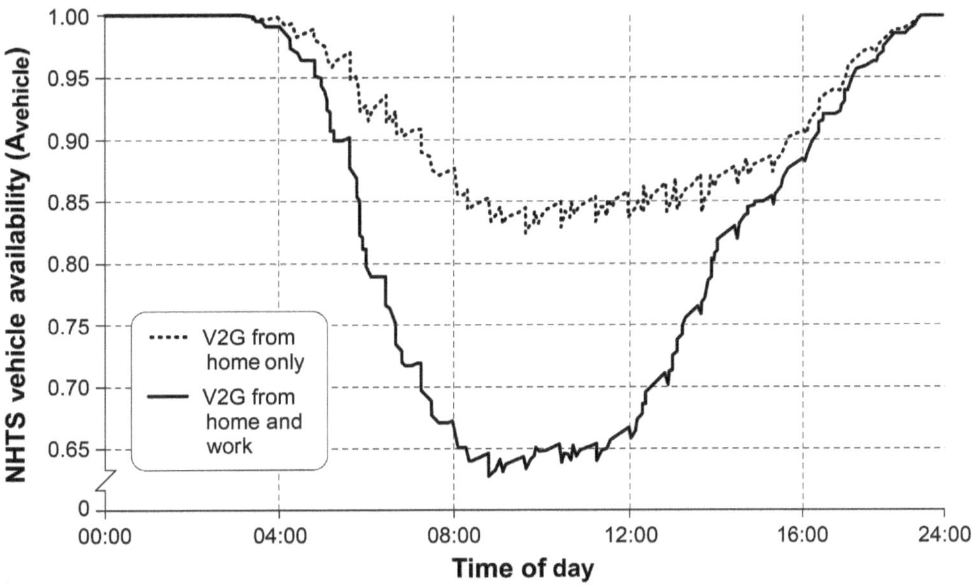

Figure 7.12 shows the availability of the vehicle-to-grid enabled vehicles as a function of the time of day for two V2G scenarios (Quinn et al., 2009).

7.12 Charging Systems for Electric and Hybrid Vehicles

Charging equipment for EVs and PHEVs is typically classified by the rate at which the vehicle batteries are charged. Charging times vary based on the state-of-charge of the battery, the type of battery, and the type of charging station also known as EVSE (electric vehicle support equipment). Vehicle charging times can range from 15 minutes to 20 hours or more (U.S. Department of Energy, 2016).

Currently there are several charging systems for hybrid and EVs. They are typically categorized as Level 1 charging, Level 2 charging, and Level 3 charging. The Society of Automotive Engineers (SAE) J1772 standard recommends that Level 1 and Level 2 EVSE be located on-board the vehicle, while Level 3 is external to the vehicle (Fig. 7.13).

- Level 1 Charging

  This charging method uses the typical 120 Vac, 15 A single phase outlet as the charging source and provides 2–5 miles of range per 1 hour of charging time. Typically, 8 hours of charging will provide 40 miles of travel range. The EV or PHEV uses the SAE J1772 standard connector.

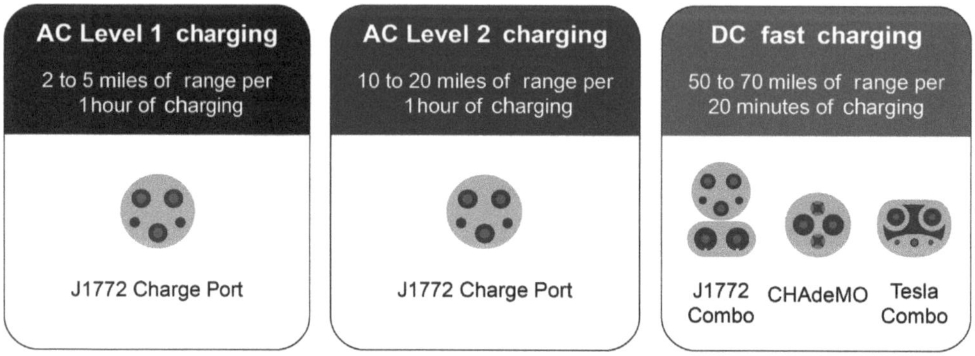

Figure 7.13 shows the charging infrastructure to charge EVs and PHEVs (U.S. Department of Energy, 2016).

- Level 2 Charging

  All commercial EVs and PHEVs have the ability to charge using Level 1 or Level 2 charging systems. Level 2 charging uses a 208 V (typical in commercial applications) or 240 Vac (typical in residential applications) to supply charging current of up to 80 A (19.2 kW) and provides 10–20 miles of range per 1 hour of charging time. Level 2 charging also uses the SAE J1772 ac charge port to charge the vehicle batteries.
- Level 3 Charging

  Level 3 charging, provides a rapid charging method to charge the vehicle batteries at installed stations. This charging method uses a 208/480 Vac, 3-phase source with a dc output to the vehicle and provides 50–70 miles of range per 20 minutes of charging time.

Table 7.1 presents the charging characteristics of some EVs and PHEV's. Fig. 7.14 presents details of the SAE J 1772 combo connector for Level 1 and Level 2 charging. Fig. 7.15 shows the different charging systems and power levels for EVs.

7.13 Hybrid Vehicle as a Miniature Power System

Currently there are a number of pilot projects where hybrid or EVs are used as a miniature power station, using high-power charging stations to bi-directionally charge and discharge EVs into a microgrid, thereby using the vehicle's batteries as a storage element for electrical energy. The vehicle is charged during low-demand hours and supplies electric power to the grid during high demand hours, thus permitting peak shaving

Table 7.1 Charging Characteristics of Some EVs and PHEVs (Yilmaz and Krein, 2013)

Vehicle	Battery Type and Energy (kWh)	All Electric Range (miles)	Connector	Level 1 Charging		Level 2 Charging		Level 3 Charging	
				Demand (kW)	*Charge Time (hours)*	*Demand (kW)*	*Charge Time (hours)*	*Demand*	*Charge Time*
Toyota Prius PHEV (2012)	Li-Ion 4.4	14	SAE J1772	1.4 (120 V)	3	3.8 (240 V)	2.5	N/A	N/A
Chevrolet Volt PHEV	Li-Ion 16	40	SAE J1772	0.96–1.4	5–8	3.8	2–3	N/A	N/A
Mitsubishi i-MiEV EV	Li-Ion 16	96	SAE J1772 JARI/TEPCO	1.5	7	3	14	50 kW	30 min
Nissan Leaf EV	Li-Ion 24	100	SAE J1772 JARI/TEPCO	1.8	12–16	3.3	6–8	50 + kW	15–30 min
Tesla Roadster EV	Li-Ion 53	245	SAE J1772	1.8	30+	9.6–16.8	4–12	N/A	N/A

Figure 7.14 SAE J1772 combo connector for ac or dc Level 1 and Level 2 charging. SAE J1772 is being revised to address dc fast charging (enabled by the two power pins at bottom) (SAE International, 2011).

of electric power on the Microgrid. Such pilot projects demonstrate the feasibility of marketing EVs and PHEVs as a renewal energy storage device and allows maximizing the use of renewable sources such as solar and wind.

An example of such an innovative experimental project to rapidly charge EVs and provide a connection to the grid is at the Los Angeles Air Force Base in El Segundo, California. This Department of Defense facility uses bi-directional charging stations based on the UL-Certified multi-port converters. A Grid-Tied Inverter and Battery Controller (GTIB-30) is used that is capable of 15 and 30 kW charging rates. This V2G pilot project provides real-world data for the charging the vehicles and for discharging the vehicle battery back into grid (Francis, 2014). Fig. 7.16 shows an EV at the Los Angeles Air Force Base.

The Princeton Power GTIB-30 Grid-Tied Inverter and Battery Controller with peak shaving functionality and demand dispatch allows a wide ac input voltage range of 208 or 480 Vac, 3-phase, with a current of 85 A at 208 Vac and 40 A at 480 Vac. The 30 kW hybrid inverter has an efficiency of 96.5%. The voltage at full power is 280–600 V with a maximum current of 120 A. It includes an automatic transfer switch, a front-panel interface, and includes performance monitoring with an operating temperature of 0°C to 50°C with derating above 40°C (Princeton Power Systems, 2014). A 100 kW hybrid inverter, the GTIB 100-G1.2 has a voltage range of 290–800 Vdc with a maximum current of 380 A, a maximum power of 110 kW, ac grid voltage of 480 V, 3-phase, 60-Hz and a peak efficiency exceeding 97% (Princeton

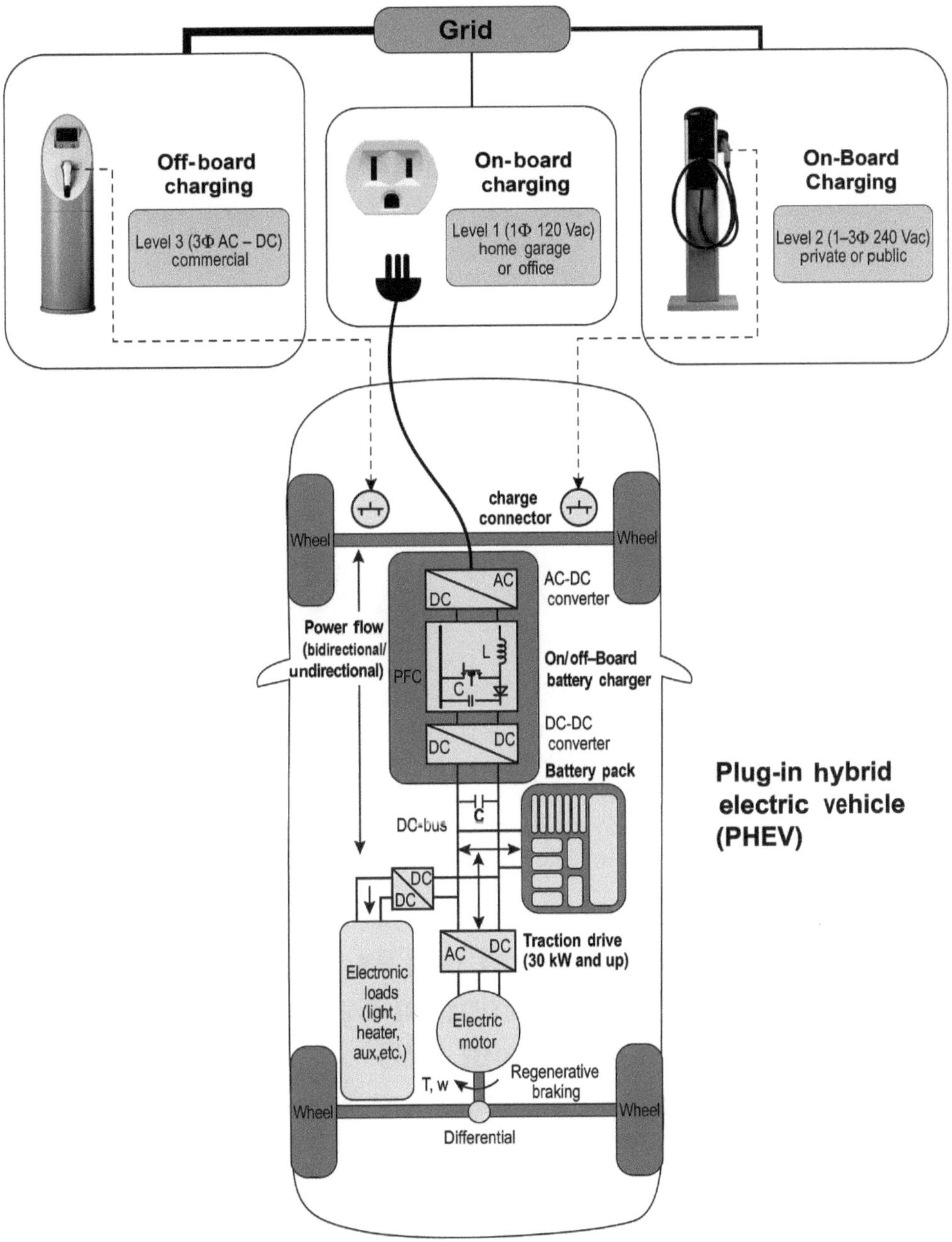

Figure 7.15 Level 1, Level 2, and Level 3 charging systems and power levels for EVs. Adapted from Yilmaz, M., Krein, P.T., 2013. Review of battery charger topologies, charging power levels and infrastructure for plug-in electric and hybrid vehicles. IEEE Trans. Power Electron. 28(5).

Figure 7.16 Electric vehicle at the Los Angeles Air Force Base in El Segundo, CA (Francis, 2014).

Power Systems, 2015). Fig. 7.17 shows the GTIB-100G1.2 100 kW hybrid inverter for battery charging.

7.14 Impact of EVs and PHEVs on the Distribution System With and Without V2G

Electric propulsion vehicles such as EVs and PHEVs need to be connected to the grid on a regular basis for charging their batteries. The electrical distribution system is the point-of-connection for the EVs and PHEVs. During charging, the specific energy consumption for a PHEV can vary between 0.15 and 0.34 kWh/mile with a median of 0.24 kWh/mile for a typical sedan (Raghavan and Khaligh, 2011).

In the existing electrical distribution system, there is little electronic communication and coordination regarding real-time load demand between the consumer and the utility.

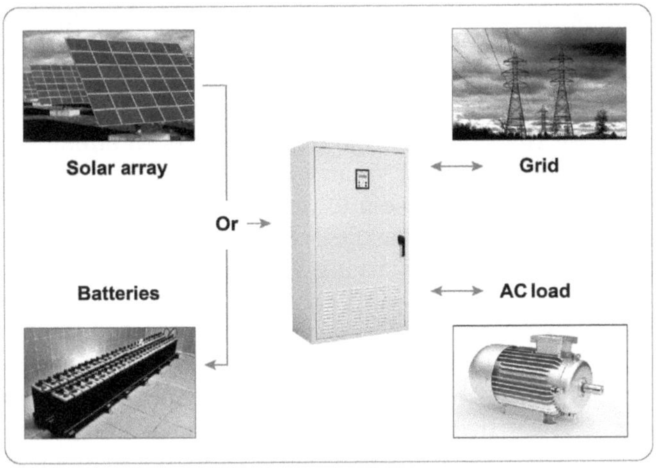

Figure 7.17 GTIB-100G1.2 connected to the grid and supporting a solar array, batteries and an ac load. Adapted from Princeton Power Systems. 2015. Grid-Tied Inverter and Battery Controller GTIB-100G1.2 Data Sheet. Princeton Power Systems, Lawrenceville, NJ.

Consequently, it is likely that this absence of an existing load demand coordination policy for the charging of these EVs and PHEVs may result in uncontrolled charging with increased stress on an aging distribution system and may result in several adverse effects such as:

- Increased demand on loaded distribution systems resulting in excessive real and reactive power losses
- Higher operating temperatures and higher hotspot temperatures on the distribution transformers
- Increased issues on the connected loads due to multiple and regular excessive high and low voltage deviations resulting in voltage swells and sags
- Unequal loading on the overall distribution system, where some distribution systems may be over loaded while other distribution systems in adjacent neighborhoods may see relatively light loading
- Increased issues in power quality including excessive harmonic distortion, and flicker
- Increased peak demand with higher voltage drops
- Imbalance in the magnitude of the phase voltages resulting in phase voltages or phase currents being outside of permissible limits

A number of computer simulation studies have been conducted to evaluate the impact and possible detrimental effect of uncontrolled charging of EVs and PHEVs using V2G power from existing local residential distribution systems. The results of one load profile study are described here. The studies indicate that in order to have a favorable demand side load regulation, it is necessary to implement a high degree of strategic real-time monitoring of the power demand parameters to effectively control the charging load of EVs and PHEVs and thus mitigate the adversarial impacts of these additional loads.

Fig. 7.18 shows the results of one study comparing the loading of uncontrolled charging with intelligent local charging optimization without V2G and charging optimization with V2G. Clearly the coordinated strategy using the V2G charging algorithms results in substantially reducing the electrical loading on the distribution system. This would correspondingly result in imposing a lower demand on the distribution transformer and the distribution lines with resulting lower losses, lower operating temperatures and increased operating life. This would also reduce the voltage deviations, harmonic distortion and issues regarding power quality. The data indicates that at the distribution level, when V2G power is used, it reduces the peak load and it also reduces the variability of the load profile. This is in addition to the previously mentioned advantages of using

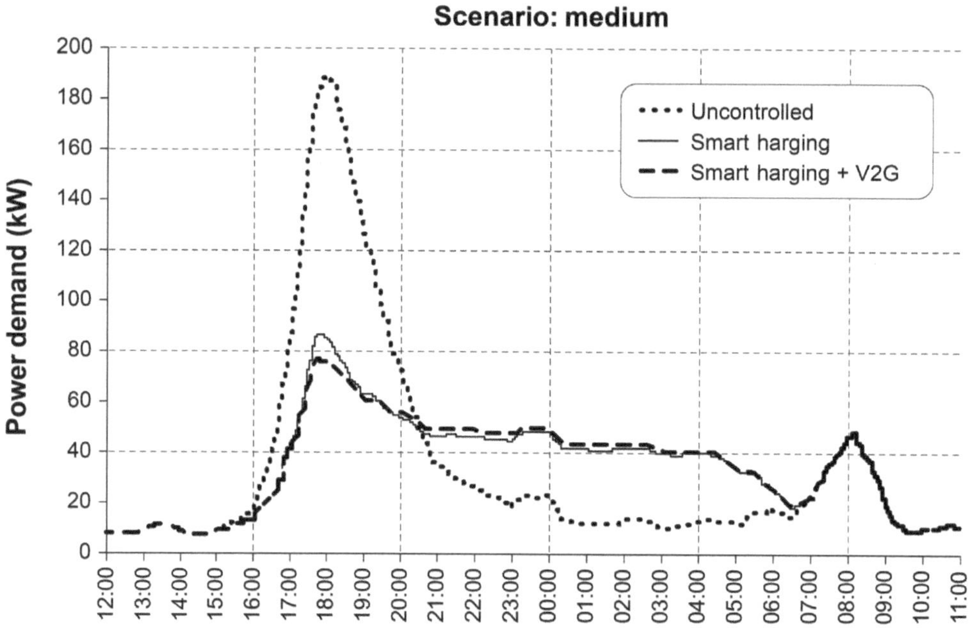

Figure 7.18 Average load profile as measured by the distribution transformer for PHEV penetration of 45% (Mets et al., 2011).

V2G power for a global country-wide smart grid network where different time zones would allow V2G power to greatly reduce the load demand on the smart grid.

7.15 V2G Complexities for Hybrid Vehicle Systems

V2G power has several advantages and provides a methodology for providing load equalizing for the smart grid. However, it does have its own share of problems and complexities for the existing hybrid vehicle system, which need to be addressed to obtain a feasible operating solution.

One of the issues to be addressed is the effect of the daily, multiple charge–discharge profiles using V2G power on the battery cycle life. Since V2G power would promote a fairly consistent, day-to-day, multiple battery charge–discharge cycles, battery manufacturers would need to meticulously investigate and judiciously determine the expected battery life for an EV or PHEV under this rigorous charge–discharge cycle life condition. In a non-V2G operation, the daily charge–discharge cycles are typically limited by the vehicle operator and are comparatively infrequent. However, in the V2G power scenario, the daily charge–discharge cycles are no longer controlled by the operator of the vehicle. In this case, the vehicle, when parked is connected to the grid and thus the control parameters in grid monitoring system primarily determine the charge–discharge cycle rate profile. This profile may be highly dependent on several variables such as the time of the day, the connected loads on the grid, and the location and number of EVs and PHEVs connected to the grid in the neighboring area.

Furthermore, safety standards committees would need to standardize the acceptable battery state-of-charge, depth-of-discharge and other critical parameters used during the V2G power operation. The parameters may also include the maximum charge and discharge rate and the maximum depth-of-discharge so that the EV or PHEV is readily available during an unexpected emergency.

7.16 Interconnection and Communication Standards for V2G Power

It is imperative that reliable and efficient operation with ease of G2V communication and interconnection be an integral part of the operation of the V2G system. To this effect, existing safety standards are currently used and others will be developed as the integration of the overall smart grid progresses. The interconnection and the diversity of the multitude of systems require a harmonization of existing and new standards. These foundational interface standards are developed by various organizations such as the American National Standards Institute (ANSI), Institute of Electrical Engineers (IEEE), Society of Automotive Engineers (SAE), German Institute of Standardization (DIN), and possibly others (Fig. 7.19).

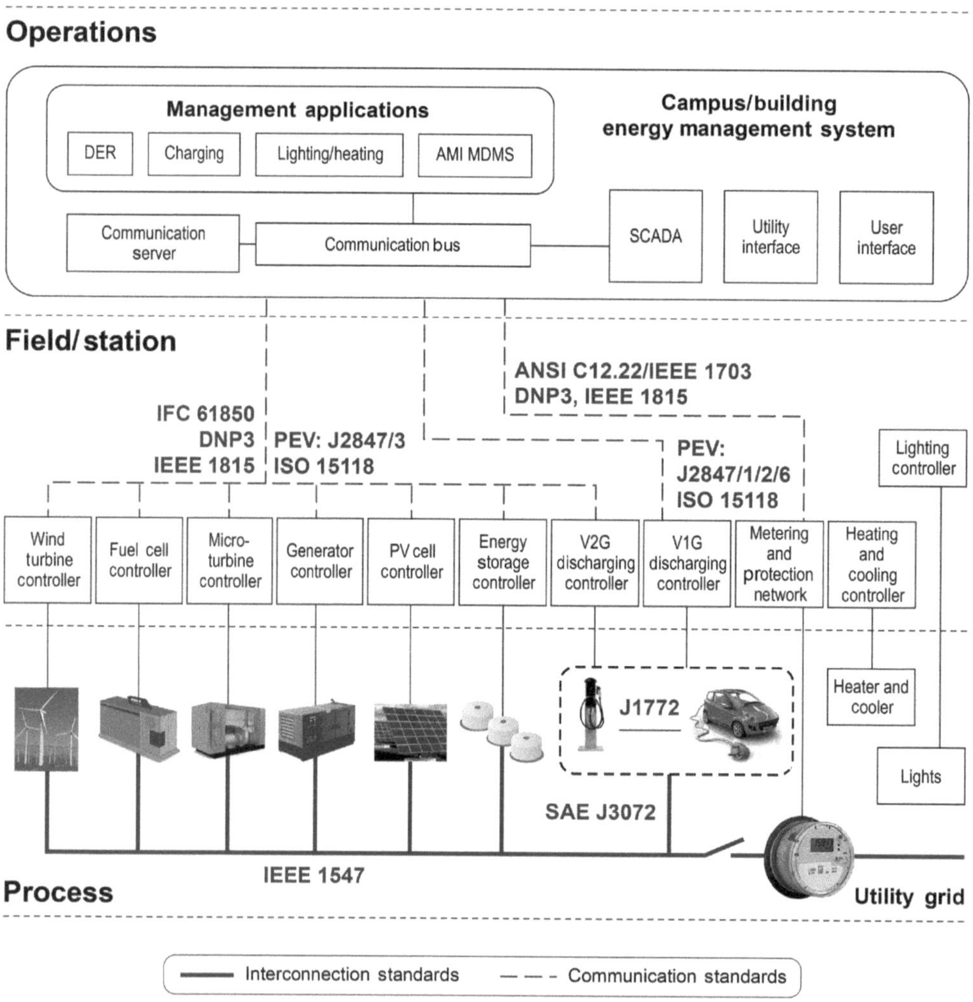

Figure 7.19 Major interconnection and communication standards for Vehicle Grid Integration (VGI) (Markel et al., 2015).

These standards include several categories such as connectivity, communications, information exchange, and cyber security: Standards related to vehicle-to-grid integration (VGI) are listed below (Markel et al., 2015):

1. Physical connectivity standards for wired charging/discharging and performance criteria for wireless charging
   a. IEEE 1547: Interconnecting distributed resources with electric power systems
   b. SAE J1772: PEV connector (charge coupler)
   c. SAE J3072: Interconnection requirements for onboard, utility-interactive inverter systems
   d. SAE J2954: Minimum performance and safety criteria for wireless charging of electric and plug-in vehicles
2. Communications standards
   a. IEC 61850, Distributed Network Protocol (DNP3), and IEEE 1815: Automation of DER
   b. SAE J2847/3 and ISO 15118: PEV as a DER (distributed energy resources)
   c. SAE J2847/1 and ISO 15118: AC PEV smart charging standards
   d. SAE J2847/2, ISO 15118, DIN 70121, and IEC 61851-24: DC PEV charging standards
   e. SAE J2847/6: Wireless PEV charging standard
   f. SAE J2847/5: PEVs and customers
   g. ANSI C12.22 and IEEE 1703: Metering
3. Information exchange standards
   a. IEC 61970, IEC 61968, and IEC 62325: Common Information Model for DER
   b. IEC 61850: Data models for DER
   c. ANSI C12.19/IEEE 1377: Metering
4. Cyber security standards
   a. IEC 62351
   b. SAE J2931/7
5. Use Cases defined by the SAE Hybrid Communication Task Force and ISO. The SAE Hybrid Communication Task Force and International Organization for Standardization have defined a series of use cases for plug-in electric vehicle (PEV) charging/discharging, listed below
   a. AE J2836/1: AC PEV smart charging use cases
   b. SAE J2836/2: DC PEV smart charging use cases
   c. SAE J2836/3: Use cases for PEV as a DER
   d. SAE J2836/5: Use cases for PEV customer communications
   e. SAE J2836/6: Use cases for wireless charging
   f. ISO 15118-1: General information and use-case definition.

7.17 Hybrid Vehicle as a Stand-Alone Emergency Power System

The multiplicity of the pilot projects using V2G power for EVs and PHEVs also provide information and useful data on the use of electric and hybrid vehicles as a standalone, sole energy storage source in case of an extended loss of grid power due to a major storm, earthquakes, or other natural calamity. It is recognized that after the major power outage caused by Hurricane Sandy, many homes experienced a loss of utility power for extended periods of time. The realization came quickly that for the typical residential owner the immediate requirements include heating, food and water, lighting and a means of electronic communication. Many homeowners used their vehicle batteries and a car-charger-phone-adapter to charge their cell phones. A strategic situation can be envisioned where a hybrid or an EV's batteries can be used to supply the necessary electric power to operate critical elements in a house such as a few florescent lights, a small refrigerator, and a small heater for a period of time (Francis, 2014).

7.18 Vehicle to Grid of the Future

The vehicle grid of the future will probably require continuous, real-time metering and communication infrastructure between utilities to ensure that the smart-grid parameters including grid voltage and frequency are within a stable operating window. A variety of factors will need to be accommodated in an optimum manner. These factors include power factor correction, power quality, voltage imbalance, base power, peak power, spinning reserve, and regulation among others. Moreover, alternate power sources such as solar, wind power, and others, though intermittent, will need to be utilized and commissioned in an efficient manner such that the smart-grid will not be unduly stressed.

This will require a multifaceted distributed communication system using complex regulation control algorithms, with wireless links to the grid operators. The system will probably implement load profile management strategies with dynamic smart power devices to integrate and accommodate real-time control of critical parameters. Fig. 7.20 shows a block diagram of a proposed layout to support the high penetration of the EV and PHEV distributed resources.

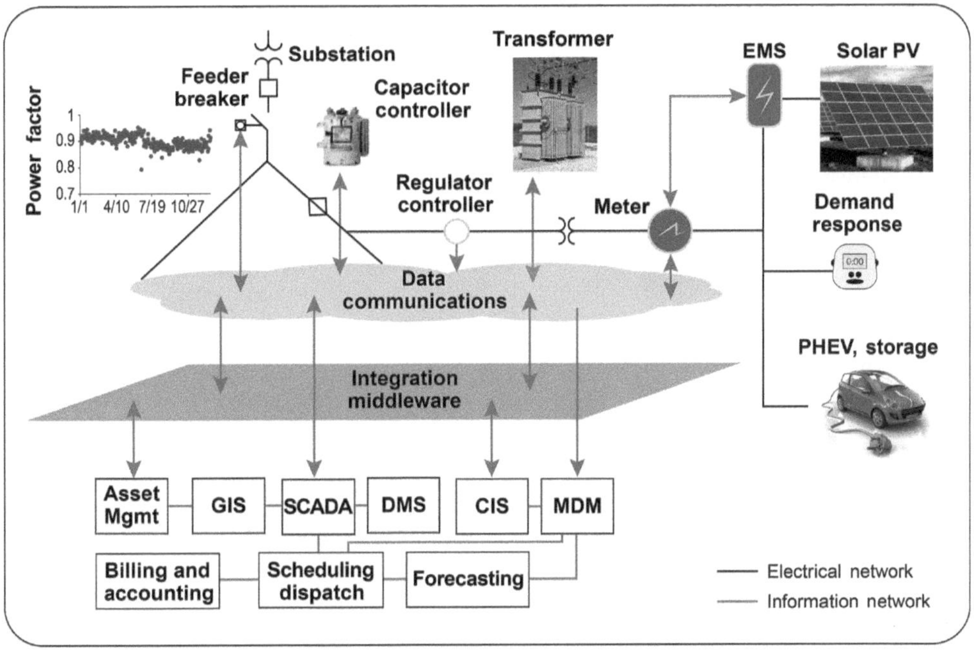

Figure 7.20 Proposed systems to support the high penetration of distributed resources (Ipakchi and Albuyeh, 2009).

A unique factor to the smart grid renewal energy storage system is the Transformerless Intelligent Power Substation (TIPS) using high-voltage silicon carbide (SiC) based IGBT's and interfacing a 13.8 kV, 3-phase 60-Hz system to a 480 V, 3-phase, 60-Hz distribution grid using a front-end dc/dc converter, a dual-active bridge powering a low voltage inverter. Some of the design challenges include the large dV/dt during switching of the high-voltage power components and the parasitic elements in the high voltage/high frequency TIPS (Bhattacharya, 2014).

Fig. 7.21 shows a conceptual drawing of the vehicle grid of the future.

7.19 Expectations for the Future

The typical utility grid of today is undergoing a significant change driven by the need for improved operational efficiency, increased reliability, modernization, cost, and reliable customer service.

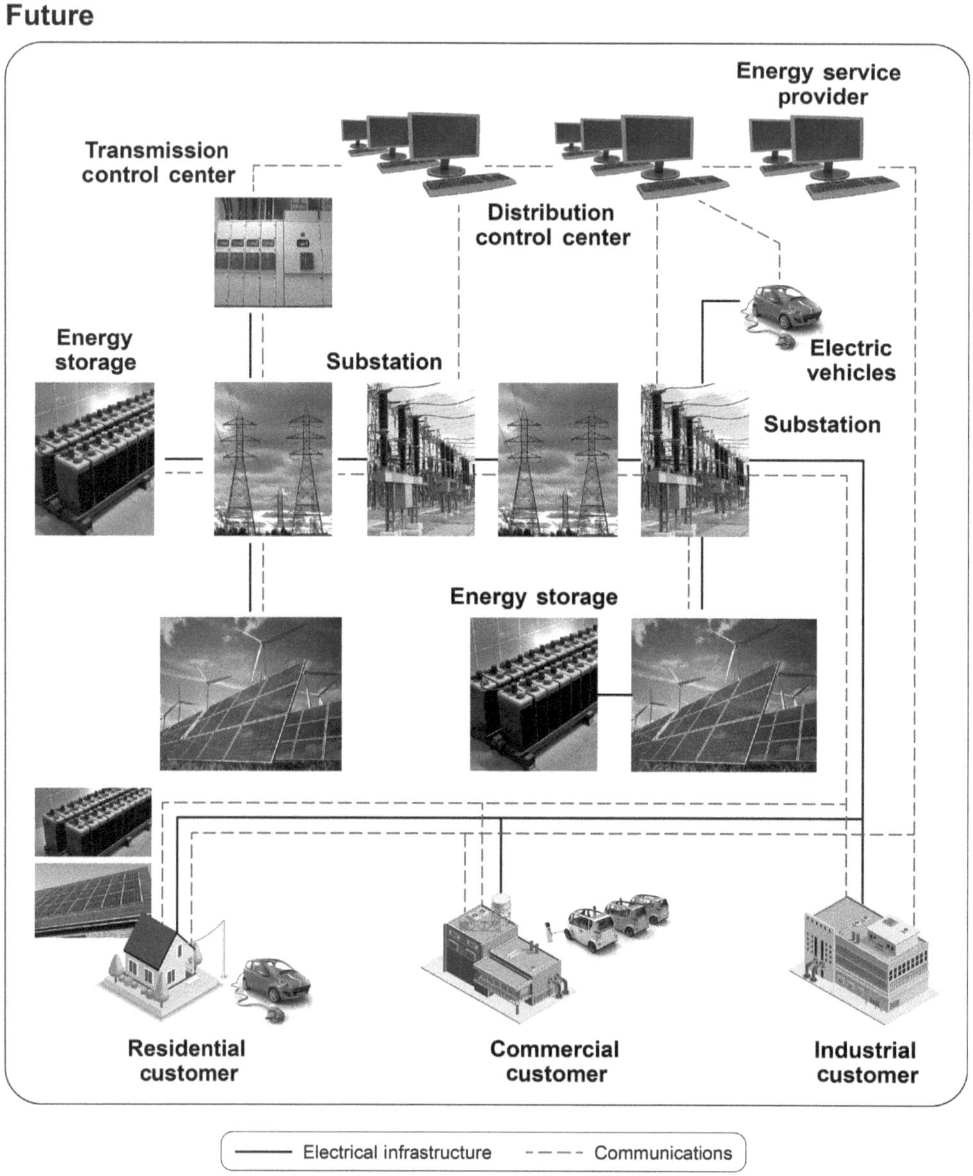

Figure 7.21 Conceptual drawing of the US electrical grid of the future. Adapted from OECD/IEA, 2011. Technology roadmap: smart grids, IEA Publishing, pp. 8 of 52.

The emerging smart grid will utilize sophisticated and intelligent new and uniquely different systems and components developed using power electronics technologies. These will probably include solid-state transformers, electronic circuit breakers, and smart transformers.

An upcoming area of considerable interest is V2G capability where the stored energy in a parked EV or PHEV is judiciously utilized as a load averaging system. Studies conducted indicate that there are major benefits realizable by using an EV or a PHEV as a load equalizing system. In such systems, EVs and PHEVs are used as a miniature power station to supply real power back to the grid during high demand periods and recharge their batteries during low demand periods. This allows maximizing the use of renewable energy sources such as solar and wind and also permits utilizing the available grid resources in an optimum manner.

However, grid modifications are necessary to allow this bi-directional power flow. Furthermore, maximum benefits can only be achieved by upgrading the existing grid to a technologically advanced grid system using the upcoming smart grid technologies.

There also needs to be a strong economic incentive to the EV or PHEV customer to permit the time-sharing use of their vehicle as a load leveling and load management device. Further, there needs to be a high level of interactive coordination between the consumer and the utility to ensure the least stress on the distribution system. Implementing an astute coordinated strategy and by intelligently using the V2G charging algorithms can result in substantially reducing the loading on the distribution system with lower operating temperatures and with increased operational life.

Additionally, during an emergency with total loss of grid power due to a major storm, the EV or the PHEV can provide emergency power for a limited period of time. This stored power could be strategically used to supply power to judiciously selected devices, such as smart phones, and operate critical elements such as a few fluorescent lights, a small refrigerator, and a small heater.

It is expected that the potentials of the smart grid will continue to grow as more efficient systems are used and the increased technological advancements are incorporated into the smart grid at all levels. Studies are being extensively conducted to improve the reliability and stability of the electric distribution grid.

There is currently discussion and ideas for not only smart grids but of smart cities (Freeman, 2016). It is estimated that in the next few years approximately 50 billion machines will be online and connected to and controlled via the Internet. This will result in large data control system handling mega terabytes of information resulting in an increased demand for electric power from the smart grid with increased requirements for efficient management, maintenance costs, reliability, sustainability, revenue generation and operational efficiency. Real-time monitoring will also be a high-level requirement with automatic notifications and supervising by smart sensors. As technology progresses, it is likely that other concepts and designs will be brought to fruition to build, support, and develop the smart grid.

References

American Wind Energy Association. 1996–2016. Wind Energy Facts at a Glance. <http://www.awea.org/Resources/Content.aspx?ItemNumber = 5059>.

Bhattacharya, S., 2014. Transformer Intelligent Power Substation. Presentation, ARPA-E ADEPT Program, 15 kV SiC IGBT Modules for Grid Scale Power Conversion, September 4, 2014. North Carolina State University.

Broussely, M., 2010. Battery requirements for HEVs, PHEVs, and EVs: an overview. In: Electric and Hybrid Vehicles, Power Sources, Models, Sustainability, Infrastructure and the Market, Chapter 13. Elsevier B.V., 307 pp.

Cheng, Y, 2011. Test Platform of Smart Grids with Renewable Energy Systems and Plug-in Hybrid Electric Vehicles. VITO–Flemish Institute for Technological Research, Unit of Energy Technology, B-2400 Mol, Belgium.

Divan, D., Moghe, R., 2014. Power Electronics at the Grid Edge, IEEE Power Electronics Magazine, December 2014.

Francis, E., 2014. Transforming EVs into mini power stations. Automotive Industries, 193 (4), (accessed 23.09.16.).

Freeman, R., 2016. On the road to smart cities, LD + A, the magazine of the Illuminating Engineering Society of North America, June 2016.

Gorguinpour, C., 2013. Executive Director, United States Department of Defense, DOD Plug-In Electric Vehicle Program "Drive Oregon," Presentation, October 24, 2013.

Hirsch, J. 2014. 253 million cars and trucks on U.S. roads; average age is 11.4 years. Article in Los Angeles Times, June 9, 2014. IHS Automotive reports. Anne Cusack/Los Angeles Times.

Ipakchi, A., Albuyeh, F., 2009. Grid of the future. IEEE Power & Energy Magazine, March/April.

Kempton, W., Tomic, J., 2005a. Vehicle-to-grid power fundamentals: calculating capacity and net revenue. J. Power Sources. 144 (1), 268–279.

Kempton, W., Tomic, J., 2005b. Vehicle-to-grid power implementation: from stabilizing the grid to supporting large-scale renewable energy. J. Power Sources. 144 (1), 280–294.

Kulshrestha, P., Wang, L., Chow, M.Y., Lukic, S., 2009. Intelligent energy management system simulator for PHEVs at a municipal parking deck in a smart grid environment. Presentation, North Carolina State University, Raleigh, NC.

Kuss, M., Markel, T., Kramer, W., 2010. Application of distribution transformer thermal life models to electrified vehicle charging loads using Monte-Carlo method. Presented at the 25th World Battery, Hybrid and Fuel Cell Electric Vehicle Symposium & Exhibition, Shenzhen, China, November 5–9, 2010.

Markel, T., Meintz, A., Hardy, K., Chen, B., Bohn, T., Smart, J., Scoffield, D., Hovsapian, R., Saxena, S., MacDonald, J., Kiliccote, S., Kahl, K., Pratt, R., 2015. Multi-Lab EV Smart Grid Integration Requirements Study. National Renewable Energy Laboratory (NREL), US Department of Energy, Technical Report NREL/TP-5400-63963 May 2015.

Medora, N.K., Kusko, A., 2012. Battery management for hybrid electric vehicles using supercapacitors as a supplementary energy storage system. 2012 IEEE 34th International Telecommunications Energy Conference, Scottsdale AZ, September 30–October 4, 2012.

Mets, K., Verschueren, T., De Turck, F., Develder, C., 2011. Exploiting V2G to optimize residential energy consumption with electrical vehicle (dis)charging. 2011 IEEE First International Workshop on Smart Grid Modeling and Simulation (SGMS) – at IEEE SmartGridComm.

OECD/IEA, 2011. Technology Roadmap: Smart Grids. IEA Publishing, Paris Cedex, France, p. 8.

Princeton Power Systems, 2014. Grid-Tied Inverter and Battery Controller GTIB-30 Data Sheet. Princeton Power Systems, Lawrenceville, NJ.

Princeton Power Systems, 2015. Grid-Tied Inverter and Battery Controller GTIB-100G1.2 Data Sheet. Princeton Power Systems, Lawrenceville, NJ.

Quinn, C., Zimmerle, D., Bradley, T., 2009. The effect of communication architecture on the availability, reliability, and economics of plug-in hybrid electric vehicle-to-grid ancillary services. J. Power Sources. 195 (5), 1500–1509.

Raghavan, S., Khaligh, A., 2011. Impact of plug-in hybrid electric vehicle charging on a distribution network in a smart grid environment. ©IEEE.

SAE International, 2011. SAE J1772 combo connector for ac and dc charging. <http://articles.sae.org/10128/> (accessed 23.09.16).

Sliverstein, A., 2011. Transmission 101. NCEP Transmission Technologies Workshop, April 20–21, 2011.

U.S. Department of Energy, 2016. Developing infrastructure to charge plug-in electric vehicles. Alternative Fuels Data Center.

US PEV Sales by Mariordo (Mario Roberto Durán Ortiz) – Own work. Licensed under CC BY-SA 3.0 via Commons. <https://commons.wikimedia.org/wiki/File:US_PEV_Sales.png#/media/File:US_PEV_Sales.png>.

Wikipedia, Electricity Sector of the Unites States. <https://en.wikipedia.org/wiki/Electricity_sector_of_the_United_States> (accessed 2016).

Yilmaz, M., Krein, P.T., 2013. Review of battery charger topologies, charging power levels and infrastructure for plug-in electric and hybrid vehicles. IEEE Trans. Power Electron. 28 (5).

8

CYBERSECURITY FOR THE SMART GRID

Adam Sorini and Ernesto Staroswiecki
Exponent Inc., Menlo Park, CA, United States

CHAPTER OUTLINE

8.1 Introduction

The term "grid" generally refers to the network of wires, substations, transformers, switches, and other utility industry equipment used to carry electricity from the source to consumers. Until recently, data gathering and management of devices in the "grid" was a manual process involving utility workers physically visiting the locations in the grid. For example, utility workers visit consumer premises to read meters, inspect for broken equipment, measure voltages, etc. Slowly, however, computerization and automation are transforming the industry. This computerization of the grid, has led to the use of the term "smart grid," much as the term "smart phone" has been used to refer to the integration of the computer and the phone.[1]

[1]http://energy.gov/oe/services/technology-development/smart-grid.

The Power Grid. DOI: http://dx.doi.org/10.1016/B978-0-12-805321-8.00008-2

Broadly speaking, the term "smart grid" refers to computer-based remote control and automation technologies that enable efficiencies in energy use for consumers. The smart grid relies on digital control, monitoring, and telecommunications to provide bi-directional flow of energy and information to various stakeholders in the energy chain, including the electricity generation plant, commercial, industrial, and residential end users. This enables a number of efficiencies such as energy management during peak hours, "microgrids," real-time pricing, mobility services, and ultimately, allows customers to better manage their consumption and even sell unused home-produced electricity back to the grid.[2]

However, with computers and connectivity arises the opportunity for mischief and attacks. Essentially, a malicious user, be it a government or another powerful entity, could potentially launch an attack that undermines the integrity of the smart grid system, or cause localized-to-widespread failures. As the smart grid becomes the predominant infrastructure to generate and distribute electricity, the impact of a widespread attack could be catastrophic to the economy of the nation and its infrastructure.

In fact, recent events underscore the potential impact of well-planned attacks to computerized energy systems. A notable attack against the U.S. Government's Office of Personnel Management was reported in 2015, in which approximately 21.5 million individual background investigation records were stolen by cyber criminals (some including fingerprint data).[3] An example (now classic) of a true cyber physical attack is the StuxNet computer worm, which is believed to have been developed in order to manipulate and destroy nuclear centrifuge systems.[4] Another well-known cyber physical attack was the recent demonstration[5] of remote manipulation of control systems in automobiles (which are mainly controlled by microprocessors on a controller area network), which resulted in a recall of 1.4 million vehicles.[6]

"The architecture of the nation's digital infrastructure, based largely upon the Internet, is not secure or resilient. Without major advances in the security of these systems or significant

[2]Aillerie, Y; Kayal, S; Mennella, J; Samani, R; Sauty, S; Schmitt, L. Smart Grid Cyber Security. McAfee White Paper. http://www.mcafee.com/in/resources/white-papers/wp-smart-grid-cyber-security.pdf.

[3]See, e.g., https://www.opm.gov/news/releases/2015/09/cyber-statement-923/.

[4]See, e.g., https://www.wired.com/2014/11/countdown-to-zero-day-stuxnet/.

[5]http://illmatics.com/Remote%20Car%20Hacking.pdf.

[6]https://www.wired.com/2015/07/jeep-hack-chrysler-recalls-1-4m-vehicles-bug-fix/.

change in how they are constructed or operated, it is doubtful that the United States can protect itself from the growing threat of cybercrime and state-sponsored intrusions and operations. Our digital infrastructure has already suffered intrusions that have allowed criminals to steal hundreds of millions of dollars and nation-states and other entities to steal intellectual property and sensitive military information. Other intrusions threaten to damage portions of our critical infrastructure. These and other risks have the potential to undermine the nation's confidence in the information systems that underlie our economic and national security interests."[7]

8.2 Cybersecurity Best Practices and Guidelines for the Power Industry

8.2.1 Risk-Based Thinking

Risk-based thinking (RBT) is the currently dominant trend in quality management practices. RBT is featured prominently in the 2015 version of the most recognized quality management standard, ISO 9001, becoming the official tag line for the standard itself.

RBT is a paradigm that emphasizes risk and risk management as concepts that are to be taken into account during every stage of a project lifecycle. Under this paradigm, risks should be identified and documented, first early on during the inception of a product or process. As a project advances, new risks are identified and the ones previously documented are updated. These risks are used throughout the development of a product to inform the decision making process at every stage.

Within the RBT framework, risk is defined as the "the effect of uncertainty on an expected result." This variance or deviation from an expected outcome does not need to be construed exclusively as a negative effect. Risk, as defined, can also lead to opportunity. Furthermore, risks, as defined, do not need to be necessarily eliminated or minimized, yet they do need to be understood as thoroughly as possible, and managed appropriately.

The energy industry is very familiar with and has been a trendsetter in risk management. Energy generation and distribution involve activities and products that are inherently risky.

[7]White House, Cyberspace Policy Review, available from http://www.whitehouse.gov/assests/documents/Cyberspace_Policy_Review_Final.pdf.

Power plants and power lines have risks relating to safety, security, quality of service, etc. Furthermore, as part of the critical infrastructure of the country, some of the most stringent standards relating to risk and quality management are applied to the energy sector.

The smart grid incorporates an entire new set of risks to those typically present for the energy sector: those involving the elements of information technology used to make the power grid a smart grid. Now it is necessary to consider risks involving cyber security, bad actors, and inadvertent or intentional software errors that could be present at any point in the grid, from the power generating plant, to the final residential consumer.

From an RBT point of view, these risks should be assessed as early as possible in the deployment of a smart grid, and enter the risk management process immediately. For example, consider the risk of a zero-day exploit being used to access metering devices at consumers' locations without authorization. After this risk is identified, a further analysis and assessment of this item would lead to an understanding of the potential effects of this event. One of these effects could be a disruption of service to the affected consumer, which would be a potential safety concern in cases when consumers depend on the electrical network. Another effect could be the loss of privacy of consumers if such unauthorized access allows hackers to track energy usage by consumers. Further analyses should shed light on other aspects of these events associated with risks, including their causes, likelihood, potential preventive measures, detection, correction after the event, notification of users, mitigation, etc. This allows for the management of these risks as is necessary to maintain the levels of services required for the energy grid.

Some of the classical tools and methods of risk and quality management, including PHA, FMEA, FTA, etc., have been used extensively by the energy sector, and in some cases even introduced or developed by this industry. With some changes, which are introduced below in following sections, they can be adapted to be used to take into consideration the cyber security concerns introduced by the deployment of the smart grid. The current trend of RBT makes these tools, and risk management in general, essential to the successful development and deployment of smart grid technologies.

8.2.2 C2M2 for Cybersecurity

The Cybersecurity Capability Maturity Model (C2M2) program was developed by the United States Department of Energy to help improve industrial cybersecurity. It is a sub-sector non-specific model that was developed based on a previous electricity-sector–specific model, the ES-C2M2 (Electricity Subsector Cybersecurity Capability Maturity Model). The C2M2 seeks to help strengthen cybersecurity capabilities of organizations and to help them effectively measure their cybersecurity capabilities. This model is applicable to organizations regardless of size or specific industry and concentrates on cybersecurity implementation and management in relation to industrial technology assets, e.g., information technology. The C2M2 supports adopting the NIST Cybersecurity Framework.[8]

The C2M2 is a generalization of an electricity subsector specific model called the ES-C2M2. It was also specialized for the oil and natural gas subsector in a model called the ONG-C2M2. At the time of the publication of this book the most recent versions of the C2M2 and ES-C2M2 are versions 1.1, which are available here: http://energy.gov/sites/prod/files/2014/03/f13/C2M2-v1-1_cor.pdf; http://energy.gov/sites/prod/files/2014/02/f7/ES-C2M2-v1-1-Feb2014.pdf.

The C2M2 model itself is described in the document "C2M2-v1-1_cor.pdf" and the recommended usage of the model is depicted in Fig. 8.1. An "initial evaluation," illustrated as the top hexagon in Fig. 8.1, is the starting point of the model and the return point for reevaluation after a cycle is complete.

In order to enter into the evaluation cycle and complete a C2M2-based self-evaluation it is important to select an effective facilitator. A facilitator is a person who is familiar with the model and who can help an organization to establish scope for the cybersecurity evaluation and help a group within the organization work through the model and achieve the model goals. The C2M2 supplies a Facilitator Guide.[9] An initial evaluation is performed (with the help of a facilitator) in a workshop-like setting and the end result of the evaluation is a scoring report that helps identify gaps in performance. The evaluation relies on participant responses to evaluation questions. The second step in the model involves analyzing the identified performance gaps to determine if the identified gaps are truly meaningful for the organization performing the evaluation. The gaps will generally relate to a specific Maturity Indicator Level (MIL) that the organization desires to achieve for a given function. MILs are

[8]See, http://energy.gov/oe/services/cybersecurity/cybersecurity-capability-maturity-model-c2m2-program; http://energy.gov/oe/services/cybersecurity/reducing-cyber-risk-critical-infrastructure-nist-framework.

[9]See, http://energy.gov/mode/795826.

Figure 8.1 Recommended usage of the C2M2 model: Evaluate a function against model practices; Identify gaps in performance; Prioritize and plan actions required to achieved desired capabilities; Implement plans to address gaps; Repeat. [See, C2M2-v1-1_cor.pdf document referenced in text at Section 4].

discussed in greater detail below. A given organization will not necessarily achieve the highest MIL in all the domains of the C2M2, but rather should seek to achieve the appropriate level in light of the organization's overall cybersecurity strategy. In the third stage of the model, the participants in the model workshop will determine how the important gaps in cybersecurity practices affect the organization's objective. The costs and benefits of implementing plans to fill the cybersecurity gaps will be evaluated and a specific plan will be developed. The fourth stage entails actually implementing the plans developed previously, tracking progress toward the planned goals, and periodic reevaluation.

The C2M2 is divided into 10 domains, which are groupings of cybersecurity objectives and practices. The 10 domains of the C2M2 are illustrated in Fig. 8.2.

Each of the 10 domains is associated with specific objectives and practices, which are achievements and activities that establish capabilities and support the domain. For example, the Risk Management domain has three objectives: (1) Establish Cybersecurity Risk Management Strategy; (2) Manage Cybersecurity Risk; (3) Management Activities. Each objective can be rated at a given MIL depending on the level of practice. For example, the three objectives in the Risk Management domain can each be described by one of four MILs: MIL0, MIL1, MIL2, and MIL3, as shown in Fig. 8.3 [See C2M2-v1-1_cor.pdf at Section 5.1].

Figure 8.2 The 10 domains of the C2M2. [See, C2M2-v1-1_cor.pdf document referenced in text at Section 5].

In general, MIL0 indicates that there is no practice; however, MIL1 can also be attained for some objectives with no practice. Typically, practices at MIL1 are ad hoc. In addition, MIL1 has been designed such that all organization should be capable of achieving at least MIL1 in all 10 domains. Each of the practices required for MIL1 in all 10 domains is illustrated in Fig. 8.4.

8.2.3 ES-C2M2 for Cybersecurity

The ES-C2M2 was developed by the United States Department of Energy to help improve cybersecurity in the electricity subsector. This electricity subsector-specific guidance was used as the basis for the subsector-independent guidance of the C2M2 discussed above.

The general purpose of the ES-C2M2 is the same to the above-described C2M2. That is, the purpose is to improve cybersecurity capabilities, allow consistent benchmarking, and so on. The 10 domains of the ES-C2M2 are also identical to the 10 domains of the C2M2. Indeed, much of the ES-C2M2 model and program described in the main body of the document "ES-C2M2-v1-1-Feb2014.pdf" (available at http://energy.gov/sites/prod/files/2014/02/f7/ES-C2M2-v1-1-Feb2014.pdf) is effectively identical to the C2M2 described above, except for Section 3

1. Establish cybersecurity risk management strategy

MIL1	No practice at MIL1
MIL2	a. There is a documented cybersecurity risk management strategy b. The strategy provides an approach for risk prioritization, including consideration of impact
MIL3	c. Organizational risk criteria (objective criteria that the organization uses for evaluating, categorizing, and prioritizing operational risks based on impact, tolerance for risk, and risk response approaches) are defined and available d. The risk management strategy is periodically updated to reflect the current threat environment e. An organization-specific risk taxonomy is documented and is used in risk management activities

2. Manage cybersecurity risk

MIL1	a. Cybersecurity risks are identified b. Identified risks are mitigated, accepted, tolerated, or transferred
MIL2	c. Risk assessments are performed to identify risks in accordance with the risk management strategy d. Identified risks are documented e. Identified risks are analyzed to prioritize response activities in accordance with the risk management strategy f. Identified risks are monitored in accordance with the risk management strategy g. Risk analysis is informed by network (IT and/or OT) architecture
MIL3	h. The risk management program defines and operates risk management policies and procedures that implement the risk management strategy i. A current cybersecurity architecture is used to inform risk analysis j. A risk register (a structured repository of identified risks) is used to support risk management activities

3. Management activities

MIL1	No practice at MIL1
MIL2	a. Documented practices are followed for risk management activities b. Stakeholders for risk management activities are identified and involved c. Adequate resources (people, funding, and tools) are provided to support risk management activities d. Standards and/or guidelines have been identified to inform risk management activities
MIL3	e. Risk management activities are guided by documented policies or other organizational directives f. Risk management policies include compliance requirements for specified standards and/or guidelines g. Risk management activities are periodically reviewed to ensure conformance with policy h. Responsibility and authority for the performance of risk management activities are assigned to personnel i. Personnel performing risk management activities have the skills and knowledge needed to perform their assigned responsibilities

Figure 8.3 Example MILs for the three objectives of the Risk Management domain of the C2M2, reproduced from the C2M2 documentation. [See, C2M2-v1-1_cor.pdf document referenced in text at Section 5.1].

of the ES-C2M2, which gives some general information regarding the electricity subsector. In particular, it is noted that the ES-C2M2 was developed with four primary electricity sector functions in mind: (1) generation; (2) transmission; (3) distribution; (4) markets. Each of these electricity sector functions is linked either directly or indirectly with the others and each function also depends on a complex information technology and operations technology infrastructure.

An additional difference between the ES-C2M2 and the C2M2 is that some threat, vulnerability, and incident reporting resources are identified in the electricity subsector specific form. For example, the Electricity Sector Information Sharing and Analysis Center (ES-ISAC, See https://www.esisac.com/) is

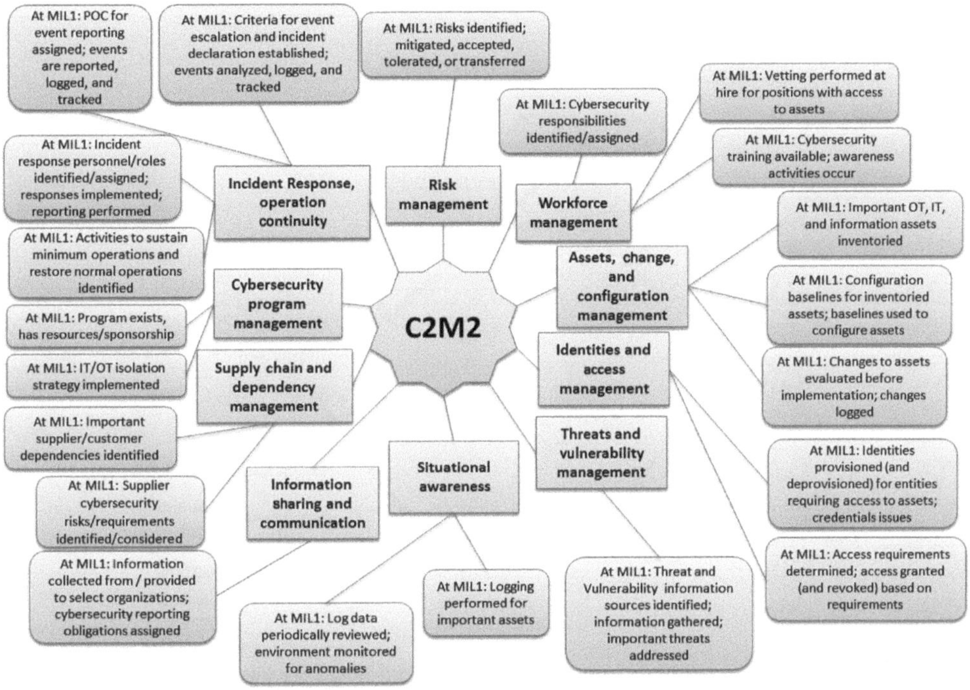

Figure 8.4 The MIL1 practices of all ten C2M2 domains. [See, C2M2-v1-1_cor.pdf document referenced in text at Section 5].

mentioned specifically in the ES-C2M2, whereas the C2M2 mentions ISACs in general. As an additional example, the ES-C2M2 explicitly mentioned the DOE Form OE-417 (See https://www.oe.netl.doe.gov/oe417.aspx) for reporting electric emergency incidents and disturbances.

8.2.4 Specific Smart Grid Related Cybersecurity Guidance and Regulations

Electrical power grid high-level domains include: generation, transmission, distribution, service providers, customers, operations, and markets.[10] Electrical energy flows between the generation, transmission, distribution, and customer domains. However, information flows among all these high-level domains as indicated in Fig. 8.5 below.

[10]See, NIST SP-1108r3 table 5-1 and figure 5-1.

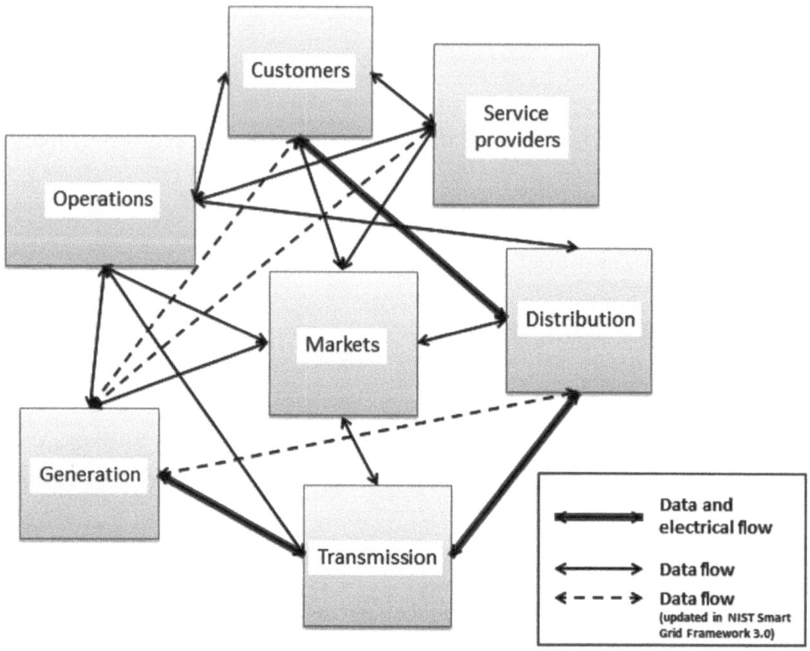

Figure 8.5 Smart Grid domains as specified in the NIST Smart Grid Framework (NIST SP 1108, rev 3).

Indeed, the "smart" grid is characterized by a fast two-way information flow among elements of the high-levels domain. For example, a "smart" meter may be able to transmit information from a customer site to a service provider computer. If this information flow is to take place wirelessly or over public networks the data channels may have to be secured. It is the introduction of new fast data flow channels that allow the grid to act "smartly" and this massive flow of data can present significant cybersecurity challenges. Of course, an increase in cybersecurity challenges affects all industries, not just the electric power industry. However, there are cybersecurity challenges that are unique to the electric power industry. For instance, since the electrical power industry employs numerous and varied high power physical control system it is subject to high impact combined cyber-physical attacks, wherein a purely "cyber" attack is combined with physical control systems to create a higher impact attack.[11]

[11]See, e.g., NIST IR 7628 at Section 1.4.

8.3 Cybersecurity Risk Assessments and Tools

8.3.1 Penetration Testing

Penetration testing is the process of simulating realistic attacks to test the functionality of security systems. The attacks are simulated in the sense that even if the vulnerability is exploited successfully by a penetration tester the underlying assets are not subsequently harmed by the tester. Rather, a typical penetration test is performed with the asset owner's permission. The purpose of the test is to gain information about vulnerabilities and to determine if and whether suspected vulnerabilities are actually exploitable.

Penetration testing (in the context of cybersecurity) is often performed using specially developed computer applications that assist in testing. For example, the network tool "nmap" is a freely available application that can be used to automatically explore computer networks.[12] Many other free and proprietary tools are available that are used by professional penetration testers. Indeed, even entire operating system distributions (e.g., Kali Linux[13]) are designed specifically for penetration testing and preloaded with numerous free "pen testing" tools. In this context, pen tests often focus on computer networks since the network is the path by which most information flows and by which access to informational assets is likely to be most easily attained surreptitiously. The concept of network penetration testing is immediately applicable to smart grid technology.

In a smart grid system, the smart components are typically connected via networks in order to make use of information being produced by grid components. The amount of information and the ease with which this information is transferred is the underlying feature (or vulnerability) of the smart grid that was not present in the "old" grid system. For example, in the "old" electrical grid system the service provider would have to be informed that the power was "out" via a land-line phone call from a residential customer. However, with a smart meter, the meter itself can sense and report a power outage to the service provider via the Internet.

The connection of smart "things" (e.g., meters, solar panels, light bulbs, cars, refrigerators, etc.) via network interfaces to the Internet is known as the "Internet of Things." This collection of

[12]https://nmap.org/.
[13]https://www.kali.org/.

nonhuman "smart" components communicates amongst themselves and with humans via the Internet. Electrical grid components on the Internet of Things are particularly interesting from a cyber-security perspective since they can control important electrical-physical resources and are subject to cyber-physical attacks.

Consider an example of a solar power generation system for residential use. A typical system consists of roof-mounted solar panels connected to inverter modules, and a distribution/communication module located in the home. A system's distribution/communication module can include built-in Power Line Communication functionality for monitoring individual solar panels/inverters. The built-in communication unit typically includes a computer that can perform a number of functions. For example, the system residential solar power system may be able to: read program/data from a "SD Card"; display information to the user; perform any number of other typical computer functions.

The typical example system described above could also be connected via Cat-5 cable to a router and it could send data about the home solar power system to a remote server for subsequent analysis via Web-based tools. Data sent to the external remote server can typically be accessed via a Web interface by the home owner. The registered system installer can also likely access system data, possibly at a deeper level (more data) than the home owner for diagnostic purposes.

Network-connected systems like the above-described solar panel communication system communicate via the Internet. If hundreds, or thousands, or millions of these systems are manufactured and set up in homes, they can present a large geographically distributed attack surface, where attack may be possible via the Internet. For example, if these types of solar energy communication systems (or any smart electrical system) accept incoming connections or display banners on any Internet socket port then they can likely be located by penetration testers or hackers on an Internet-of-Things search engine such as shodan.io.[14] For example, a search for "solar" on Shodan.io turns up hundreds of hits in the United States alone. Similarly, a search for "electric" turns up thousands of hits, some of which are shown below in Fig. 8.6.

Penetration testing of smart-grids and other distributed cyber-physical systems can be augmented by IOT search engines as described above. Realistic penetration testing of the

[14]https://www.shodan.io/.

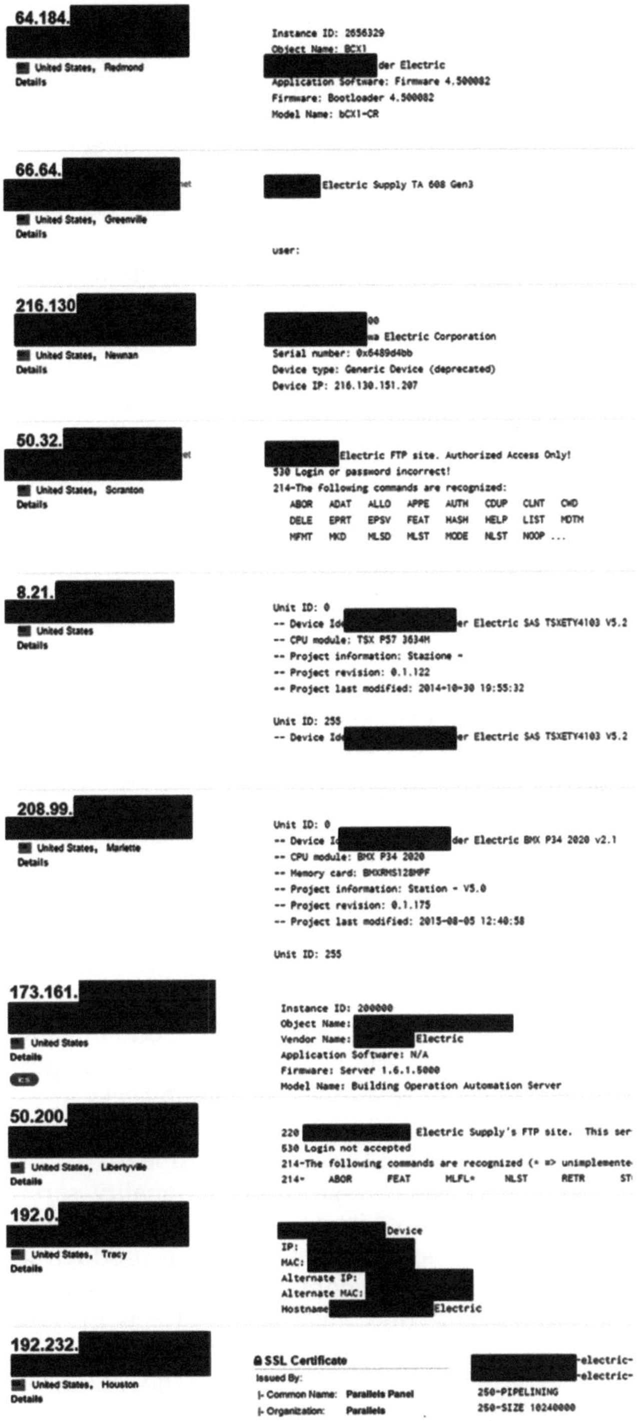

Figure 8.6 Screen shot of an Internet-of-Things search result on "shodan.io" for the search term "electric," showing the first six search results with IP addresses located in the United States. Annotation in the form of black bars has been added to obfuscate the IP addresses and other information regarding the "smart" devices (modified from shodan.io).

grid will include similar types of attacks to test the functionality of the system. The results of a penetration test will include useful data on how to better secure the grid system against real-world attackers.

A penetration test performed on the smart grid components for a given company could involve testing the entire set of smart components the company has produced and distributed to geographically disparate locations. If any of the smart devices are running old operating systems with known vulnerabilities, one starting point for penetration testing could be to try all known exploits of the operating systems using automated techniques.[15]

8.3.2 Preliminary Hazards Analysis

An electrical power grid cybersecurity Preliminary Hazards Analysis (PHA) is a top–down risk assessment process that is used for identifying assets, threats, and vulnerabilities of power grid components. The analysis is conducted via a team approach where the analysis team is made up of experts from different disciplines including power grid technologies, cybersecurity principles, and other diverse aspects of the specific electrical power components under investigation. The goal is exhaustive analysis, and this is more easily achieved when a multidisciplinary team attacks the problem from a number of different perspectives.

The first step in the PHA is to identify the assets that need to be protected. Although "asset discovery" may seem simple, it is essential because if there are no assets to secure then there is no need for security. Therefore one must identify all the relevant assets. The valuation of assets is not performed at this initial stage. Rather, initially, the team seeks to exhaustively identify all assets that require protection. As part of the asset discovery phase the PHA team needs to learn as much as possible about the specific power grid functionality, service, or product that is the subject of investigation. The team may have general expertise, but specific details of the product/service are required in order to have a comprehensive understanding of all the assets.

The second stage of the PHA is the "threat discovery phase" in which a catalog of all the possible threats to the previously identified assets are listed. The diverse make-up of the PHA team adds to the effectiveness of threat discovery "brain storming." Again, at this stage, the threats do not have to be ranked,

[15]For example, using Metasploit (see https://www.metasploit.com/index.jsp) or other penetration testing suites.

but rather the goal of threat discovery is to catalog all the possible threats to the power grid technology under investigation.

After the asset-discover and threat-discovery phases, the next step is to estimate the severity level and the likelihood for each of the identified threats. Severity level refers to the amount of damage that would occur if the threat were to be successfully carried out. The severity level scale is typical a qualitative one such as: "low," "medium," and "high." Likelihood refers to the probability of the threat actually being carried out. Typically this scale will also be qualitative (e.g., "low," "medium," "high") or semi-quantitative. In general estimates of likelihood and severity are subject to uncertainty and quibbling over specific values is probably not useful. When in doubt go with the more conservative estimate. Establishing these severities and likelihoods for each identified threat is important for having a well-defined system for ranking threats. The risk associated with the threat is a function of both the severity and the likelihood and can be taken as the product of severity and likelihood in a semi-quantitative model.

As illustrated in Fig. 8.7, this relationship models the fact that a high severity does not necessarily result in a high risk if the likelihood is low enough (and vice-versa). For example, the power grid system under consideration is certainly vulnerable to destruction by a massive asteroid from space, but since this particular threat is of very low likelihood the risk is low enough that it need not even be considered.

After the likelihood and severity have been determined for each threat, the risk associated with all the identified threats

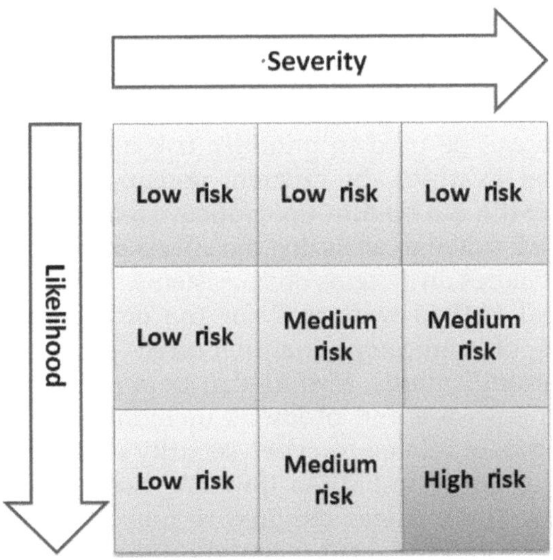

Figure 8.7 Qualitative "Risk Matrix" illustration of the relationship between Severity and Likelihood.

can be established. A key deliverable of the PHA is a risk-ranked catalog of threats along with the associated documentation used to establish risk, threat, and asset information. This provides a complete picture of the security environment and can be used as a launching point for subsequently instituting risk mitigation strategies in a focused manner.

8.3.3 Failure Modes and Effects Analysis

One of the most commonly used tools for risk management is the Failure Modes and Effects Analysis (FMEA). FMEA is one of the earliest systematic techniques developed for failure analysis, and has been part of the basic toolset of reliability and quality engineers since the 1950s. A closely related analysis is the Failure Modes, Effects, and Criticality Analysis, where the criticality analysis is undertaken in more detail.

FMEAs are widely used in almost all stages of a product's lifecycle, and there are different variations of traditional FMEAs, including Process, Design, and Functional FMEAs. In all cases, generating a successful FMEA requires first identifying all potential failure causes and the failure modes they generate for in every system, subsystem, and component of the product being developed. A FMEA, typically requires a worksheet with an entry for each failure mode identified. On each entry the potential effects and causes of the failure are recorded. Following that, an FMEA requires an assessment of each failure's probability (P) (also called likelihood or occurrence), severity (S), and detectability (D) (also called observability). Typically, each of these variables, P, S, and D, are assigned a value between 1 and 10, with 1 being the least probable, least severe, and most detectable value, and 10 being the most likely, most severe, and least detectable value. The product P*S is called the criticality (C) of the failure, and the product P*S*D is the Risk Priority Number. These values can be used to prioritize risk management activities (i.e., the order in which the different potential failures should be addressed). FMEA is a bottom-up, inductive reasoning analysis, and is specially well suited to analyzing the effects of a single function or component failures on systems or subsystems.

Traditional FMEAs work well for the design and other lifecycle stages of traditional manufactured products. However, with some modifications, FMEAs can be an effective and valuable tool for software and products including software, including assessing risks related to cyber-security concerns.

The use of Software-FMEAs (SFMEA) has been proposed as early as 1979. The changes required to make the FMEA process useful for software products are based on the aspects of software development that differ from hardware product manufacturing.

Hardware products are susceptible to manufacturing defects (i.e., defects resulting from the copying of a design into many products) as well as wear from use. This is not the case for most software products or software components of products, where the defects, or bugs, are introduced during the development of the software (i.e., the design or coding of the software product), a stage typically associated with the design stage of a hardware product lifecycle. Furthermore, it is understood that SFMEA are more useful if they are undertaken early on in the software development lifecycle.

Additionally, the traditional hardware product FMEA does not capture intentional actions by bad actors, yet this is necessary in order to take into account threats, vulnerabilities, and other cyber-security considerations. FMEA-based approaches (e.g., Intrusion Modes, Effects, and Criticality Analysis, or IMECA; Failure Modes, Vulnerabilities, and Effects Analysis, or FMVEA) exist that use the structure and methods of FMEAs, and incorporate traditional failure modes caused by intrusions that have been enabled by system vulnerabilities.

As an example, FMVEAs are created using the basic structure of an FMEA. An early step of an FMEA is to identify the failure causes involved in the system under analysis. The FMVEA approach additionally requires the identification of the potential vulnerabilities and threats combinations related to the same system as causes of failures. The next step involves the identification of the failure modes (in the case of failure causes) or threat modes (in the case of vulnerabilities and threats) produced by the causes of failures. The following stages are common to FMEAs and FMVEAs: evaluate the effects of the failure or threat modes, assess the severity of said effects, and finally estimate or measure the probability of each failure or threat mode, allowing at this point an estimation of the criticality of each failure or threat mode. Lastly, incorporating the observability of each failure or threat mode to the analysis results in a more complete picture of the risks associated with the system analyzed.

Finally, while Software FMEAs and FMEA-based security assessment frameworks are extremely valuable tools for risk management involving software products and cyber-security concerns, given the high level of uncertainty in evaluating the likelihood of software failures and cyber-security intrusions, these tools have found their highest level of usefulness as qualitative tools that are well suited for comparisons and prioritization of risk mitigation or elimination activities.

8.3.4 Fault Tree Analysis

Another widely used technique used for safety and reliability analysis is the Fault Tree Analysis (FTA). FTA was developed in the early 1960s, and it is a top–down, deductive reasoning analysis, making it complementary to FMEA. While FMEAs are intended to analyze the effects of a single function or component failures on systems or subsystems, FTAs are better suited to analyzing the resistance of a system to single or multiple initiating faults. Given their contrasting nature, typically both FMEA and FTA analyses are carried out during a product's lifecycle to evaluate associated risks, reliability, quality, and safety of said product.

The FTA is a procedure used to determine which events, actions, and faults, and which combinations of events, actions, and faults can lead to a system failure or other undesired event. The process begins with a conclusion, the top-level failure or fault condition to be analyzed. The next step in an FTA is to construct a multi-level logic diagram (called a fault tree) of the potential causes for the analyzed failure, starting in a very general sense for the causes at the second level, and increasing in specificity as the levels increase. For example, consider a top-level failure of a "computer crash." Appropriate second-level causes include "hardware failure," and "software failure." Third-level causes for "software failure" could include "operating system failure," "driver failure," and "application program failure." Each of these failures is considered a node, and each can in turn have causes.

The causes of a node are related to said node through a logical operation: for example, if either one of several causes is enough to generate the failure described in the node they relate to, the node failure is the result of an OR operation with the causes as inputs; if all of several causes need to occur to generate the failure described in the node they relate to, the node failure is the result of an AND operation with the causes as inputs. In many cases it is possible to envision a failure node as the result of a Boolean-logic operation of the causes considered for said node. However, there are extensions to Boolean logic used for FTAs to allow considerations for the order in which causes would need to occur to generate the failure described in the node.

FTAs are typically used to identify potential causes of a system failure before it occurs, and to guide testing and failure analysis efforts once an undesirable event has taken place. If the failure rates or probabilities of all lower level nodes are known, FTAs can also provide estimates of the probability of the top level even to occur. As with many other quantitative and qualitative analysis techniques, there are software tools that facilitate their application. Within the energy sector, one of the most widely FTA software tools used is EPRI's CAFTA (Computer Aided Fault Tree Analysis).

The application of FTAs to software products is called a Software FTA (SFTA), and was first described in the early 1980s.

As with other techniques, the modification required to make FTAs useful for software are related to the different nature of software faults from hardware faults: software faults are the result of a defect, typically incorporated into the product during the design stages; unlike hardware faults that can be the result of wear or manufacturing problems. Therefore software faults are not probabilistic in the same way that a hardware fault is: a software fault is either present in the software (and in every copy of it) or not. So while SFTAs are still powerful tools to understand system failures, their quantitative aspects are limited compared to hardware FTAs.

In contrast with FMEAs, FTAs are well suited to consider external causes of failure in its analysis. These external causes may include actions of users or bad actors. An FTA-based technique focused on considering top-level undesirable events resulting from attacks to a system in conjunction to the system vulnerabilities is called an Attack Tree Analysis (ATA) or a Threat Tree Analysis. ATAs have been described sin the 1990s, and are useful tools to understand the potential security threats to a system, the protective measures already in place, and the risks remaining. They can also be used to prioritize these risks, and to evaluate the impact of mitigating activities to the overall security of the system being analyzed.

8.4 Challenges and Conclusions

We have presented a number of cybersecurity practices and tools for protecting the smart grid. In order to be affective these practices and tools must be properly applied and maintained. To this end, employee awareness and education regarding cybersecurity is essential. The cybersecurity capabilities of the grid must be tested periodically via "penetration testing" technique. Periodic threat and risk assessment is also important. Finally, even the most secure system will eventually be broken. As such, a disaster recovery plan in the event of a breach of security should also be implemented and tested.

There are many resources available to stakeholders including education and training from the International Information System Security Certification Consortium $(ISC)^2$, or the International Council of E-Commerce Consultants (EC-Council); literature from a variety of sources including the American Bar Association associations books, *The ABA Cybersecurity Handbook* and *A Playbook for Cyber Events*; and a plethora of cybersecurity conferences throughout the world.

Further Reading

American Bar Association Standing Committee on Law and National Security. 2014. Cybersecurity working group. *A Playbook for Cyber Events*, second ed.

Greenberg, A., 2015. *After jeep hack, Chrysler recalls 1.4M vehicles for bug fix*, Wired. Available from: <https://www.wired.com/2015/07/jeep-hack-chrysler-recalls-1-4m-vehicles-bug-fix/> (accessed 30.08.16).

Kali, 2016. *Our most advanced penetration testing distribution, ever*, Kali Linux. Available from: <https://www.kali.org/> (accessed 30.08.16).

McAfee, 2013. *Smart grid cyber security*, Alstom Grid, Intel, McAfee. Available from: <http://www.mcafee.com/in/resources/white-papers/wp-smart-grid-cyber-security.pdf> (accessed 30.08.16).

Metasploit, n.d. *Metasploit home page*. Available from: <https://www.metasploit.com/index.jsp> (accessed 30.08.16).

Miller, C., Valasek, C., 2015. *Remote exploitation of an unaltered passenger vehicle*, Illmatics. Available from: <http://illmatics.com/Remote%20Car%20Hacking.pdf> (accessed 30.08.16).

National Institute for Standards and Technology, 2010. *Introduction to NISTIR 7628 guidelines for smart grid cyber security*, US Department of Commerce. Available from: <http://www.nist.gov/smartgrid/upload/nistir-7628_total.pdf> (accessed 31.08.16).

National Institute for Standards and Technology, 2014. *NIST framework and roadmap for smart grid interoperability standards, release 3.0*, US Department of Commerce. Available from: <http://www.nist.gov/smartgrid/upload/NIST-SP-1108r3.pdf> (accessed 31.08.16).

NMAP.org, n.d. *NMAP home page*. Available from: <https://nmap.org/> (accessed 30.08.16).

OPM.Gov, 2015. *Statement by OPM press secretary Sam Schumach on background investigations incident*, US Government. Available from: <https://www.opm.gov/news/releases/2015/09/cyber-statement-923/> (accessed 30.08.16).

Rhodes, J.D., Polley, V.I., 2013. The ABA cybersecurity handbook: a resource for attorneys, law firms, and business professionals. American Bar Association.

Shodan, 2016. *Shodan home page*. Available from: <https://www.shodan.io/> (accessed 30.08.16).

U.S. Department of Energy, n.d. *Cybersecurity capability maturity model (C2M2) program*, US Government. Available from: <http://energy.gov/oe/services/cybersecurity/cybersecurity-capability-maturity-model-c2m2-program> (accessed 30.08.16).

U.S. Department of Energy, n.d. *Reducing cyber risk to critical infrastructure: NIST framework*, US Government. Available from: <http://energy.gov/oe/services/cybersecurity/reducing-cyber-risk-critical-infrastructure-nist-framework> (accessed 30.08.16).

U.S. Department of Energy, n.d. *Smart grid*, US Government. Available from: <http://energy.gov/oe/services/technology-development/smart-grid> (accessed 30.08.16).

Zetter, K., 2014. *An unprecedented look at Stuxnet, the world's first digital weapon*, Wired. Available from: <https://www.wired.com/2014/11/countdown-to-zero-day-stuxnet/> (accessed 30.08.16).

9

BIG DATA AND MONITORING THE GRID

Sonal K. Phan[1] and Cathy Chen[2]
[1]Exponent Inc., Atlanta, GA, United States [2]Exponent Inc., New York, NY, United States

CHAPTER OUTLINE

9.1 Introduction

An electrical power grid is a complex network of transmission lines, substations, transformers, and other infrastructure that delivers electricity from generation sites like traditional power plants to locations where it is distributed to, and consumed by consumers (also known as loads). The power grid in the United States was initially started in the 1890s, and has

The Power Grid. DOI: http://dx.doi.org/10.1016/B978-0-12-805321-8.00009-4

been expanded and upgraded over time, as technology has changed. The power grid now consists of over 9200 generating units that produce more than one million megawatts of generating capacity that are connected by more than 300,000 miles of transmission lines (US Department of Energy). In today's power grid, an electrical disruption, like a blackout, can have a cascading, domino-like effect (US Department of Energy). An excellent example of this cascading effect is the North American blackout of 2003, in which a series of failures of equipment in the greater Ohio area led to a blackout of most of the Northeastern United States and Canada. While there were many issues and violations that led to this blackout, one contributing factor was a lack of information transfer and coordination between equipment and utilities (US-Canada Power System Outage Task Force, 2004). In order to meet increasing power demand (energy consumption in the United States increased from 39,109 trillion BTUs in 2012 to 40,394 trillion BTUs in 2016) (US Energy Information Administration, 2016), and intelligently incorporate new equipment, while also preventing blackouts and power losses, the power grid must evolve. Technological changes are being implemented throughout the entirety of the power grid that facilitates the evolution of the grid into the next-generation that is needed in the 21st century (US Department of Energy). There is now a need for a power grid that can automate, cope with increasing complexity, adapt, and actively manage the needs of consumers in real time, without the constant interference of humans; there is a need for the grid to be smarter.

There are many changes occurring to the legacy power grid to transition it closer to a smart grid. For example, renewable energy sources, like geothermal, solar, and wind power, are now viable sources of generation. These renewable energy sources, however, which are often affected by weather patterns, add unpredictability and volatility to the legacy grid that it is not equipped to handle. In addition to more forms of generation, many parts of the grid have digital technology that allow for two-way communication between the utility and customers through the use of sensors (e.g., pervasive sensors, actuators, smart meters, and data networks) (US Department of Energy). In long-distance transmission networks, phasor measurement units (PMUs) are being added to detect grid instability within milliseconds and taking corrective actions to prevent cascading failures (like the 2003 blackout in the Northeastern United States and Canada). This two-way communication allows the smart grid to automatically reroute power when equipment fails or outages occur. It also can better predict loads, potential outages, and give utilities greater insight into the health of the power grid (US Department of Energy). The power grid is expected to become more dynamic and require constant real-time adjustments based on the measured status of the system at any given point in time (Yin et al., 2011). These digital technologies and their respective algorithms are at the heart of the smart grid.

These new technologies in the smart grid have led to the generation of massive amounts of data. This data set is too large and at times too complex to compute with traditional data analytics, and consequently is known as a big data challenge. The big data generated by the smart grid offers many possibilities, specifically, new insight into dynamic energy management that the information from PMUs, smart meters, renewable energy sources, and the like, can provide. Dynamic energy management in conventional power grids has been well investigated; it is, however, not well understood in smart grids. Energy Management Systems, or Energy Portfolio Management, includes a number of systems, data, technologies, and analysis techniques that will be discussed in this chapter including:

- Wide-area awareness in real-time grid status
- Information extraction
- Demand response algorithms
- Supervisory control through computer-based systems
- Forecasting and economic dispatch

The smart grid aims to provide steady availability of power, energy sustainability, environmental protection, prevention of large-scale failures, and optimized operational expenses of power generation and distribution, while reducing future capital expenses for thermal generators and transmission networks (Diamantoulakis et al., 2015). Traditional computer techniques cannot handle the fast data processing needed in the smart grid for real-time monitoring, dynamic energy management, and power flow optimization. As a result, high performance computing techniques, like virtualization, and in-memory computing, can effectively reduce compute times (Diamantoulakis et al., 2015). In this chapter, big data challenges that smart grids create will be explored, and methods for analyzing and visualizing this large data set will be analyzed. The different elements of the smart grid work together to achieve better balance than legacy power grids by utilizing high speed controllers that forecast loads, optimize portfolios, trade energy, and analyze usage.

9.2 Smart Grid Big Data Challenge

As illustrated in Fig. 9.1, the smart grid is becoming more complicated, with a number of different sources of generation, various types of loads, and vast amounts of monitoring data. The smart grid consists of a variety of sensing equipment and tools for measuring its overall health. The ongoing development in micro-generation is resulting in an increase in small generators (including renewables like solar and wind) distributed geographically throughout a region, as opposed to the present model of centralized power generation. Moreover, the genesis of the "prosumer"—the customer that is also a producer—adds to the resources associated with the smart grid (IBM, 2012). Preparing for a potential surge in electric vehicles and their charging stations, which can both charge and discharge from the smart grid, adds additional complexities and opportunities (IBM, 2012). As more intelligent systems are added to the grid, data are generated at a faster rate, creating a unique big data challenge. Data must be processed and analyzed in order to best utilize the information generated by the smart grid. While these processes and analytics will be discussed, we must first

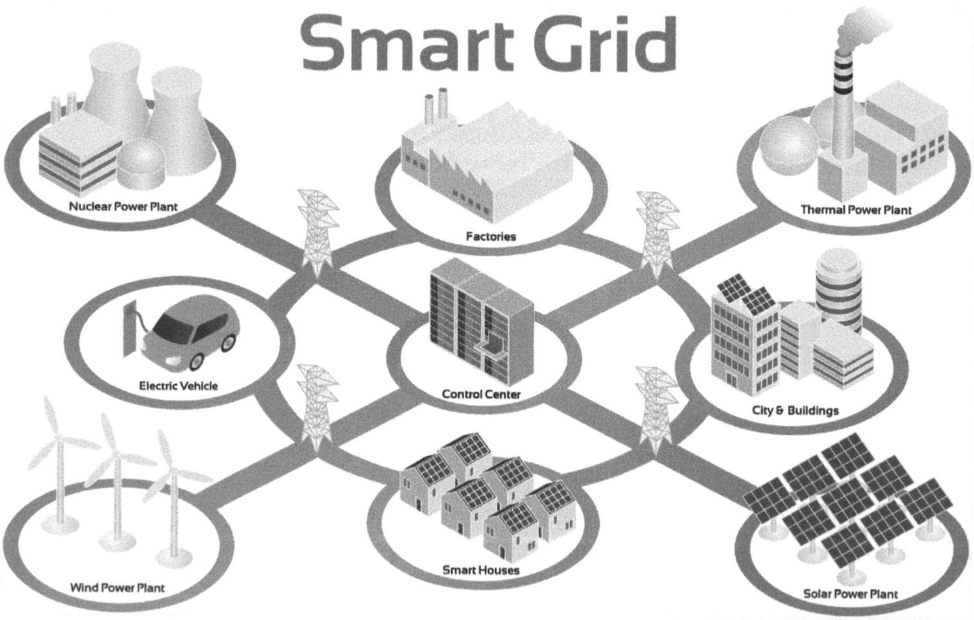

Figure 9.1 The smart grid is growing in sources for data, consumption, and generation.

discuss and understand some of the sources of this big data. Two of the sources of this information boom are the use of supervisory control and data acquisitions (SCADAs) systems and the increasing use of smart meters.

9.2.1 Supervisory Control and Data Acquisition Systems (SCADA)

Industrial control systems (ICS) are computer-based systems that control and monitor industrial processes. The ICS used in the smart grid is the SCADA system. SCADA systems consist of a collection of equipment that provides a remote operator with sufficient information to determine the status of a particular piece of equipment or process and determines if any action is required in response to that information (Thomas & McDonald, 2015). SCADA systems allow for data acquisition, remote control, historical data analysis, human/machine interface, and report writing using coded signals over communication channels (Thomas & McDonald, 2015). As defined by Thomas and McDonald (2015), the SCADA system is an integrated technology traditionally composed of four major components:

1. *Remote Terminal Unit (RTU)*—control hardware that acquires all field data from different field devices, and communicates the relevant data to the master station. Concurrently, the RTU distributes the control signals received from the master station to the field devices. Today, intelligent electronic devices have begun to replace RTUs, as discussed below.
2. *Data Communication Systems*—the remote locations of RTUs impose many restraints on the system because data communication over such long distances involves third-party media, such as telephone lines or cellular towers. These communications are necessary, but often have limited bandwidth.
3. *Master Terminal Unit/Master Station (MTU)*—the group of input, and output devices, computers, and peripherals that allows operators to observe the state of the power grid and control it.
4. *Human/Machine Interface*—the interface required for the interaction between the MTU and the operators or users of the SCADA system.

SCADA systems are generally event driven; cover large geographic areas; are suited to multiple, independent systems; have poor data quality and media reliability; and must contain power efficient hardware, often focused on binary signal detection (Galloway & Hancke, 2012).

SCADA systems are used at all levels of the power system: generation, transmission, and distribution. SCADA system functions include data acquisition, remote control of equipment in the grid, analytics, data display to humans, and report writing for consumption at different levels of management and from different departments of the utility (Thomas & McDonald, 2015). In the smart grid, the big data collected by SCADA systems needs to be analyzed, displayed, and reported on, a process that traditional analytics cannot handle. New techniques and emerging research into analysis of smart grid big data will be discussed later in this chapter.

9.2.2 Smart Meters

In addition to the data the SCADA generates, the inclusion of more and smarter RTUs has exacerbated the big data problem. One such example of this is the smart meter. Smart meters are being deployed to consumer premises to monitor real time power consumption and communicate back to the utility. These meters can also receive signals from the utility that provide incentives for reducing loads at a particular time of high demand, as illustrated in Fig. 9.2 above (See also Demand Response later in this chapter) (Simmhan et al., 2013a). The data gathered from smart meters can provide better understanding of customer segmentation, behavior, and how pricing influences usage. For example, time-of-use pricing encourages cost-savvy retail customers to run their washing machines, dryers, and dishwashers at off-peak times. The customer not

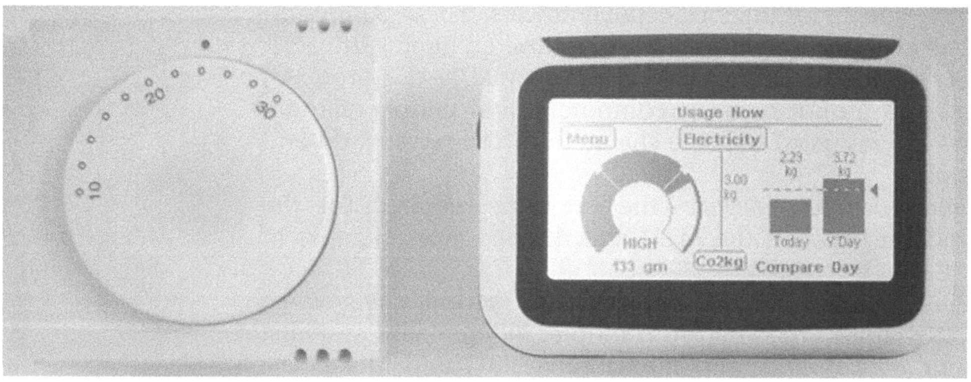

Figure 9.2 Smart meters monitor real time power consumptions and communicate them back to the utility. The meter can also receive signals from the utility that provide incentives for reducing loads.

only saves money, but also requires less generation capacity from energy providers, meaning lower capital outlay for new generation and overall greater operational efficiency for utilities (IBM, 2012). Going from one (or fewer) meter reading each month to a meter reading every 15 minutes equates to 96 million reads per day for every 1 million meters, resulting in a 3000-fold increase in data, which can be overwhelming if not properly managed and processed (IBM, 2012).

9.2.3 Intelligent Electronic Devices

As mentioned above, RTUs in SCADA systems are slowly being replaced with intelligent electronic devices. An intelligent electronic device performs electrical protection functions, advanced local control intelligence, has the ability to monitor processes, and can communicate directly to a SCADA system (Subnet Solutions Inc.). In addition to other devices that measure physical quantities and report those back to the SCADA system in real time, other examples of intelligent electronic devices include electronic temperature monitors, on-line dissolved-gas analyzers, and electronic tap position monitors (Harlow, 2012). Individual intelligent electronic devices provide sensors, methods of communicating data, and software to analyze the data monitored to provide alarms when any of the monitored parameters goes beyond preset limits (Harlow, 2012).

Data from smart meters and intelligent electronic devices need to be reliably transmitted on communications channels to the MTUs where the data are processed, digested, and visualized in a way that can be meaningful to the human operator reading the reports. These data are then used in the smart grid in multiple parts of the system, from feature extraction to event monitoring to forecasting techniques like demand response (DR) and price forecasting. These methods and applications are discussed in the following sections.

9.3 Informative Feature Extraction

Smart grids continuously generate data such as the voltage, current, phase angle, power (kW, kVA, kVAR), temperature, type of grid connectivity, and statistics on supply and demand. In their native form, such raw data signals can be noisy, redundant, and require large amounts of memory to store. Hence, these data signals need to be converted into informative features for meaningful and timely analysis. In this section, various feature extraction methods are discussed.

9.3.1 Waveform Characteristics

Power Quality (PQ) characterizes the deviations in the Alternating Current (AC) transmission waveform from a standard sinusoid alternating at 60 Hertz. Deviations from this waveform, such as those caused by harmonics, can cause equipment to operate incorrectly or ultimately fail. Most automated monitoring systems extract parameters such as energy, entropy, minimum, maximum, standard deviation, mean, and root-mean-square (rms) values from the AC transmission waveforms. These parameters are extracted from either the original waveform or a time-frequency transformation of the original waveform (Khokhar et al., 2015; Mittal et al., 2012). The following time-frequency transformations are commonly used for feature extraction.

9.3.1.1 Short-Time Fourier Transform

Fourier transform decomposes a signal into a weighted linear combination of sinusoids at different frequencies. The Fourier coefficients (i.e., weight of sinusoids) constitute a spectrum for the signal. The spectrum will be affected with the presence of a long-term disturbance in a power signal and can be used to identify long-term disturbances, but it may not be able to resolve short-term disturbances. Moreover, simple Fourier transform does not provide any time information about the disturbance. Hence, a modified form of Fourier transform, called short-time Fourier transform (STFT), is used for extracting informative features from power signals. STFT analyzes small overlapping sections of a signal. A section is extracted using a moving window function. Figure 9.3 illustrates a STFT using a sinusoidal waveform and a sinusoidal waveform with a transient power quality disturbance. It is a challenge to select an appropriate window size for STFT because of a compromise between time and frequency information. A large window size leads to good frequency resolution and bad time resolution, while a small window size leads to good time resolution and bad frequency resolution (Khokhar et al., 2015; Mahela et al., 2015).

9.3.1.2 Wavelet Transform

Wavelet transform decomposes a signal into a weighted linear combination of a set of scaling functions and wavelet

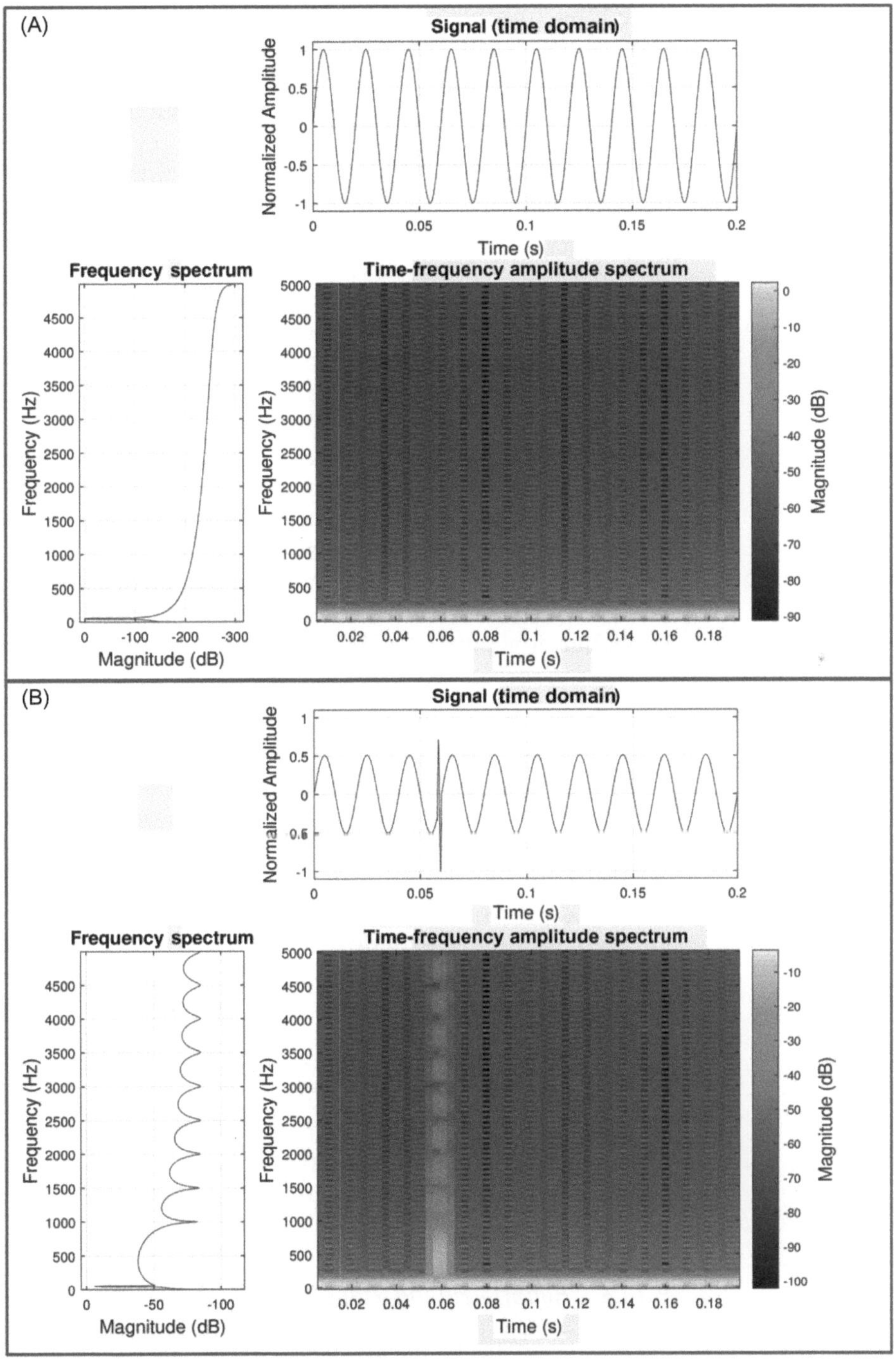

Figure 9.3 Short-Time Fourier Transform of (A) a pure sinusoidal waveform and (B) a sinusoidal waveform with a transient power quality disturbance.

functions (mother wavelet). Scaling and wavelet functions are orthogonal functions that divide the function space into a series of orthogonal high and low frequency spaces. Coefficients (weights) associated with the scaling function, called approximation coefficients, capture low frequency information, while coefficients associated with wavelet function, called detail coefficients, capture high-frequency information. Unlike Fourier transform, wavelet transform provides local information (in both time and frequency) of a given signal, which makes this transform very useful for extracting disturbance information from power signals. Wavelet transform can be performed using different types of wavelet functions. For power signal analysis, the fourth order Daubechies wavelet (db4) was found to be appropriate because of the similarity of its shape to power disturbance signals (Ekici, 2009). The number of levels for multiresolution analysis is another parameter for this analysis. After a certain number of levels, detail coefficients are susceptible to noise. Features extracted from higher level coefficients often are correlated to features extracted from lower level coefficients and do not add much value to the classification system (Khokhar et al., 2015; Mahela et al., 2015; Figure 9.4).

9.3.1.3 S-Transform

S-transform is time-frequency analysis method, similar to STFT and wavelet transform. A limitation of STFT is that it uses a fixed, preselected window width for analysis of signals containing low and high frequency signals. Because of fixed window width, it is difficult to analyze low frequency power signals affected by high frequency power signals. On the other hand, wavelet transform has fixed decomposition frequency bands, low resolution for low frequency components, and high sensitivity to noise. S-transform is a hybrid of two methods. Similar to STFT, S-transform uses a window function to extract features for small segments but unlike STFT, S-transform scales the window function with frequency analogous to wavelet transform. Instead of a rectangular window function, S-transform uses a Gaussian window function (Mittal et al., 2012).

9.3.1.4 Hilbert–Huang Transform

Hilbert–Huang transform (HHT) is a two-step method for analysis of nonlinear and nonstationary signals. The first step is empirical mode decomposition (EMD) that decomposes the

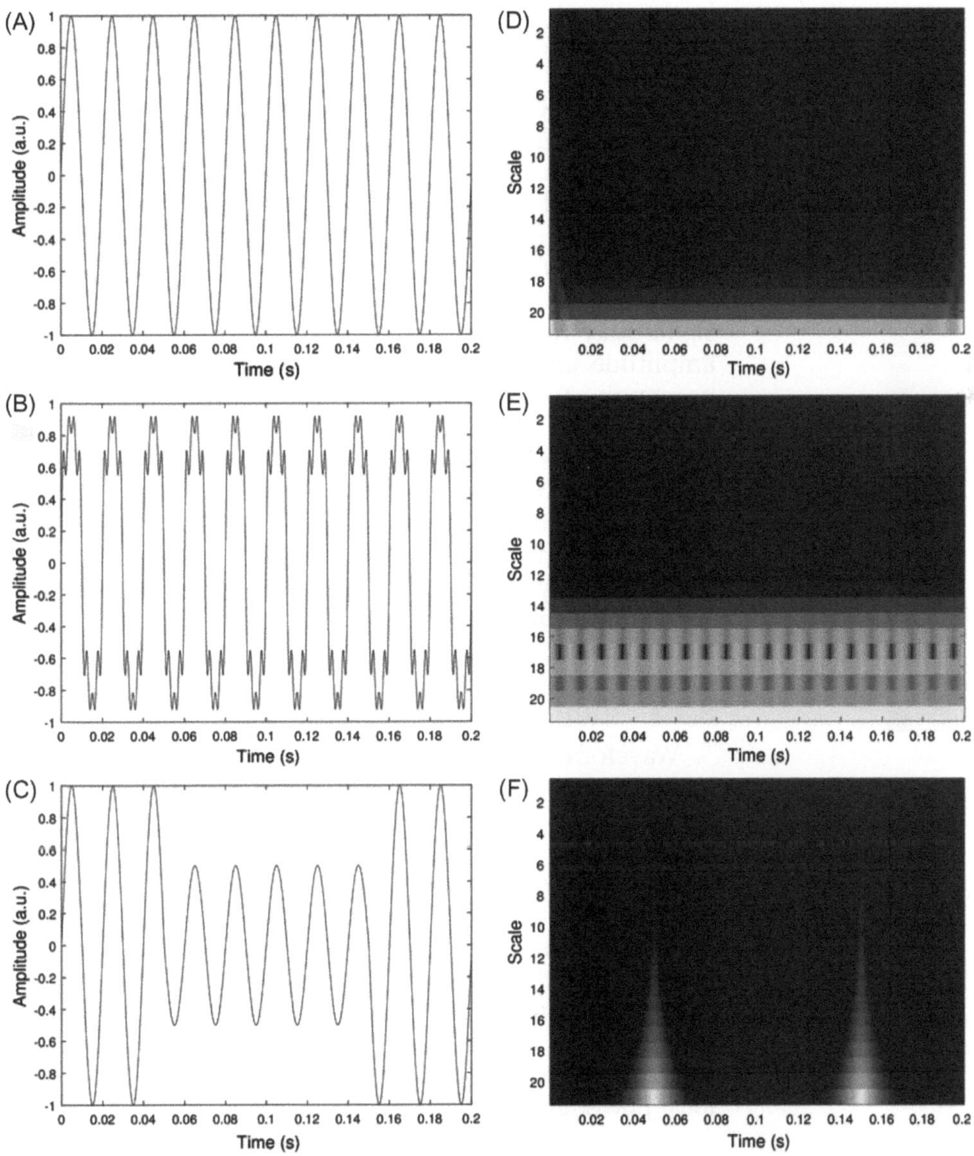

Figure 9.4 Wavelet transform using Morlet wavelet. Output coefficients corresponding to scales 1 to 21 for three input sinusoidal signals (A) pure 50 Hertz, (B) with harmonics, and (C) with a sag in amplitude are shown in (D), (E), and (F), respectively (Torrence & Compo).

original signal into a finite number of intrinsic mode functions (IMFs). IMFs are time-varying mono-component (single frequency) functions. The signal is decomposed into IMFs in such a manner that the highest frequency component of each event in the signal is captured by the first IMF. The second step of HHT is Hilbert transform that produces an orthogonal pair for each IMF that is phase shifted by 90 degrees. Each IMF set and its orthogonal pair can be used to estimate the instantaneous variation in magnitude and frequency of the IMF with respect to time. Thus, HHT is very useful to extract information from nonlinear and nonstationary time series data such as power signals with PQ events. Over the years, this simple HHT has been further modified to better address the challenges posed by power signals. For example, a transition or discontinuity in the signal may result in overshoot and ambiguity in instantaneous amplitude and frequency response because of the Gibbs phenomenon. To overcome this, a windowing technique, similar to STFT, was developed. Also, arbitrarily close harmonics in power signals are not accurately decomposed by the EMD process and they result in IMFs with mixed frequency and amplitude information and incorrect instantons amplitude and frequency information. An iterative HHT was proposed to identify accurately decomposed harmonic components in the PQ signal (Drummond & Sutanto, 2010).

9.3.2 Consumption-Related Features

Waveform-based features primarily capture information related to PQ, which can help prevent power grid failures, as discussed in the next section. Understanding and modeling local demand and supply patterns, however, is a key to effective management of power systems. Demand-side management technologies help in an economical operation of power systems by regulating power generation and adjusting the pricing schemes based on the demand. The demand for a region is often modeled using historical time-series data with the real-time data being used to predict future demand. Environmental factors and site-specific factors play an important role in capturing demand for a local power grid.

9.3.2.1 Environmental Factors

Many demand forecasting models include parameters or metrics for capturing change in environmental conditions of the region that can affect power supply and demand. Environmental factors include population, gross domestic product, median household income, climate, and renewable energy sources (Kialashaki & Reisel, 2013). The amount of electrical power consumption has been correlated to the population of a region with differences in consumption in rural versus urban areas (Amaral et al., 2005; Shi et al., 2014). Also, population changes in a local region have been a strong predictor of power demand (Hor et al., 2005). A high correlation has been observed between daily temperature and daily electricity demand (Franco & Sanstad, 2008). Furthermore, inclusion of the effects of relative humidity improved the demand forecast during the summer months in California (Hor et al., 2005). The demand on the power grid is also reduced with an increase in renewable power production in a local grid (Aghaei & Alizadeh, 2013).

9.3.2.2 Site-Specific Factors

Besides environmental factors that affect demand in a region, demand in a particular building can be further modeled using some site-specific parameters. Demand is affected by the type of building: industrial, office, residential, or college campus (Simmhan et al., 2013b). Demand is further affected by the usage patterns of individuals inhabiting these buildings including their working, sleeping, and cleaning hours (Muratori et al., 2013). The age of a building, type of insulation, use of energy efficient appliances, heating, ventilation, and air-conditioning settings, and number of electrical vehicles, if any, are some other factors (Simmhan et al., 2013b). As compared to a legacy grid, in smart grids, smart meters continuously collect data related to site-specific usage patterns and send these data back to the centralized data processing units. Mining these large volumes of data will be a key for effective energy forecasting, as discussed in a later section of this chapter.

9.4 Event Monitoring

9.4.1 Power Quality Disturbance

In a modern business environment with the increased use of computers, digital controllers, and electronic switching components, power systems are expected to provide uninterrupted, high-quality power. Deviation in the voltage, current, and frequency from the standard rating may be considered a PQ disturbance based on their magnitude and may result in the failure of electrical and electronic equipment. For example, PQ disturbances can lead to cascading failures thereby increasing the risk of blackouts. In a smart grid, integration of unpredictable renewable energy sources and the use of modern control equipment (with non-linear, electronic switches) have exaggerated the PQ problem. PQ events may impose increased expenses on customers with costs related to supporting ancillary services, such as reactive power, re-dispatch, and load curtailment. Therefore, it is essential to monitor and mitigate PQ events. As compared to a legacy grid, smart grids collect data at a much more granular level and as such, a large amount of data needs to be monitored and analyzed to detect PQ events. Therefore, automated classifiers are developed to monitor data and detect PQ events (Mahela et al., 2015). Classifiers can successfully detect the following PQ events using the waveform features discussed in previously (Ekici, 2009; Monedero et al., 2007).

1. *Sag*—a short duration undervoltage event that causes a decrease in rms voltage to 10–90% of rated value at power frequency for durations of a half cycle to 1 minute. It is caused by a sudden increase in load impedance, source impedance, short circuits, or ground faults (Fig. 9.5B).
2. *Swell*—a short duration overvoltage event that causes an increase in rms voltage of 10–90% over the rated value at power frequency for durations of a half cycle to 1 minute. It is caused by a sudden reduction in load on a circuit with a poor or damaged voltage regulator (Fig. 9.5C).
3. *Momentary interruption*—a drop of 90–100% of the rated voltage for durations of a half cycle to one minute. It is caused by faults, broken conductors, or operator intervention (Fig. 9.5D).
4. *Switching transient*—a short-lived oscillation event in a system. It is caused by a sudden change of voltage, current, or load due to the operation of a switching device. A similar event, an impulsive transient (i.e., a fast-voltage event), is mainly caused by lightning (Fig. 9.5E).
5. *Harmonic distortion*—the deviation of the supply voltage waveform from an ideal sinusoidal waveform. It is caused by the interaction of distorted customer loads with the impedance of the supply network. In the presence of some harmonic distortion, power factor correction capacitors can further increase the distortion (Fig. 9.5F).

Several existing models can classify PQ events with high accuracy (>90%), given sufficient processing time. The successful application of these methods in the field, however, requires a short processing time. Thus, most of the ongoing research is focused on real-time detection and mitigation. Also, the majority of existing classifiers can only detect and classify single, isolated PQ events; however, multiple PQ events can occur simultaneously due to several sources of disturbances in a power grid. With the presence of multiple PQ events, the compound signal can be significantly different.

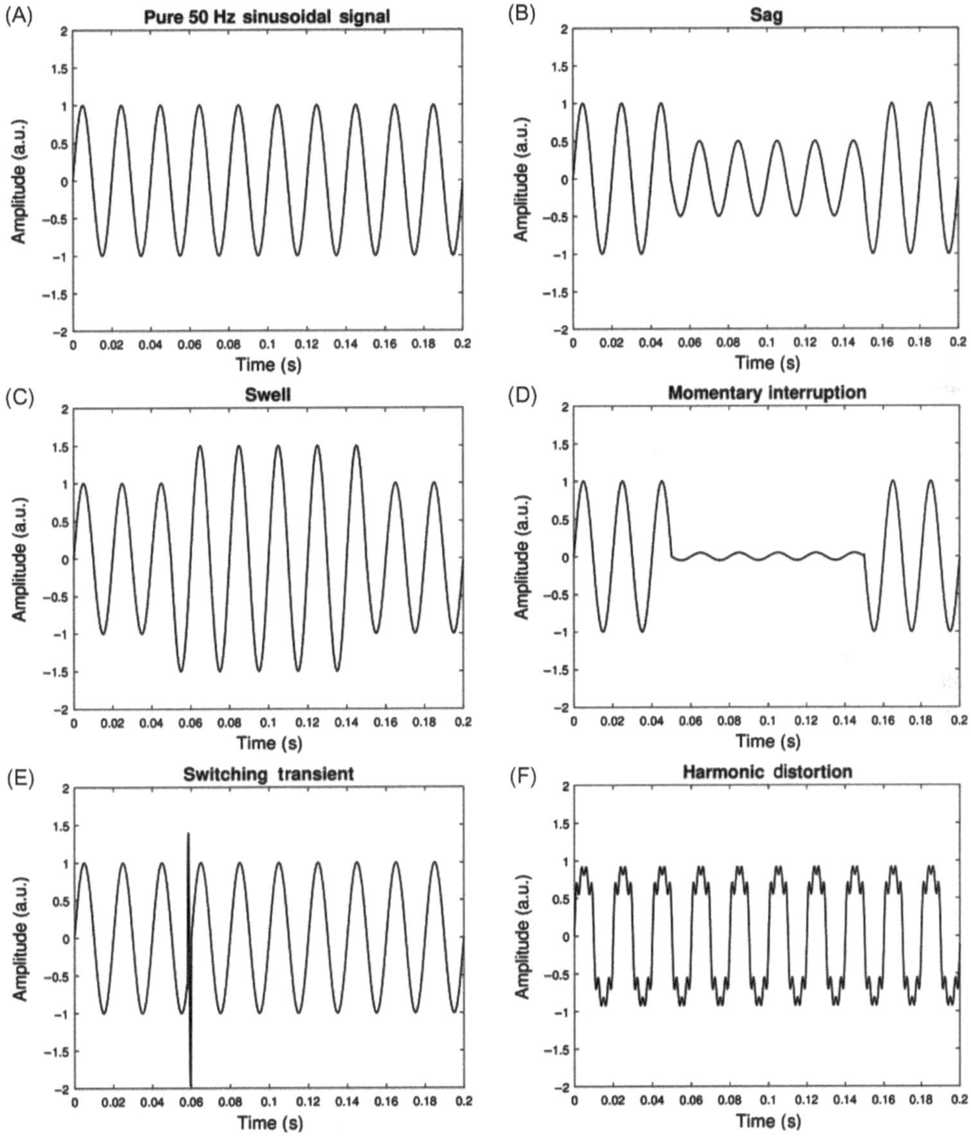

Figure 9.5 Signal waveforms during various power quality disturbance events: (A) pure 50 Hertz sinusoidal signal, (B) sag, (C) swell, (D) momentary interruption, (E) switching transient, and (F) harmonic distortion (Moravej et al., 2009).

9.4.2 Intrusion

One of the key components of a smart grid is a two-way communication network between energy suppliers and their customers that allows for real-time pricing, load management, and integration of renewable resources to the grid. This

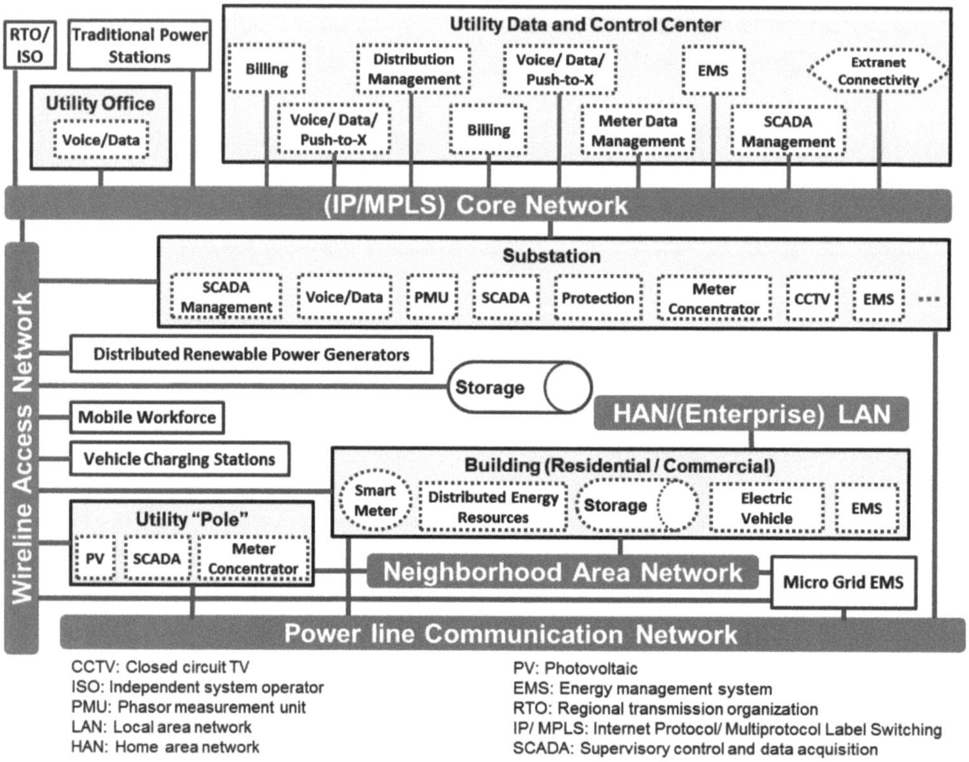

Figure 9.6 Physical networking architecture of a smart grid with various potential access points for intrusion (Budka et al., 2010).

end-to-end, Internet-like communications network along with smart devices introduces cyber threats to the security of the smart grid (Fig. 9.6). A smart grid is vulnerable to cyber-attacks at several locations including the advanced metering infrastructure (AMI) and unsecured wireless mesh network communications. In AMI, an attack can modify the memory of a device by inserting malicious software or disconnecting the meters, which would block the transmission of metering information to the grid. An attack on the network can damage the integrity of configuration, routing, and communication traffic that can lead to alteration of data, unintentional dissemination of consumer data, and denial of service. An attack could also provide false data, which could then be used to change system configurations for purposed of financial gain or can shut down a power system completely. If a SCADA system at a substation is left unsecured, it can provide multiple access points allowing the attacker to reprogram the measuring and controlling programs. (Zhang et al., 2011).

In order to prevent such attacks, it is essential that the networks at generation, transmission, and distribution levels of the grid are secured using adequate encryption, authentication, and access-control protocols. The goal of encryption is to ensure security of data in an unsecured network domain while minimizing latency involved in data delivery. Developing strong and adaptable authentication mechanisms is imperative to prevent unauthorized access to the systems. Maintaining authentication (keys and passwords) in a smart grid poses a challenge because of long-term deployments of a system with limited system update capabilities. While encryption and authentication protects the network from external attacks, access-control protocols limit damage to the system after a malicious user succeeds in getting some access. Access control significantly reduces repercussions of an attack by disabling unnecessary functionalities from software and protocols used on grid systems. In addition to securing the network, it is also essential to secure devices such as smart meters. Smart meters are intentionally designed in such a manner that it is possible to remotely update their firmware as needed, but this capability also opens them attack. To avoid these attacks, only cryptographically signed firmware should be loaded on a smart meter. While mechanisms discussed above are essential preventive measures to secure the smart grid, these measures may fail in restricting access of a malicious user. Therefore, machine learning techniques are used to detect and mitigate any intrusions. (Sridhar et al., 2012).

The intrusion detection functions have two core functionalities: collecting audit data in binary or human-readable files or databases and analyzing the data to either categorize them into binary—good and suspicious—processes or assigning a probabilistic measure to each process. Audit data used for intrusion detection can be broadly classified into two categories: host- and network-based data. Host-based data capture logs, such as file-system details, for each node/device in the grid. Host-based analysis is attractive because it allows distributive analysis in large grids and it is relatively straightforward given that host-specific knowledge is well defined. Disadvantages of host-based analysis include additional work for each node in the grid and potential modification of audit data at each node by a sophisticated attacker, rendering the intrusion detection methods useless. Network-based data capture network activity such as traffic and transmitted packets. The advantage of network-based auditing is that individual nodes are not responsible for collecting logs while the major disadvantage is that visibility of nodes collecting data diminishes the effectiveness of the method and, as such, it is challenging to set up effective data collection sensors. (Mitchell & Chen, 2014; Zhang et al., 2013).

Data analysis for intrusion detection can be broadly classified into knowledge- and behavior-based techniques. Knowledge-based techniques look for runtime patterns that match a specific pattern of misbehavior. These techniques use supervised learning methods to train a classifier that can identify suspicious patterns. These classifiers usually have a very low false positive rate because they react to known bad behavior. One main limitation of knowledge-based methods is that they only look for known, bad-behavior patterns and their success depends on the extent of the dictionary used for training purposes. Behavior-based techniques look for patterns that are not ordinary. Ordinary patterns are defined by analyzing the history of patterns for repetitive patterns. These techniques use unsupervised learning techniques to isolate suspicious patterns. Because the unsupervised techniques are not trained to look for specific patterns, rather they just isolate out of the ordinary patterns, these techniques are very good at protecting the grid against new unknown attacks. This method, however, is highly susceptible to false positives, especially during the initial stages of setting up the intrusion detection system, when there is only a limited history of patterns. One modification of behavior-based unsupervised techniques is behavior-specification techniques. In behavior-specification techniques, experts define a behavior model for various components of the grid. The models include range of allowable values for various system parameters. Any deviation from these allowable ranges raises a red flag. Behavior-specification methods have low false positive rates and they do not need any empirical training data; however, it can be challenging to define formal specifications for various components (Mitchell & Chen, 2014; Berthier et al., 2010).

Intrusion detection techniques pose a big data challenge with thousands of samples collected every day. Information systems have to tackle not only storing and retrieving these terabytes of data, but also analyzing them in real-time to assist with system protection (Zuech et al., 2015).

9.4.3 Islanding

Islanding refers to a condition in which a part of the electric power distribution network is disconnected from the power grid but remains energized by local distributed power generation resources, such as wind turbines, solar panels, and other local generators. Islanding can lead to several problems related to PQ, safety, and operation. The current industry standards recommend disconnecting these local distributed resources immediately after islanding occurs. Disconnecting these local distributed resources, however, results in wasted power, and therefore research is being conducted on secure islanding of distribution networks. Regardless of the practice of disconnecting local distributed resources or secure islanding of a distribution network, the first step is to detect the islanding phenomenon. Islanding detection techniques can be broadly classified as remote and local (Raza et al., 2015; El-Arroudi & Joos, 2007).

9.4.3.1 Remote Islanding Detection Techniques

Remote islanding detection techniques work on the principle of communication between the utility and distributed power generation resources. Remote techniques monitor either the parameters (such as voltage and frequency) for the entire distribution system or the status of circuit breakers that could island a distribution system. Remote techniques often collect the required data via multiple SCADA systems distributed across the grid. When a disconnection is detected, a trip signal is sent to the distributed power generation resource. These techniques are fast and reliable. They have negligible impact on PQ and system transients; however, they require real-time monitoring of all distributed power generation sources in a smart grid, and as such, they are very expensive to implement. (Khamis et al., 2013)

9.4.3.2 Local Islanding Detection Techniques

In contrast, local islanding detection techniques can be employed more economically. Local techniques detect islanding at the distributed generation site by measuring the variations in system parameters such as voltage, frequency, phase angle, impedance, and harmonic distortion, collected via a SCADA system or other means. Local techniques are further categorized into passive and active techniques. Passive techniques monitor the system parameters and compare them to preset threshold values to detect islanding. Active techniques interact with the power systems by introducing minor disturbances and monitoring their impact on the system. While in the presence of islanding, these minor disturbances would be significant, but in the absence of islanding, the impact of these minor disturbances is negligible. Passive techniques are fast, cost efficient, and they have no impact on PQ, but they are not very sensitive if the load and generation are closely matched in the islanded system. Also, passive techniques require selection threshold values for decision making. Active techniques increase the sensitivity of island detection, but they inject disturbances in the power system; as such they are more complex and reduce PQ (Raza et al., 2015). Some hybrid techniques also were developed that apply passive techniques to monitor the distribution system continuously and if islanding is suspected, active techniques are used for confirmation (Khamis et al., 2013).

As discussed above, passive techniques traditionally compare system parameters to threshold values, but the selection of threshold values is quite a difficult task. If the threshold is too high, it will result in a false negative and if the threshold is too low, it will result in a false positive. Moreover, analyzing raw system parameters can be time consuming and susceptible to noise. These challenges can be addressed by extracting information on waveform-based features, and training and optimizing classification models for decision making. Similar to power disturbance detection, islanding detection can be done using different feature selection and classification methods (Raza et al., 2015).

9.5 Energy Consumption Forecasting

One of the prime objectives of smart grid is to allow better control and balance of energy supply and demand. Hence, in order to get the best return on smart grid infrastructure investments, it is crucial to develop an ability to extract system knowledge and compare it to historical observations in real-time. Smart meter readings allow for insight into usage patterns that are unique to each premise and each aggregate level of the distribution hierarchy. Large volumes of data collected by smart meters can be used to improve the energy forecasting process by providing improved input data when the ISO performs its economic dispatch and unit commitment processes. Big data can improve the accuracy in many of the ISO's markets such as the hour-ahead market, the day-ahead market, and even the month- or year-ahead markets (Mirowski et al., 2014; Hedman et al., 2010; Puttre, 2015).

9.5.1 Forecasting Techniques

Traditionally, energy consumption and load forecasting have been applied on the user side, where information about energy consumption was collected and plotted against some variable (i.e., production activity or historial energy consumption data). The data were then used as a basis for a simple interpolation of the attained data (Shandilya et al., 2016). Energy consumption forecasting has several distinct facets: short-term load forecasting (STLF), long-term load forecasting, spatial-load forecasting, and price forecasting (Hong, 2014). Nowadays, these interpolation-based forecasting techniques have been replaced by a number of varying techniques. Smart grid investments have brought new challenges to the energy consumption forecasting field, such as DR forecasting (Hong, 2014).

STLF is essential for securing an inexpensive supply of reliable energy, as mandated by the US Energy Policy Act of 2005. This type of forecasting can be used to predict consumption demands up to one week ahead (Soliman & Al-Kandari, 2010). The pros and cons of the three most dominant STLF methods, as outlined by Hong (2014) are illustrated in Table 9.1 below:

Table 9.1 Short Term Load Forecasting (STLF) Techniques

Method	Pros	Cons
Multiple Regression	• Interpretable • Easy to implement • Easy to update • Easy to automate • Good accuracy	• Needs an explanatory variable • Needs a designated functional form • Needs at least two years of history
Autoregressive Integrated Moving Average (Box–Jenkins model)	• Few parameters to estimate • Does not require a long history • Good accuracy in a very short term	• Low accuracy in the longer term • High cost to implement • High cost of update • High cost to automate • Difficult to interpret
Artificial Neural Network	• Minimum statistical knowledge required • Minimum domain knowledge required • Good accuracy during normal weather conditions	• Heavy computation • Over parameterization • Difficult to interpret • Low accuracy during extreme weather conditions

Table 9.2 Spatial Load Forecasting Techniques

Method	Pros	Cons
Trending	• Ease of use • Simplicity • Short-range response to recent load-growth trends	• Low accuracy in the longer term • Overfitting of high-ordered polynomial
Simulation	• Fair short-range accuracy • Good long-range usefulness for planning	• Expensive development • Training costs
Hybrid	• Responds to recent trends of load history • Keeps long-range defensibility provided by simulation methods	• Requires little skill • Requires little interaction from the user

Different factors are the driving forces for forecasting, depending on the forecasting length. While in the short-term, forecasts are based on variation of electricity load and generation dispatches, in long-term load forecasting (traditionally defined as any forecasting beyond one week), other factors, as discussed below, become more dominant. For example, in midterm to 1-year forecasting, the major market forces affecting loads and forecasts are seasonal weather, annual generation, and planned outages (Chan et al., 2012).

Spatial load forecasting aims to pinpoint when, where, and how much load growth would occur, often forecasting anywhere from several years to several decades (Willis, 2002). These forecasts have been widely used in transmission and distribution planning (Hong, 2014). The pros and cons of the three most dominant spatial load forecasting methods, as outlined by Hong (2014), are illustrated in Table 9.2.

9.5.1.1 Demand Response

With the installation of millions of smart meters, sensors, and communication devices that make up the smart grid, the increase in data available offers a unique opportunity to make smart grids greener, more reliable, and cheaper. A way to achieve these goals is through DR, which is defined by the Federal Energy Regulatory Commission as:

> *Changes in electric usage by end-use customers from their normal consumption patterns in response to changes in the price of electricity over time, or to incentive payments designed to induce lower electricity use at times of high wholesale market prices or when system reliability is jeopardized (Federal Energy Regulatory Commission, 2009).*

It is possible for the smart grid to become more adaptive to the increased penetration of dynamic energy response, encouraging users' participation in energy savings, and cooperation through DR mechanisms due to the bi-directional flow of information in the smart grid (Diamantoulakis et al., 2015). It is estimated that the capacity to meet demand during the top 100 hours in 1 year accounts for between 10% and 20% of electricity costs (Wells, 2004), as generation and transmission capacity is provided to meet peak demand that occurs infrequently (Arnold, 2011). DR allows for a utility to mitigate these costs. In order to effectively design and implement DR, utilities have to perform a series of tasks including forecasting electricity prices, forecasting load at household levels with or without DR programs, and after-the-fact estimation of the consumption pattern changes due to various DR programs. DR can be applied to both residential and industrial loads and includes three different concepts (Diamantoulakis et al., 2015):

- Energy consumption reduction;
- Energy consumption (or production) shifting to periods of low (or high) demand, and;
- Efficient utilization of storage systems.

A crucial issue in smart grids is how to manage DR in order to reduce peak electricity load, while at the same time utilizing renewables and storage systems more efficiently. The effectiveness of DR algorithms depends critically on demand, price, load, and renewable energy forecasting, which all need sophisticated signal processing techniques (Diamantoulakis et al., 2015). The major challenge in this area is the estimation of normal (baseline) consumption patterns, with over one hundred methods available to perform this estimation (Hong, 2014). Most are derived from simple averages of load profiles, however, and have been shown to be insufficient at the household level. Research into estimation of baseline consumptions is an active field of research (Hong, 2014).

9.5.1.2 Price Forecasting

Effective DR depends critically on load energy forecasting on the utilities' side and price forecasting at the market and customers' side (Chan et al., 2012). Price forecasting, which is used to determine how much energy a utility should buy or sell, has become an integral part of energy consumption forecasting (Hong, 2014). With the inclusion of more variable energy sources, like renewables, forecasting tools now need to provide confidence intervals in addition to point estimations. This has to be done in much shorter intervals of half an hour, or even ten minutes due to the variable nature of renewable energy (Chan et al., 2012). Common behavior of electricity prices can be summarized generally into four aspects:

- The strongly seasonal nature of prices;
- The mean reversion nature of prices;
- Price-dependent volatilities and spikes, and;
- The correlation between electricity price and load.

Using these aspects, price forecasting in general follows two methodologies: simulation-based methods and statistical and artificial intelligence (AI)-based methods (Hong, 2014). Simulation-based methods create a mathematical model of the electricity market, load forecasts, outage information, and bids from market participants (Chan et al., 2012). The results of the simulation methods yield the optimal economic dispatch for the system by solving the unit commitment mathematical problem, setting the locational marginal prices, and solving the optimal power flow. Price forecasting accuracy using simulation methods is highly dependent on the quality of the input data. Statistical, or AI, methods, conversely, do not require a comprehensive knowledge of market procedures. Statistical methods use historical prices, weather, outages, and load to forecast future prices. Statistical methods cost less to implement, but are less accurate in analyzing price spikes that are caused by congestion in the transmission network (Hong, 2014).

9.6 Visualization

Large volumes of data generated by smart grid can overwhelm power grid data analysts and decision makers. A well-designed visualization can aid in data interpretation and decision making by systematically representing a large amount of data. Visualization tools are key components in the daily operation of the power grid, as well as in fault diagnosis and forecasting. Researchers have developed visualization tools for both unprocessed state variables of a power system, such as voltage and phase angles, and informative metrics derived from these state variables (He et al., 2015; Nga et al., 2012). Information visualization tools developed especially for the power grid often include common line, bar, and pie charts. The usual intent of visualization tools is to show variation of one or several state variables over time and region. Researchers also have shown the value of multivariate plots such as scatter plot matrices and parallel coordinates to study multiple variables in conjunction

(Mikkelsen et al., 2011). In addition to these commonly used univariate and multivariate tools, three types of custom visualizations are used to analyze information related to power grids: maps, networks, and virtual grids.

9.6.1 Maps

Map-based visualization associates data with geographic locations and facilitates geo-spatial data interpretation. Maps are a powerful tool for multilevel understanding of a power grid and for studying cascading effects of an event (Yan et al., 2013). When completely zoomed out, the user can visualize one or multiple variables across the complete power grid in a country. When zoomed in, the user can pan through a geographical region and study the variation in a local substation. In a map-based visualization, the map is used as a base layer and different visualization symbols, networks, and heat maps are overlaid on top of it (Fig. 9.7).

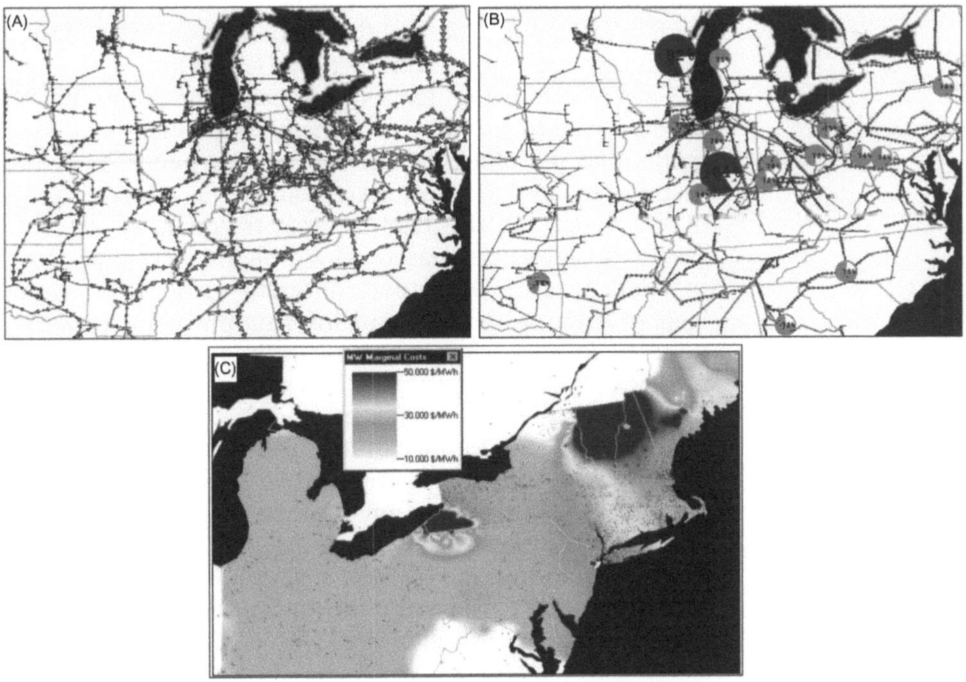

Figure 9.7 Map visualization of power grip data: (A) line diagram of the high-voltage transmission system in the Eastern Interconnect in North America, (B) pie charts superimposed on the figure in (A) indicating loading on each transmission line, (C) color contouring to indicate locational marginal prices for northeast United States. Reprinted with permission from IEEE.

In order to study the state of a transmission system, it is crucial to understand the current flows and percentage loading of various transmission lines. One of the commonly used techniques for transmission system representation is a one-line diagram imposed on maps. In a zoomed out view, one-line diagrams only include a small number of high-voltage buses and transmission lines. Small arrows are plotted on the lines to indicate the direction of real power flow (in megawatts). The arrows point toward the direction of flow and the arrow size is proportional to the amount of flow. This visualization is useful when the number of transmission lines is limited, but it is not an efficient tool to monitor a large number of transmission lines simultaneously (Overbye & Weber, 2000).

Pie-charts imposed on maps also have proven to be a powerful visualization tool. Dynamically sized pie charts have been used to indicate the loading on each transmission line. The loading on a line is indicated by the percentage fill of the pie. The color and size of pie is dynamically changed if the loading exceeds a limit, drawing the attention of an operator. There is a limit on the number of pie charts, however, that can be represented on a map before the visualization gets cluttered and difficult to interpret (Overbye & Weber, 2000).

Color contouring techniques can overcome the challenge posed by large number of transmission lines on one-line diagrams and pie charts. Color contouring paints a region on the map based on the accumulated data from all transmission lines in the region. In the regions with no data, an algorithm estimates virtual values based on a weighted average of neighboring data points. Of course, different weighting functions can lead to different results and information from these visualizations has to be interpreted accordingly. Contour visualization has been used successfully to visualize bus voltages and the marginal cost of electricity. Instead of contouring a whole map region, color contours can be used to visualize data around lines in a one-line diagram (Overbye & Weber, 2000).

Map-based visualization can be used for spatio-temporal data interpretation by using animation. Animations are useful for understanding the change in state variables at fault locations over time and to see how the faults cascade through the grid (Nga et al., 2012; Yan et al., 2013).

9.6.2 Networks

As discussed, a map-based power grid representation can become cluttered when analyzing a large number of transmission lines. Also, the focus of map-based visualization is to associate variables such as voltage, loading, and pricing with a geographic location, but it fails to show the electrical connectivity view of the electric power system. A network-based visualization can help view this inner working independent of geographic locations. A network-based visualization is composed of nodes and links. While visualizing a power grid, nodes can represent a power generator or transformer, while links represent transmission lines. The nodes and links are weighted to reinforce the fundamental concept of a force-directed model. In a force-directed model, an attractive spring force is applied on the links to maintain the connectivity between nodes while a repulsive force is applied on nodes to preserve the global structure of a graph (Wong et al., 2009).

A network-based visualization with a force-directed model can be used to study various phenomena on a power grid. For example, if the repulsive force is linked to the voltage angle at each bus while the attractive force is linked to transmission line impedance, then the visualization shows an impedance representation, where system nodes are clustered into various groups. These groups are formed by electrical buses and generators with a strong synchronous coupling. The links between these strongly coupled groups show weak coupling (i.e., the lines where an unstable system is most likely to separate). The visualization evolves as the node and link weighting is changed. Nodes can be weighted by variables such as voltage magnitude, voltage angle, real power injection, reactive power injection, real load, and reactive load. Links can be weighted by variables such as resistance, reactance, impedance, real power flow, reactive power flow, real losses, and reactive losses (Wong et al., 2009). For example, in Fig. 9.8 the base voltages of the Western Electricity Coordinating Council's (WECC) buses are visualized using both a geographic layout and a force-directed network layout. On the geographical layout, the Northern California and Southern California portions of WECC are represented by small regions on the visualization. In reality, California represents a much larger share of the power system, which is accurately depicted in the network layout.

9.6.3 Virtual Grid

Virtual environments use computer graphics to simulate a three-dimensional reality. This three-dimensional reality is used as a base on top of which layers of informative variables are plotted for data interpretation. The three-dimensional rendering

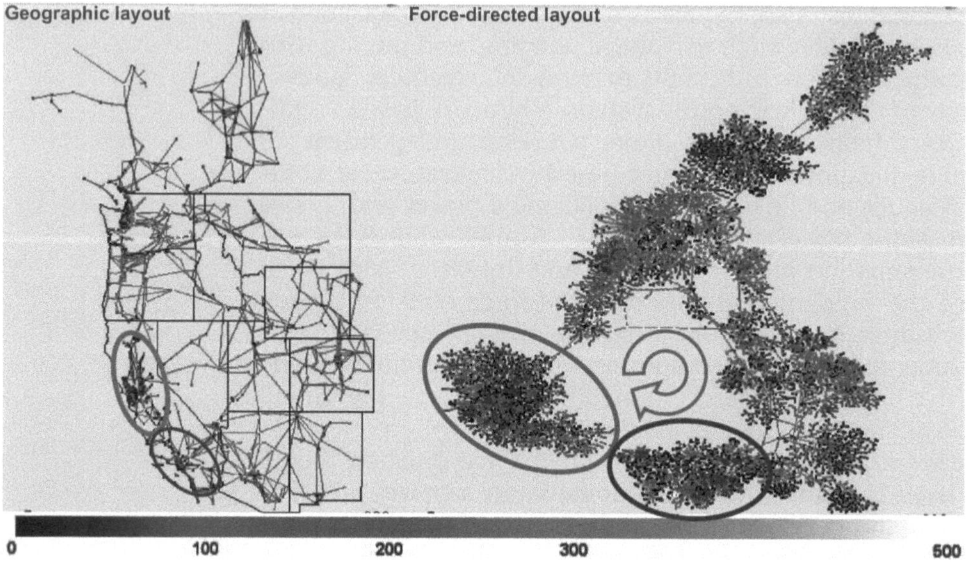

Figure 9.8 Geographical and force-directed network layout of a heat map visualization illustrating the base voltages of the WECC buses. The color in the heat map maps to the base voltage. The Northern and Southern California portion of the WECC are marked using ovals. The "ring path" of WECC is shown in purple. Reprinted with permission from IEEE.

is kept lightweight to allow for real-time updates, so the user can catch any new events quickly (Overbye et al, 1999). The user also can navigate the virtual environment and review the state variables at a specific location and time. To simulate this environment, researchers have designed a virtual grid using common one-line representations, such as the IEEE 118-bus test system. A one-line representation, which is a two-dimensional line plot, forms one plane of the virtual environment. The remaining third dimension is often used to plot the state variables via three-dimensional bars. In addition to bars, variables can be represented on the one-line representation plane itself using thickness, line color, contour lines, and pie charts. One of the advantages of virtual grid is that it allows one to visualize multiple variables along the grid. For example, in a single view, stacked bars can show current reactive power output and reactive power reserves, and color contours can show bus voltage (Overbye & Weber, 2000). While the operator is navigating the virtual environments, it is crucial that the operator can interpret the data quickly and notify authorities about any events. Data interpretation can be much easier if the visualizations show discrete bar lengths and color coding rather than continuous values (Mikkelsen et al., 2012). Bars and locations in virtual grid visualization can be linked easily with an alarm or event list. If an operator notices an alarm, he or she can open a table listing the incident that led to the alarm (Mikkelsen et al., 2011).

9.7 Big Data Infrastructure

Making the power grid smart depends on the ability to understand the unprecedented influx of sensing data to draw insights into the system's behavior and automate the available controls. This is a "Big Data Challenge," requiring advanced informatics techniques and cyberinfrastructure (Simmhan et al., 2013a). Data are acquired from thousands of equipment sensors and smart meters, including instantaneous data that is sampled and streamed periodically, in addition to historical data aggregated over several years (Simmhan et al., 2013a). Unique processing is required in order to discern intelligent decisions in the smart grid from Big Data. Two such platforms for processing Big Data are the Apache Hadoop parallel and distributed computing platform (Hadoop) and cloud computing.

Hadoop is an open-source software for reliable, scalable distributed computing. It is a framework for distributed processing of large data sets across clusters of computers (The Apache Software Foundation, 2015). The University of Southern California's Los Angeles Smart Grid Project requires data-driven models that are trained using historical data on micro-grid behaviors, utilizing a large dimension of features that are direct and indirect indicators of energy demand or reduction. The model training is computationally costly and data intensive. The OpenPlanet regression tree learning application uses the Hadoop MapReduce platform for performing analytics; it is optimized to scale on private cloud infrastructure (Simmhan et al., 2013a). In the Tennessee Valley Authority's smart grid, active PMUs send global positioning system stamped data 30 times per second. In order to reliably store this data and make the data available and reliable at all times, the Tennessee Valley Authority also utilized Hadoop (Bisciglia, 2009).

9.8 Future Trends

As the century progresses, smart grids will continue to grow and will need to cope with larger-scale integration of distributed energy resources (such as renewables), cyber-physical components, and devices. With this growth, the volumes of data collected by smart grids will also continuously increase. As a result, new architectures, concepts, algorithms, and procedures will be necessary (Strasser et al., 2015). As more renewable energy sources are added to the smart grid, the intermittent behavior of these sources will provide additional challenges in the planning and operation of smart grids. Since many of these renewable energy sources are related to weather patterns (wind, solar), meteorologists will have to work with energy forecasters to process large amounts of data. The volatility of renewable power generation is a major challenge to system operators and energy traders, and major advances need to be made in renewable power generation forecasting by meteorologists and forecasters (Hong, 2014).

On the demand side of the power grid, intelligent appliances, increased adoption of electric vehicles, and rooftop solar panels make the consumer load profile variable (Simmhan et al., 2013a). Consequently, new algorithms will have to take into account forecasting of energy consumption at electric vehicle charging stations, and variable loads from consumers when they become electricity generators due to their rooftop solar panels or their household battery storage. These evolving algorithms may need to leverage machine learning and time-series predictions, which can be executed in a cell phone environment (Strasser et al., 2015). The future smart grid will continue to adjust to these changes and become smarter, more efficient, and more adaptable.

References

Aghaei, J., Alizadeh, M.-I., 2013. Demand response in smart electricity grids equipped with renewable energy sources: a review. Renewable Sustainable Energy Rev. 64–72.

Amaral, S., Câmara, G., Monteiro, A.M., Quintanilha, J.A., Elvidge, C.D., 2005. Estimating population and energy consumption in Brazilian Amazonia using DMSP night-time satellite data. Comput. Environ. Urban. Syst. 179–195.

Arnold, G., 2011. Challenges and opporutnities in smart grid: a position article. Proc. IEEE. 99 (6), 922–927.

Berthier, R., Sanders, W.H., Khurana, H., 2010. Intrusion detection for advanced metering infrastructures: requirements and architectural directions. IEEE International Conference on Smart Grid Communications. IEEE, Gaithersburg, MD, pp. 350–355.

Bisciglia, C., 2009. *The smart grid: hadoop at the tennessee valley authority (TVA)*. Retrieved 1 12, 2016, from <http://blog.cloudera.com/blog/2009/06/smart-grid-hadoop-tennessee-valley-authority-tva/>.

Budka, K.C., Deshpande, J.G., Doumi, T.L., Madden, M., Mew, T., 2010. Communication network architecture and design principles for smart grids. Bell Labs Tech. J. 15 (2), 205–227.

Chan, S., Tsui, K., Wu, H., Hou, Y., Wu, Y., Wu, F., 2012. Load/price forecasting and management of demand response for smart grids: methodologies and challenges. IEEE Signal Processing Magazine, Special Issue on Signal Processing Techniques for Smart Grid. 68–85.

Diamantoulakis, P.D., Kapinas, V.M., Karagiannidis, G.K., 2015. Big data analytics for dynamic energy management in smart grids. Big Data Res. 2 (3), 94–101.

Drummond, C., Sutanto, D., 2010. Classification of power quality disturbances using the iterative Hilbert Huang transform. International Conference on Harmonics and Quality of Power (ICHQP). IEEE, Bergamo, pp. 1–7.

Ekici, S., 2009. Classification of power system disturbances using support vector machines. Expert Systems with Applications. 36 (6), 9859–9868.

El-Arroudi, K., Joos, G., 2007. Data mining approach to threshold settings of islanding relays in distributed generation. IEEE Trans. Power Syst. 22 (3), 1112–1119.

Federal Energy Regulatory Commission, 2009. The Natiuonal Assessment of Demand Response Potential. Department of Energy, Washington, DC.

Franco, G., Sanstad, A.H., 2008. Climate change and electricity demand in California. Clim. Change. 87 (1), 139–151.

Galloway, B., Hancke, G.P., 2012. Introduction to industrial control networks. IEEE Communications Surveys & Tutorials. 860–880.

Harlow, J.H., 2012. Electric Power Transformer Engineering. *third ed.* CRC Press Taylor & Francis Group, Boca Raton, FL.

He, X., Ai, Q., Ni, J., Piao, L., Xu, Y., Xu, X., 2015. 3D power-map for smart grids—an integration of high-dimensional analysis and visualization. arXiv preprint arXiv, 1503.00463.

Hedman, K.W., Ferris, M.C., O'Neill, R.P., Fisher, E.B., 2010. Co-optimization of generation unit commitment and transmission switching with N-1 reliability. IEEE Trans. Power Syst. 25 (2), 1052–1063.

Hong, T., 2014. Energy forecasting: past, present, and future. Foresight: Int. J. Appl. Forecasting. 43–48.

Hor, C.-L., Watson, S.J., Majithia, S., 2005. Analyzing the impact of weather variables on monthly electricity demand. IEEE Trans. Power Syst. 20 (4), 2078–2085.

IBM, 2012. Managing Big Data for Smart Grids and Smart Meters. IBM, Somers, NY.

Khamis, A., Shareef, H., Bizkevelci, E., Khatib, T., 2013. A review of islanding detection techniques for renewable distributed generation systems. Renewable Sustainable Energy Rev. 28, 483–493.

Khokhar, S., Zin, A.A., Mokhtar, A.S., Pesaran, M., 2015. A comprehensive overview on signal processing and artificial intelligence techniques applications in classification of power quality disturbances. Renewable Sustainable Energy Rev. 51, 1650–1663.

Kialashaki, A., Reisel, J.R., 2013. Modeling of the energy demand of the residential sector in the United States using regression models and artificial neural networks. Appl. Energy. 108, 271–280.

Mahela, O.P., Shaik, A.G., Gupta, N., 2015. A critical review of detection and classification of power quality events. Renewable Sustainable Energy Rev. 41, 495–505.

Mikkelsen, C., Johansson, J., Cooper, M., 2012. Visualization of power system data on situation overview displays. Sixteenth International Conference on Information Visualisation (IV). IEEE, Montpellier, pp. 188–197.

Mikkelsen, C., Johansson, J., Rissanen, M., 2011. Interactive information visualization for sensemaking in power grid supervisory systems. Fifteenth

International Conference on Information Visualisation (IV). IEEE, London, pp. 119–126.
Mirowski, P., Chen, S., Ho, T.K., Yu, C.-N., 2014. Demand forecasting in smart grids. Bell Labs Tech. J. 135–158.
Mitchell, R., Chen, I.-R., 2014. A survey of intrusion detection techniques for cyber-physical systems. ACM Computing Surveys. 55.
Mittal, D., Mahela, O.P., Jain, R., 2012. Classification of power quality disturbances in electric power system: a review. IOSR J. Electr. Electron. Eng. 3 (5), 6–14.
Monedero, I., Leon, C., Ropero, J., Garcia, A., Elena, J.M., Montano, J.C., 2007. Classification of electrical disturbances in real time using neural network. IEEE Trans. Power Deliv. 22 (3), 1288–1296.
Moravej, Z., Abdoos, A.A., Pazoki, M., 2009. Detection and classification of power quality disturbances using wavelet transform and support vector machines. Electric Power Components and Systems. 182–196.
Muratori, M., Roberts, M.C., Sioshansi, R., Marano, V., Rizzoni, G., 2013. A highly resolved modeling technique to simulate residential power demand. Appl. Energy. 465–473.
Nga, D.V., See, O.H., Xuen, C.Y., Chee, L.L., 2012. Smart grid and renewable energy. Visualization Techniques in Smart Grid. 3, 175–185.
Overbye, T.J., Weber, J.D., 2000. Visualization of power system data. Thirty three Annual Hawaii International Conference on System Sciences. IEEE, Hawaii, p. 7.
Overbye, T.J., Klump, R.P., Weber, J.D., 1999. A virtual environment for interactive visualization of power system economic and security information. IEEE Power Engineering Society Summer Meeting. IEEE, Edmonton, Alta, pp. 846–851.
Puttre, M., 2015, January. Analytics and Big Data Are Changing The Energy Market Map. *Solar Industry.*
Raza, S., Mokhlis, H., Arof, H., Laghari, J., Wang, L., 2015. Application of signal processing techniques for islanding detection of distributed generation in distribution network: a review. Energy Convers. Manage. 96, 613–624.
Shandilya, S., Shandilya, S., Thakur, T., Nagar, A.K., 2016. Handbook of Research on Emerging Technologies for Electrical Power Planning, Analysis, and Optimization. Engineering scicne Reference, An Imprint of IGI Global, Liverpool, UK.
Shi, K., Yu, B., Huang, Y., Hu, Y., Yin, B., Chen, Z., et al., 2014. Evaluating the ability of NPP-VIIRS nighttime light data to estimate the gross domestic product and the electric power consumption of China at multiple scales: a comparison with DMSP-OLS data. Remote Sensing. 6 (2), 1705–1724.
Simmhan, Y., Aman, S., Kumbhare, A., Liu, R., Stevens, S., Zhou, Q., et al., 2013a. Cloud-based software platform for big data analytics in smart grids. Comput. Sci. Eng. 15 (4), 38–47.
Simmhan, Y., Aman, S., Kumbhare, A., Liu, R., Stevens, S., Zhou, Q., et al., 2013b. Cloud-based software platform for data-driven smart grid management. IEEE/AIP Comput. Sci. Eng. 1–6.
Soliman, S. A.-h, Al-Kandari, A.M., 2010. Electrical Load Forecasting: Modeling and Model Construction. Elsevier, Burlinton, MA.
Sridhar, S., Hahn, A., Govindarasu, M., 2012. Cyber–physical system security for the electric power grid. Proc. IEEE. 210–224.
Strasser, T., Siano, P., Vyatkin, V., 2015. Guest editorial—new trends in intelligent energy systems-An industrial informatics point of view. IEEE Trans. Ind. Inf. 11 (1), 207–209.

Subnet Solutions Inc. n.d. *Dictionary: glossary of terms.* Retrieved 05 12, 2016, from Subnet Solutions, Inc Making Power Grides More Intelligent: <http://www.subnet.com/resources/dictionary/intelligent-electronic-device.aspx>.

The Apache Software Foundation, 2015. *Welcome to Apache Hadoop.* Retrieved 1 12, 2016, from <http://hadoop.apache.org/>.

Thomas, M.S., McDonald, J.D., 2015. Power System SCADA and Smart Grids. CRC Press, Boca Raton, FL.

Torrence, C., Compo, G.P., n.d. Retrieved 09 16, 2016, from A Practical Guide to Wavelet Analysis: <http://atoc.colorado.edu/research/wavelets/>.

US-Canada Power System Outage Task Force, 2004. Final Report on the August 14, 2003 Blackout in the United STates and Canada: Causes and Recommendations. US-Canada Power System Outage Task Force.

US Department of Energy, n.d. *What is the smart grid?* Retrieved 3 21, 2016, from smartgrid.gov: <https://www.smartgrid.gov/the_smart_grid/smart_grid.html>.

US Energy Information Administration. (2016, 08 26). eia. Retrieved 09 01, 2016, from Consumption and Efficiency: http://www.eia.gov/consumption/.

Wells, J., 2004. Electricity Markets-Consumers Could Benefit From Demand Programs, But Challenges Remain. Government Accountability Office, Washington, DC.

Willis, H., 2002. Spatial Electric Load Forecasting. Marcel Dekker, New York.

Wong, P.C., Schneider, K., Mackey, P., Foote, H., Chin Jr, G., Guttromson, R., et al., 2009. A novel visualization technique for electric power grid analytics. IEEE. Trans. Vis. Comput. Graph. 15 (3), 410–423.

Yan, J., Zhu, Y., He, H., Sun, Y.L., 2013. Multi-contingency cascading analysis of smart grid based on self-organizing map. IEEE Trans. Information Forensics Security. 646–656.

Yin, J., Kulkarni, A., Purohit, S., Gorton, I., Akyol, B., 2011. Scalable real time data management for smart grid. Proceedings of the Middleware 2011 Industry Track Workshop. ACM, Lisbon, Portugal, pp. 1–6.

Zhang, Y., Wang, L., Sun, W., 2013. Trust system design optimization in smart grid network infrastructure. IEEE Trans. Smart Grid. 4 (1), 184–195.

Zhang, Y., Wang, L., Sun, W., Green II, R.C., Alam, M., 2011. Distributed intrusion detection system in a multi-layer network architecture of smart grids. IEEE Trans. Smart Grid. 796–808.

Zuech, R., Khoshgoftaar, T.M., Wald, R., 2015. Intrusion detection and big heterogeneous data: a survey. J. Big Data. 1–41.

10

PROGNOSTICS FOR THE POWER INDUSTRY

Daniel Shai[1], Cathy Chen[1], Matthew Pooley[1] and Brian W. D'Andrade[2]

[1]Exponent Inc., New York, NY, United States [2]Exponent Inc., Washington, DC, United States

CHAPTER OUTLINE

10.1 Introduction

On Thursday, August 14, 2003, just after 4:10 p.m. EDT, the Northeast Blackout (see Fig. 10.1) collapsed the grid in just over 9 seconds. This blackout affected approximately 50 million people and 61,800 megawatts (MW) of electric load in the United States and Canada (US–Canada Power System Outage Task Force, 2004). The noticeably darkened areas in the American northeast, shown in Fig. 10.1 taken from the United States National Oceanic and Atmospheric Administration (NOAA)

The Power Grid. DOI: http://dx.doi.org/10.1016/B978-0-12-805321-8.00010-0

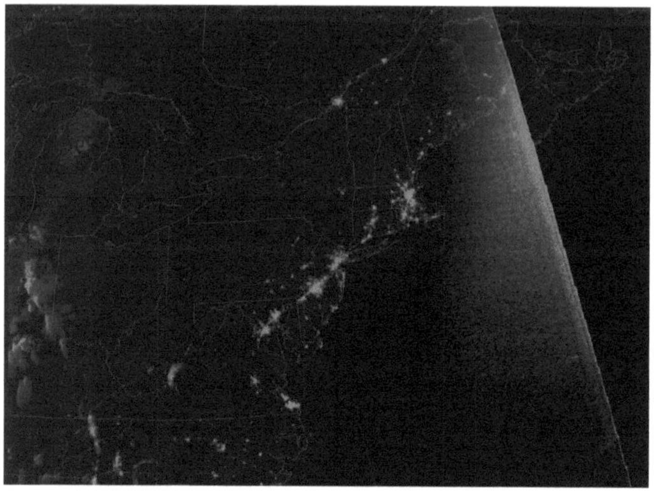

Figure 10.1 Composite image of the 2003 Northeast blackout. NOAA/DMSP.

of the American northeast, illustrate the magnitude of the Northeast Blackout of 2003.

Following the blackout, there was a rekindled public interest in the American power infrastructure that resulted in increased media attention of the United States electrical system. At that time, the United States electrical system was called "antiquated" and likened to that of a third-world nation (LaCommare and Eto, 2004). The 2003 Northeast Blackout and subsequent media coverage has in part catalyzed discussions about the aging power grid infrastructure and its modernization (LaCommare and Eto, 2004).

10.1.1 Aging Electrical Infrastructure

Infrastructure is the collective system of public works of a country, state, or region, including civil works, telecommunication systems, water delivery, and electrical power delivery. Electrical infrastructure typically comprises poles, ducts, underground vaults, substations, transformers, circuit breakers, voltage regulators, monitoring and control systems, service transformers, meters, transmission towers, and miles of electrical conductors, all of which work together to supply users with reliable power (Willis and Schrieber, 2013).

In general, power delivery relies on an interconnected system that consists of three stages: generation, transmission, and distribution. Generation facilities create power and consist of

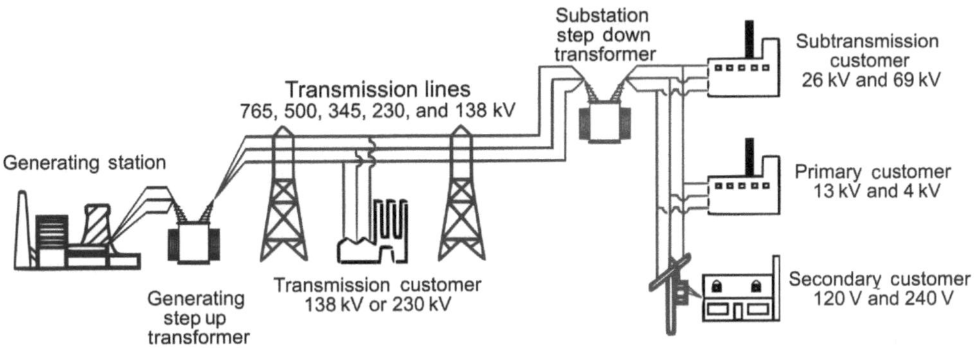

Figure 10.2 Illustration of three main parts of power infrastructure.

approximately 5800 major power plants and smaller generation facilities in the United States. High-voltage transmission lines, a network of over 450,000 miles of electrical conduit, connect generation facilities with major population centers. Local distribution networks then bring the electric power into homes and businesses via overhead lines or underground cables (ASCE, 2011). A schematic diagram illustrating key electrical infrastructure components is presented in Fig. 10.2.

The power grid in America is slowly aging. The American Society of Civil Engineers (ASCE) issues a report card every four years for the state of different parts of the United States infrastructure. An "A" is considered a rating of exceptional, while an "F" is failing. In its most recent report in 2013 (ASCE, 2013), the ASCE gave the American energy infrastructure a "D +," stating, in part, that

> *"America relies on an aging electrical grid and pipeline distribution system, some of which originated in the 1880s. Investment in power transmission has increased since 2005, but ongoing permitting issues, weather events, and limited maintenance have contributed to an increasing number of failures and power interruptions."*
>
> **(ASCE, 2013, p.8).**

Centralized electric generating plants with local distribution networks were started in the 1880s, and the grid of interconnected transmission lines was started in the 1920s. Many of the transmission and distribution systems in existence today consist of equipment and facilities that date from the economic boom of the 1940s, following World War II, and the subsequent growth period of the 1950s and 1960s. Much of the equipment installed during those periods are still in service today (Willis and Schrieber, 2013). For example, 51% of the generating capacity of the United States is in plants that were at least 30 years old at the end of 2010, while 73% of all coal-fired capacity is 30 years or older. Seventy percent of transmission lines and power transformers are 25 years or older and 70% of circuit breakers are more than 30 years old (ASCE, 2011). As equipment get older, the wear and tear of constant use, environmental conditions, disasters, and accidents enable failures.

This combination of mismatched equipment leads to an operating problem for electrical utilities. The largest portion of their infrastructure, despite spending decades in the field, is still functional enough to provide satisfactory service for perhaps a decade or more to come (Willis and Schrieber, 2013). A small portion of the system however is badly deteriorated but not to the extent that it shows visible signs of imminent failure (Willis and Schrieber, 2013). One means to possibly find these deteriorated systems may be through the use of prognostics.

10.1.2 Outages and Safety Hazards

Aging equipment has led to an increasing number of intermittent power disruptions due in part to failures that are not predicted, and in part due to the complex procedure for replacing energy sources (ASCE, 2013). In the United States, the number of significant power outages increased from 76 in 2007 to 307 in 2011 (ASCE, 2013). Some reliability issues are also beginning to emerge due to the multifaceted procedure of exchanging in new energy sources and decommissioning older infrastructure (ASCE, 2013).

Leading risk management firms agree that a blackout lasting more than several days would have cascading problems and could be considered a "worst case" scenario. Most critical systems, including hospitals, first responders, water, and sewage systems only have backup power generation that lasts anywhere between several hours to a few days. The 2003 blackout illustrated how after 3–6 hours, fuel stations without power had to shut down due to pumps being inoperable, thus leaving the public and more importantly critical systems like first responders without fuel for vehicles and backup generators.

After an extended time without electricity, water, and transportation, emergency services and critical manufacturing entities will shut down. Without water and functioning emergency services, a cascade effect could occur, leading to more danger and hazards. An unpowered house in winter will cool to freezing temperatures in a day or two, possibly leading to fires being set and fire hazards in locations with a handicapped emergency response structure (Bruch et al., 2011). Scenarios such as these underscore the importance of preventative strategies in the power industry.

10.1.3 Maintenance Strategies

In order to prolong the life of the power infrastructure, and minimize the time to failure, a number of maintenance strategies are common practice. Current maintenance strategies loosely fall into two categories: strategies where a component is replaced by a new or "good as new" (refurbished) part, or strategies where a component is maintained by a less costly activity that results in limited improvement of its condition. Within these two strategies, there are differences in implementation timing because some utilities choose to perform maintenance at fixed intervals, and others perform maintenance out as needed (Endrenyi et al., 2001).

In a survey of maintenance practices in varying countries, Table 10.1 was generated regarding common practices by utilities in Austria, Canada, Germany, Italy, Saudi Arabia, and the United States.

There are new strategies for predicting failures and most effective maintenance, several of which will be discussed in this chapter. These strategies generally include comparing models and statistics to determine where the most cost-effective for sustaining equipment reliability option lies.

10.1.4 Cost of Outages

The beginning of the 21st century has seen two of the costliest power blackouts in human history, the 2003 North American blackout and the 2011 Japan blackout (Bruch et al., 2011). The 2003 blackout in North America cost the United States between

Table 10.1 Reported Frequent Maintenance Intervals (Endrenyi et al., 2001)

	A: Generators		B: Transformers		C: Breakers	
	Interval	Duration	Interval	Duration	Interval	Duration
Minor maintenance	1 year	1–2 week	1 year	1 day	1 year	1 day
Minor overhaul	5 year	4–5 week	5 year	3 day	5 year	3 days
Major overhaul	8–10 year	6–8 week	7 year	4–8 week	8–10 year	2 week

$4.5 billion dollars and $8.2 billion dollars, with only a fraction of those losses for the utilities. A breakdown of these losses including private and government sectors includes (ELCON, 2004):

- $4.2 billion in lost income to workers and investors
- $380–940 million in costs associated with lost or spoiled commodities
- $15–100 million in extra costs to government agencies
- $1–2 billion in costs to affected utilities

Specific examples of private sector losses include the Ford Motor Company, which reported that 23 of its 44 plants in North America were shutdown. At a plant in Ohio, the blackout caused molten metal to cool, requiring a week to clean and rebuild the furnace. Furthermore, Duane Reade Inc., the largest drug store chain in the New York City metropolitan area reported that the blackout forced the shutdown of all 237 stores at the time, resulting in an approximate loss in revenue of $3.3 million (ELCON, 2004).

Costs do not only arise when there is a large-scale blackout, smaller outages during a year, known as annual power outages, are quickly adding up. The number of annual power outages that hit the American power grid was 307 in 2011 (ASCE, 2013). One estimate for the annual costs of outages for the US electrical grid is $79 billion, based on the base-case estimate of customer losses put forth by the US Department of Energy. Fig. 10.3 illustrates how the costs are split.

Given the risks previously outlined, this chapter explores large-scale grid health monitoring and asset management plans for new infrastructure in the smart grid. It reviews the

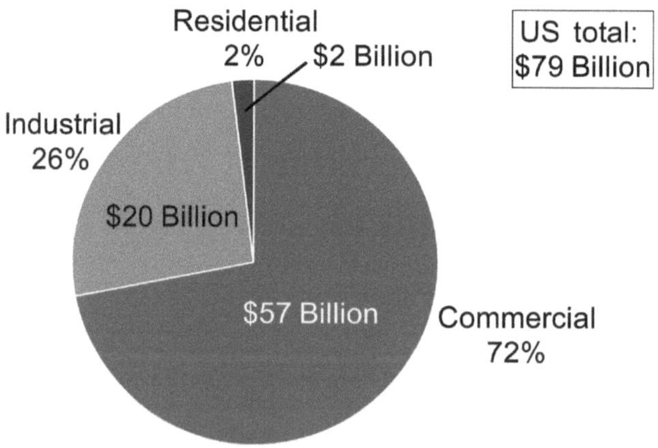

Figure 10.3 Costs of power outages annually (LaCommare and Eto, 2004).

optimization and efficient use of resources in maintaining the current power grid, the prediction and prevention of failures in the power industry, presents statistical reliability, and prognostics as a solution to current system failure predictors.

10.2 Statistical Reliability and Prognostics

In this section, existing statistical reliability methodology is discussed and the concept of prognostics is introduced. This section concentrates on reliability modeling applied to individual network locations; full network reliability monitoring is discussed later in Section 10.4.

As is presented in this chapter, statistical reliability methodology involves the use of data collected from past failure events, as well as data on the condition of components obtained via maintenance inspections, to model the likelihood of future failures. The resulting reliability models created from this methodology are then used to plan future maintenance activities to minimize the probability of significant power disruptions using a given amount of maintenance resources.

In contrast, prognostic approaches involve actively monitoring the condition of components in real time, or at least on time scales significantly shorter than the mean-time-to-failure, to dynamically concentrate maintenance activities on components that show signs of worsening condition. When using prognostics it is possible, in principle, to reduce maintenance activity while ensuring that components are not allowed to degrade to a state from which predictable failures can occur.

10.2.1 Existing Statistical Reliability Techniques

Statistical reliability techniques are widely used in existing power distribution networks as metrics for forecasting future failures based on past experience. For example, reliability data, such as mean-time-to-failure, for network components can be extracted from the recorded lifetimes and failures of all similar components. In addition, inspections performed during routine maintenance, or in response to particular events such as harsh weather, can be used to derive empirical estimates for the degradation in condition that can be expected for given categories of components due to the passage of time or the occurrence of environmental factors such as high winds or extreme temperatures. Models developed from past reliability data can be used to estimate the probability that a component, in a given condition, fails in a given time interval. The estimated effects of weather events or other sporadic detrimental incidents, such as power overloads caused by adjacent network failures, on the condition of components can be used to amend reliability models and make decisions regarding where to allocate maintenance resources after such unexpected events.

10.2.1.1 Failure Events

In order to quantify past failure events and construct reliability models, it is necessary to define various categories of failure events, along with appropriate metrics to assess the severity of such events. Precise definitions of failure modes are specific to particular categories of components. However, for the purposes of discussing statistical reliability modeling, it is sufficient to categorize failures into general categories (Dodson et al., 2009).

10.2.1.1.1 Catastrophic Failure

Catastrophic failures describe the sudden and total failure of a given component. Relevant examples include breaks in power distribution lines that cause open circuits, burnout of insulating dielectric material due to electrical arc discharge caused by voltage surges, and faults to ground that melt/permanently damage conducting cables.

Some catastrophic failures are unpredictable; one major source of catastrophic failure events is animal-related disruption. For example, large animals can damage guy-wires that hold up electric poles and cause failures, or smaller animals such as birds or squirrels can cause short circuits between phase conductors and each other or ground. Vehicular accidents or construction work that accidentally severs electrical cables are other major causes of unpredictable catastrophic failure. While the probability of such unpredictable events cannot be precisely calculated for all categories of components, it is possible to assign some qualitative adjustments to reliability models for specific components that are at risk from unpredictable events, such as exposed cables that are in close proximity to animal herds. It is often possible to mitigate these risks by taking preventative action, such as fitting appropriate guards around exposed cables.

Many catastrophic failures are preceded by a change in condition that can be assessed during maintenance inspections. For example, moderate electrostatic discharge damage can be visible during inspection and can indicate that insulating layers have degraded such that they are no longer sufficient at peak loads. The goal of reliability modeling is to estimate the probability of catastrophic failure given the last confirmed condition of a component and the duration since the condition was last verified.

10.2.1.1.2 Degradation Failure

Degradation failures are characterized by a gradual decrease in the functionality of a component that is irreversible without maintenance intervention. Examples include: reduction in stored charge density in batteries due to repeated charge cycling; degradation of dielectric layers that reduce the efficiency of transformers; and degradation of breaker fuses such that the maximum current is reduced. This mode of failure can develop into a catastrophic failure if action is not taken to mitigate the performance reduction before a threshold of degradation is reached. Reliability modeling can be used to estimate the rate at which various modes of degradation occur and to include the probability of associated failures in the corresponding failure predictions.

10.2.1.1.3 Intermittent Failure

Intermittent failures result in the sudden loss of performance, or a sudden drop in performance such that it is below the minimum acceptable level, followed by a sudden spontaneous restoration of normal performance without maintenance intervention. A common form of intermittent failure in individual components are short circuits caused by debris that temporarily create a ground fault until the debris is burnt away by the resulting fault current. Intermittent failures can also be caused by contamination of component housings by ionic particles that can form conducting filaments that are destroyed by the current they induce. As with degradation failures, reliability modeling can be used to account for the probability of intermittent failures and to make informed decisions regarding the allocation of maintenance resources.

10.2.1.1.4 Drift Failure

Drift failure modes allow a component to continue to function, but cause a gradual drift of the component operational output that can lead to values that fall outside of normal operational ranges. Unlike degradation failures, drift failures are not irreversible and normal operation may be restored remotely, e.g., by power cycling the component or temporally reducing operational load. A relevant example includes progressive ionization of insulating material between two conductors. Over time an ionized channel can be induced in an insulating layer; this is more likely to occur in older materials that have already undergone some degree of degradation (e.g., moisture ingression) and can reduce the effective thickness of the insulating material leading to performance variations in the component. When the component is deactivated the ionization dissipates and normal operation is restored upon reactivation. Reliability modeling can be used to predict the severity and expected frequency of drift failures and prioritize maintenance efforts to permanently remedy such failure modes.

10.2.1.2 Reliability Modeling

Reliability modeling for particular components allows the failure probability for a particular category of component to be estimated as a function of operational time. This section outlines the general principles of such reliability modeling. For a discussion of how the reliability of components can be incorporated into reliability models of networks please see Section 10.4.

10.2.1.3 Reliability as a Function of Time

The probability of failure for an individual component varies over time. Perhaps surprisingly, there are two time intervals where failure is most likely: immediately after the component is installed, and after the component has served its operational lifetime. This results in a characteristic "bathtub" curve for the failure probability as a function of time (Brown, 2008), as shown in Fig. 10.4.

The high probability of failure immediately after installation represents the potential for manufacturing errors, damage during transportation, and incorrect installation. The region of high failure probability soon after installation is known as the "break-in" period. After this break-in period, there is a region of consistently low failure probability known as the "operational lifetime" of the component. Following the operational lifetime, the probability of failure increases exponentially with time, representing the fact that after this time expected degradation caused by normal operation is likely to have degraded the condition of the component sufficiently to make catastrophic failure increasingly probable.

In practice, a component need not always be replaced once it reaches its operational lifetime. Often it is sufficient to perform maintenance servicing to recondition the component for continued operation. For example, perishable parts can be replaced without replacing the entire unit. The diagram in Fig. 10.5 shows the expected failure probability as a function of

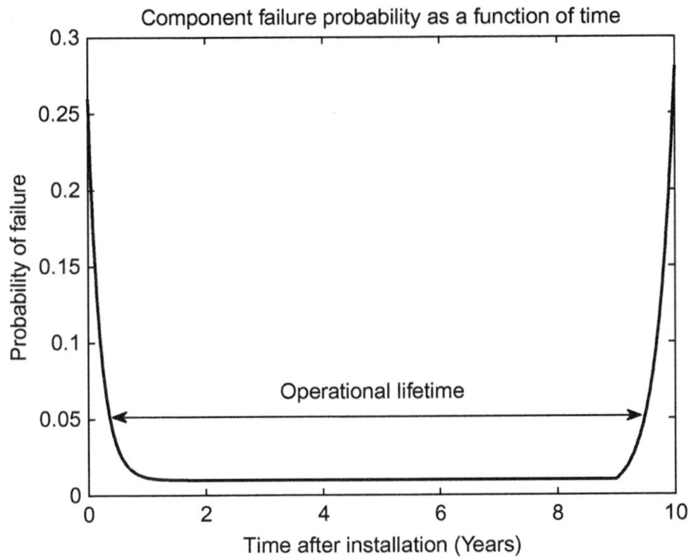

Figure 10.4 Diagrammatic representation of a typical variation in failure probability as a function of time for a component after installation.

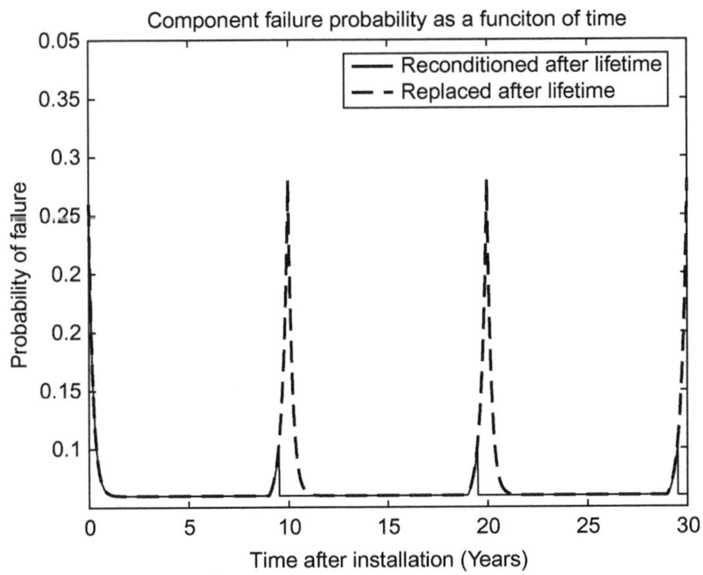

Figure 10.5 Diagrammatic representation of the failure probability as a function of time for a component that undergoes periodic reconditioning (solid line), compared with the probability of failure if the component is entirely replaced during the periodic maintenance (dashed line).

time for a typical component that undergoes periodic maintenance reconditioning. Shortly after the end of the operational lifetime, when the failure probability has begun to rise, maintenance servicing returns the component to a reliable operating condition and reduces the probability of failure back to the typical normal operational level. Because of the large probability of failure immediately after a new component is installed, performing this kind of reconditioning maintenance often minimizes the total probability of failure compared with replacing the component entirely, as shown in the figure.

Reliability modeling to calculate the probability of failure over time for a given component is based on empirical failure data, such as the expected mean-time-to-failure, collected from similar components. Such models are often used to decide on the interval between maintenance servicing. These models can be adjusted for the environmental and operational conditions in which the component will be functioning. For example, an external component that is subjected to large temperature variations may have a significantly decreased operational lifetime compared to the same component housed in a temperature controlled environment. In addition, the effects of extreme weather events, such as storms, can be incorporated into reliability models by reducing the effective operational lifetime of components following such events. This kind of adjustment and fine-tuning of reliability models can be based on an understanding of the failure modes to which a particular component is susceptible, but they are often simply based on the subjective experience of the modeling engineer. An example of how degradation effects can be modeled is given below.

10.2.1.3.1 Modeling the Expected Lifetime of Power Transformers

One factor that limits the operational lifetime of a transformer is the time it takes for winding insulation to degrade. A typical metric by which this mode of degradation can be quantified is the time it takes for the mechanical strength of the winding insulation to degrade to 50% of its original level (Brown, 2008). The rate at which the winding insulation degrades is highly dependent upon temperature, such that the insulation lifetime (in hours), L, may be reasonably described by

$$L = 10^{(K_1/T)+K_2}$$

where T, is the temperature of the insulation in degrees Kelvin, and the material-dependent constants K_1 and K_2 have been experimentally determined and are well documented in standard guides (IEEE Std C57.91, 1996). The operating temperature of the insulation, T, is dependent upon the current load of the transformer and the ambient temperature in which the transformer operates. Thus, the operational lifetime of a transformer can be estimated based on its expected operating conditions. Fig. 10.6 shows the estimated operational lifetime of

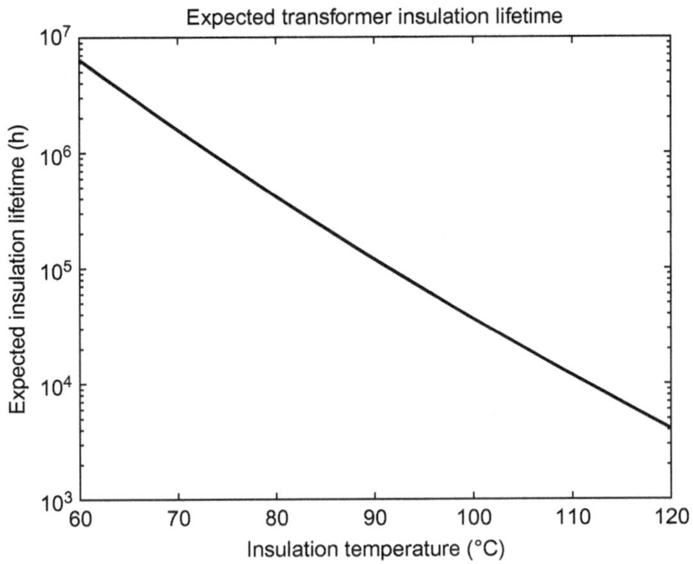

Figure 10.6 Estimated operational lifetime of typical transformer winding insulation as a function of operational temperature.

typical winding insulation within a transformer as a function of operating temperature.

This examples serves to demonstrate how an understanding of common failure modes for components, derived from empirical data extracted from collated failure data for past components, can be used to estimate the operational lifetime of components and construct reliability models. However, it also serves to highlight some limitations of this approach.

For example, the expected lifetime of the insulation at 60°C is over 700 years. Clearly it is unlikely that a transformer will actually operate normally for that length of time without intervention, even if it the insulation can be maintained at a constant low temperature. In reality, a transformer does not operate at a constant temperature because the current load through the transformer, and the ambient temperature, varies. Thus it is not sufficient to simply assume a constant temperature throughout the entire operational life of the transformer. Even relatively short durations of high temperatures can significantly degrade the insulations, and so reduce the lifetime substantially from an estimate based solely on the average operating conditions. Furthermore, even if the temperature could be maintained at 60°C, other failure mechanisms are likely to end the transformer's operational life prior to the 700-year insulation expected lifetime. In essence, statistical reliability predictions are only as accurate as the assumptions on which the underlying models are based. Prediction accuracy can be improved by incorporating more potential failure mechanisms and accounting for the expected variation in operation conditions throughout the component lifetime. However, maintenance inspections are still required to periodically verify that components are aging at the predicted rate.

10.2.2 Benefits of Prognostics

Reliability modeling is a powerful tool for improving maintenance efficiency and optimizing investment in component upgrades. However, reliability modeling is a reactive form of disruption mitigation: it is based on analysis of historical data and predicts the probability of failures based on this data, rather than actual knowledge of the real-time condition of components. Even when operational lifetimes are based on known failure mechanisms that develop according to well-understood processes, the actual condition of components cannot be verified between maintenance inspection and this can lead to unexpected failures when degradation occurs faster than expected.

Crucially, reliability models are by necessity based on aggregated data collected from a large sample of components with a range of different operational histories. Because of this, lifetime predictions often cannot adequately take into account a particular component's precise operating conditions. Also, adjustments to the probability of failure calculations following unexpected events, such as extreme weather, are not tailored to the actual change in condition of the particular components after they have endured said events. Thus, while reliability modeling is useful for predicting the expected number of failures of a given component in a given time interval, it is not ideal for identifying precisely in which individual components failures will occur.

Prognostic methodologies enable a more proactive approach to component maintenance. The general concept of prognostics is to monitor the condition of components in real-time for signs that degradation has occurred or that a failure event is likely. In principle, by monitoring the condition of components on a timescale far shorter than the time it takes for failures to develop, predictable failures can be drastically reduced by attending to degradation before the condition of components approaches a regime where failures are likely. In addition, by removing the requirement for routine maintenance inspections to determine the condition of components, maintenance costs are reduced while simultaneously improving the reliability of power systems.

10.3 Prognostics for Utility Infrastructure

In this section we will describe how maintenance and reliability programs have advanced beyond the traditional reliability model described above. In Section 10.3.1, a general overview is given describing more advanced maintenance activities that actively account for the condition of individual pieces of equipment. These additional considerations can be used to make a prognostic determination of a component's remaining life and ultimate failure mode. In Section 10.3.2, application of these new principles are discussed in the context of the power transformer that was considered previously in Section 10.2.

10.3.1 Beyond Traditional Reliability

A distinction is often made in discussions of component reliability between time-based maintenance (TBM) and condition-based maintenance (CBM). TBM, which is most closely aligned with the principles described above in Section 10.2, specifies that preventative maintenance should be performed according to a fixed schedule that is typically specified by a component manufacturer, by a standard, or is developed according to historical failure rates for specific use cases. The schedule is set so that maintenance is performed before failure rates increase near the end of equipment life (i.e., before the second upturn in the bathtub curve is reached). This approach, while simplest to implement, can often lead to financial and equipment losses through multiple channels. Effort can be wasted on equipment that does not necessarily require maintenance, while equipment with a shorter-than-average operational lifetime can fail before maintenance is performed. Furthermore, TBM often cannot account for complex, larger problems that may form as the result of multiple component failures.

One of the basic principles of using TBM is the assumption that component failure rates increase over time. However, this is not necessarily the case. For example, in the 1960s United Airlines conducted a study of the time-dependent failure rates for various nonstructural avionic components (Nowlan and Heap, 1978; Smith and Hinchcliffe, 2004). It was found that while some components did in fact obey the so-called bathtub curve described above, these cases constituted a small percentage of the overall conclusion of the study.

In Fig. 10.7, six different types of reliability curves from this study are schematically illustrated, and their individual

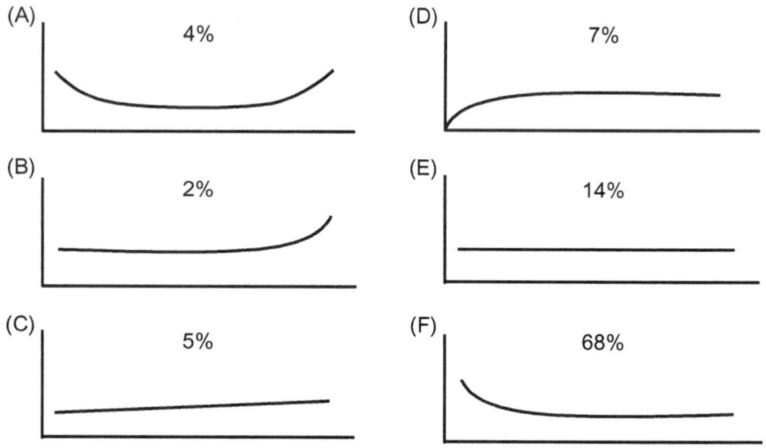

Figure 10.7 Schematic drawings of six different types of reliability curves based on a survey of aircraft equipment. Each depicts failure probability as a function of time, and their individual percentage of occurrence is indicated (Nowlan and Heap, 1978).

percentage of occurrence is indicated. The traditional bathtub curve (A) accounts for only 4% of components studied and only 11% (A–C) have failure rates that increase with time after an initial breaking in period. For the remaining 89% of components having failure rates that are either constant or that decrease with time beyond the early stages of their lifecycle, TBM does not provide an optimal maintenance strategy and therefore represents potentially wasted effort.

These shortcomings of TBM have led to the implementation of more targeted maintenance strategies such as CBM. CBM is based on measurement indicators of an individual piece of equipment's need for maintenance (Vachtsevanos et al., 2006). While more complex to plan and implement, CBM has several key advantages over TBM. First, there is less unnecessary maintenance because equipment that is operating normally will not indicate a condition that requires maintenance. Furthermore, because CBM involves a more active role in monitoring equipment health, developing problems often can be identified before they grow to become catastrophic failures. This approach can lead to significant cost savings and higher system reliability that defray the additional cost and effort for implementing such a system.

A generalized failure progression timeline is illustrated in Fig. 10.8. After a new piece of equipment is installed, it typically operates for some time before outwardly exhibiting indications of pending failure and then ultimately reaching a state of system failure. Diagnostic testing such as the activities performed with CBM would indicate the imminent failure and the problem would be addressed.

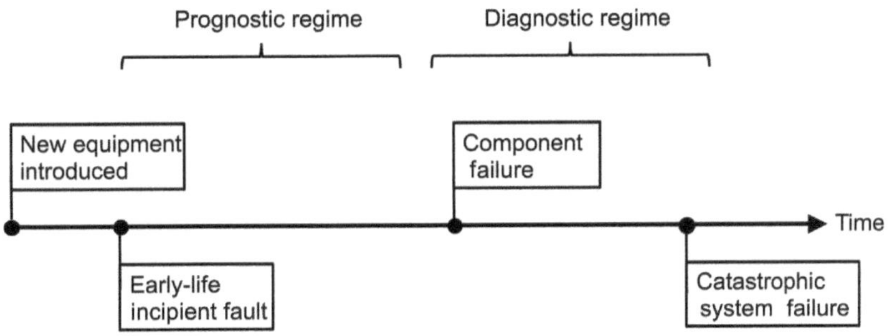

Figure 10.8 Failure progression timeline indicating difference in prognostic and diagnostic regimes (Vachtsevanos et al., 2006).

CMB thus enables corrective action to be taken in the diagnostic regime (see Fig. 10.8) by detecting component failures that occur before a catastrophic system failure. However, component failure is often a manifestation of an incipient fault that occurs much earlier in the component lifecycle, and a prognostic approach enables detection of such early-life fault events to be combined with predictive modeling to estimate the remaining useful life (RUL) of a system. In addition, monitoring in the prognostic regime grants insight into the most likely ultimate failure modes for equipment and may enable further targeting of diagnostic regimes resources.

The concept of prognostics is derived from the medical industry, where conditions are monitored in order to predict a likely outcome of a disease or ailment, and prognostics has long ago made its way to engineering disciplines as a term indicating failure prediction. A prognostics approach has long been used in the aircraft industry (Vachtsevanos et al., 2006) and is becoming more widespread with electric utilities.

Prognostics may be based on previous experience with similar components, may rely on data-driven classification methods, or in some cases may incorporate physics-based failure models. Regardless of the specific method used, the goal of prognostics is to predict the RUL and ultimate failure mode of a component given early-life data. Because it is a predictive approach, prognostic-related maintenance activities are classified as predictive maintenance, rather than the preventative or reactive maintenance described thus far.

The spectrum of maintenance activities discussed in this section is summarized diagrammatically in Fig. 10.9. On the left, TBM represents the traditional reliability approach to equipment maintenance. CBM is an intermediate approach,

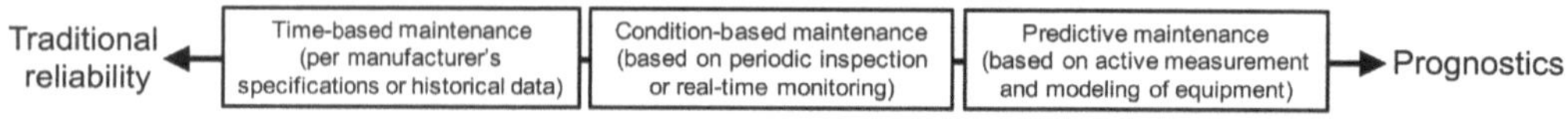

Figure 10.9 Spectrum of maintenance strategies ranging from traditional reliability strategies to prognostics.

employing equipment condition monitoring on a periodic schedule. Finally, on the far right of the spectrum, predictive maintenance approaches use both active measurement and modeling to detect and address developing problems before they result in system failure.

10.3.2 Prognostics of a Power Transformer

Continuing the example of the power transformer discussed in Section 10.2, it is worthwhile to discuss the differences between the traditional reliability based approach and the more targeted prognostics-based approach when applied to transformer maintenance. The behavior described in Section 10.2, known as Arrhenius behavior, is used to make a RUL prediction, but it has a number of shortcomings. First, as identified above, the RUL is calculated assuming the entire transformer operates in a uniform thermal environment with average temperature T_1. However, inhomogeneity in winding manufacturing may result in hotspots that significantly deviate from the average temperature and could produce a fault in a period much shorter than the RUL predicted by the Arrhenius equation. Furthermore, this approach assumes that temperature is the only ageing factor while ignoring other factors such as historical and future loading, humidity, oxidation, and corrosion.

More advanced approaches rely on testing methods to assess the current state of a transformer, and from these tests a more accurate prediction of the RUL can be made. A popular form of power transformer testing is dissolved gas analysis (DGA) (Hamrick, 2010). As a transformer ages, or when a fault occurs, gases are released from transformer components such as winding insulation and are dissolved into the transformer's oil reservoir. Periodic testing of the oil composition is performed using gas chromatography and the type and concentration of dissolved gas can be used to assess a transformer's condition and history. DGA is particularly useful for diagnosing electrical disturbances and thermal decomposition.

DGA has traditionally been conducted as a form of periodic test, where a sample is taken, often when the transformer is offline, and then analyzed off-site in a laboratory. More recently, advancements in sensor technology have made online, real-time analysis a possibility. With online testing, samples are acquired while the transformer is in service and are analyzed by a field-deployed gas chromatograph. These values can then be compared against standards such as IEEE C57.104 *IEEE Guide for the Interpretation of Gases Generated in Oil-Immersed Transformers* to assess whether dissolved gases are present at an abnormal level.

A common method to interpret DGA is to use Duval's triangle, illustrated in Fig. 10.10. Duval's triangle is a tri-axis plot with each axis representing the relative percentage of one of three hydrocarbon gases measured in a transformer oil sample.

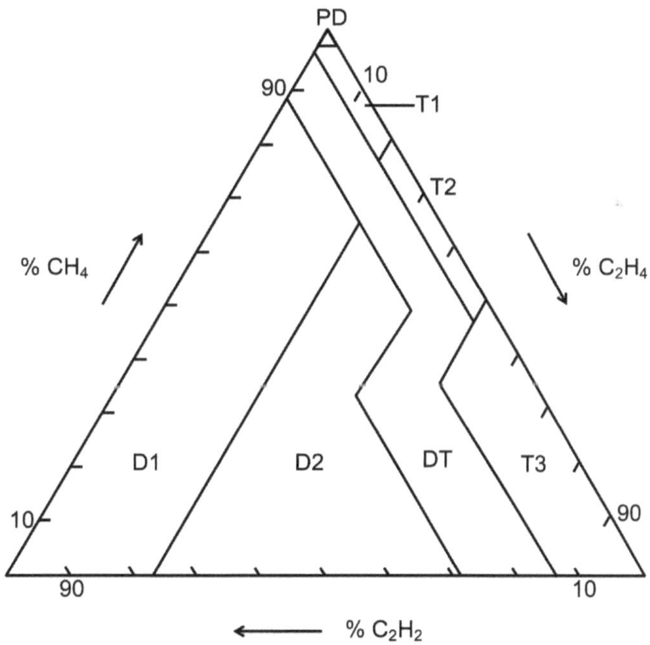

PD	Partial discharge
T1	Thermal fault < 300 C
T2	Thermal fault 300 – 700 C
T3	Thermal fault > 700 C
D1	Low energy discharge (sparking)
D2	High energy discharge (arcing)
DT	Mixture of thermal and electric faults

Figure 10.10 Duval's triangle for interpreting DGA data.

The relative concentrations coordinate a sector of the triangle, which indicates what type of fault is likely to have occurred. For example, the pinnacle of the triangle in Fig. 10.10 represents a partial discharge (PD) event, which has a signature of 100% methane (CH_4) production when compared to the relative concentrations of the other two gases (C_2H_2 and C_2H_4).

Often it is useful to determine how these quantities change over time. To facilitate this, transformer management systems are being developed (Dias et al., 2005) and in some cases have been integrated into commercial transformers such as those produced by Siemens AG (Siemens, 2010) and General Electric (General Electric, 2010), both of which advertise transformer management systems with DGA capabilities.

While methods such as Duval's triangle give an indication of an incipient fault, they occasionally fail, for example, when multiple incipient faults form simultaneously. Furthermore, they do not quantitatively predict the RUL. Rather, they only alert the asset owner of a forming problem.

Recent efforts have been made to develop intelligent monitoring systems to help characterize and diagnose incipient faults using DGA. Many of these methods use advances in machine learning classification techniques to make accurate predictions of the type of incipient fault and RUL left in a power transformer. In the recent work by Sica et al. (Sica et al., 2015), a cognitive system for transformer fault prognosis was developed and trained on over 150 historical records. This system was shown to accurately (within the calculated uncertainty) predict the failure time of two test cases. Systems such as these will likely go hand-in-hand with the integration of other smart devices involved in asset management.

The techniques described in this section represent only a small subset of maintenance and monitoring activities being implemented in the management of power transformers. Historically DGA has been only one method for power transformer condition assessment, while other activities include electrical testing of the insulation, windings, and core; vibration analysis; and IR temperature analysis (Department of the Interior, 2003). Furthermore, maintenance activities such as these are not only limited to power transformers but can be applied to a range of other equipment commonly used in utilities.

Whereas in this section we have discussed prognostic activities applied to individual pieces of equipment, in the next section we will discuss how these principles are applied to the grid as a whole.

10.4 Large Scale Grid Health Monitoring

Beyond the advances in prognostics for individual asset management discussed in the previous section, it is worth discussing the efforts and advances being made in terms of monitoring the grid, or at least large portions of the grid, as a whole. Reliable operation of the power generation and delivery systems across the United States and Canada is a monumental undertaking, owing in part to the fact that the grid is composed of many semiautonomous parts which need to work together harmoniously despite not necessarily having aligned financial interests.

In his article entitled "What is wrong with the electric grid?," Eric Lerner discusses the problems with prognostics as experienced by utilities in the past few decades. The problems arose, at least in part, due to the deregulatory acts of the government in the 1990s, where utility providers were divided into distinct entities responsible for generation and transmission separately (Lerner, 2003). Prior to the 1990s, local utilities had governance essentially all aspects of the system depicted schematically in Fig. 10.2. Power transmission between regional entities was minimal, and, consequently, the problem of identifying developing problems in the expansive modern electric grid was not the large, interconnected undertaking that it is today.

The power grid today behaves differently, where now a distribution provider can purchase power from a variety of physically remote sources, which can lead to increased and unpredictable load on the nation's transmission infrastructure. This behavior requires an enhanced communication infrastructure as the power grid of the past was not designed for such volatile load predictability (Hauser et al., 2005). This unpredictable load and lack of real-time situation awareness has contributed to failures such as the 2003 northeast blackout described in Section 10.1.

10.4.1 Phasor Measurement Units

Partly as a response to this changing power ecosystem, a new technology has emerged, allowing for faster and detailed measurements of grid conditions, and consequently allowing for more detailed analysis and understanding of grid operation both in real-time and offline. This technology is rooted in a device called a phasor measurement unit (PMU), explained below.

PMUs are derived from the concept of a phasor, which is a way of representing a sinusoidal waveform in a polar coordinate system developed over 100 years ago (Steinmetz, 1893). A PMU enables the acquisition of phasor data at a high sampling rate of tens of samples per second. This data is then combined with an extremely precise timestamp, and the resulting bundle (phasor + timestamp) is known as a synchrophasor. The necessary electronics to measure a synchrophasor are then combined with communications capabilities which can transmit these measurements to a centralized location where data from many locations can be analyzed together to achieve a holistic understand of grid conditions.

Each individual measurement location is a PMU, and currently a network of PMUs is being deployed in the US–Canadian system, as well as other locations in the word. A schematic overview of a synchrophasor network topology is shown in Fig. 10.11. Multiple PMUs are connected to a phasor data concentrator (PDC) that serves as a data aggregator. In some cases, where multiple PMUs are deployed at a single substation, a local PDC may be present to aggregate data before sending it to a corporate PDC that aggregates data from multiple substations owned by the transmission company. This data may then be exchanged with a regional PDC to enable regional overview across utilities.

Between 2009 and 2014 a drastic increase in synchrophasor popularity occurred, due in part to significant federal financial aid (NASPI, 2014a). In 2009, 200 research-grade PMUs were deployed whereas near the end of 2014 that number grew to as

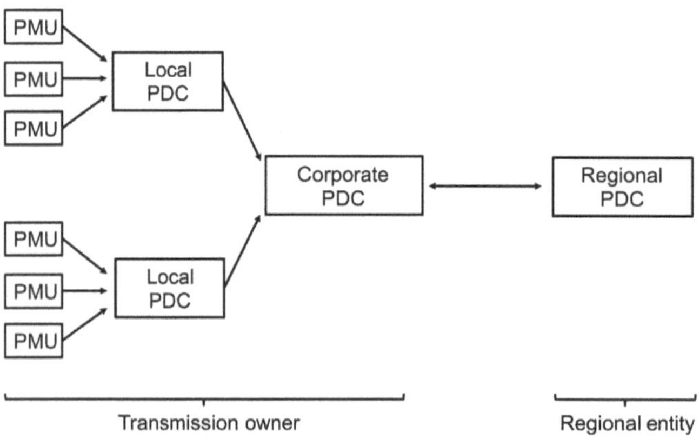

Figure 10.11 Synchrophasor network architecture as specified in IEEE C37.118.2.

many as 1700 production grade PMUs having been deployed. The boom in PMU and underlying data network deployment has led to the introduction of IEEE C37.118 (IEEE, 2011) for standardizing the format of data generated and transmitted by PMUs.

10.4.2 Prognostic Capabilities Enabled by PMUs

The concept of large scale grid conditioning monitoring is not new, and has in fact been implemented using older technologies such as Supervisory Control And Data Acquisition (SCADA). Rather it is the richness of the acquired data and the robustness of the communications network that is novel with the synchrophasor network. Synchrophasor analysis provides new capabilities in many areas, but two are particularly key for prognostics.

First, it enables real-time situational awareness. The communications network being implemented enables high speed transmission and aggregation of data from sources covering a vast geographical area and software has been developed to present this information to grid operators in useful ways and in real-time. With the nearly real-time data provided by PMUs will come enhancements such as better alarm and visualization tools, stability and loading assessment (Patel et al., 2010).

The lack of wide-area situational awareness has historically played a tangible role in most significant North American blackouts (NASPI, 2014b). Returning to the 2003 blackout, e.g., in the hour between approximately 3:00 and 4:00 PM on August 14, 2003, the phase angle between Cleveland and western Michigan steadily increased from 12 degrees to approximately 50 degrees. In the intervening minute between 4:09 and 4:10, the phase angle diverged to more than 115 degrees (Patel et al., 2010), ultimately leading to the cascade of failures leading to the blackout discussed previously. An abrupt increase in phase angle difference is an indicator of stress on the grid because phase angle is related to power transmission between the two locations with differing phases. In the case of the Cleveland–Michigan phase difference in 2003, Ohio's power draw from Michigan increased to roughly 2 GW around 4:09 PM (Lerner, 2003). Had this information been accessible to grid operators in real time, the onset of a problem could have been detected earlier and preventative actions taken. In fact, inadequate situational awareness was cited as one of the four contributing factors to the 2003 blackout (US–Canada Power System Outage Task Force, 2004).

Secondly, operation of PMUs enable the accumulation and storage of a wealth of data, both in normal state and during times of grid stress which allows for more sophisticated analysis and understanding of grid conditions and behavior. PMUs acquire between 30 and 60 measurements per second, whereas older technologies like SCADA typically acquire a sample every 2–4 seconds.

Such high sampling rates lead to an enormous database that can be drawn upon to build predictive models that enable prognostic approaches to identifying future failures. For example, as of 2014 the Bonneville Power Administration (BPA) maintains a synchrophasor data archive of 40 terabytes of synchrophasor data and adds approximately three terabytes of new data each month (BPA, 2015a). In light of this vast database, BPA has been investing in programs to improve event detection, characterization, and prediction (BPA, 2015b). It is expected that as PMUs become more prevalent, their use in prognostics will grow substantially.

10.5 Prognostics for Residences

Utility asset and grid management are important because failures of systems serving thousands of customers have obvious negative consequences. Individual residences, however, can also benefit from a smart grid to enable efficient energy usage and improved safety. The latter is interesting from a prognostics perspective because the smart grid can enable preemptive action against residential power systems failure events that affect thousands of people in North America by enabling processes utilizing data from the grid and at the point of service to residential customers. Of interest is the correlation between residential fires and the current meter sockets as well as the ability of smart meters to prevent residential fires, report on potential problems, and predict outages.

In 2013, John Hall wrote a report, published by the National Fire Protection Association, on electrical fires (Hall, 2013), which is defined by Hall as a structure fire that involved some type of electrical failure or malfunction as a factor contributing to ignition. The report stated, "In 2011, an estimated 47,700 home structure fires reported to US fire departments involved some type of electrical failure or malfunction as a factor contributing to ignition. These fires resulted in 418 civilian deaths, 1570 civilian injuries, and $1.4 billion in direct property damage.

In 2007–11, home electrical fires represented 13% of total home structure fires, 18% of associated civilian deaths, 11% of associated civilian injuries, and 20% of associated direct property damage," and "Electrical distribution or lighting equipment accounted for 6% of 2007–11 home structure fires, ranking fourth among major causes behind cooking equipment, heating equipment, and intentional."

In a separate study, BC Hydro in Canada commissioned a report on the frequency of electrical related fires in residential structures during the period July 2010 and June 2014. The report written by the University of the Fraser Valley presents data regarding fire rates before BC Hydro installed new residential smart meters as well as after BC Hydro installed smart meters. The report stated that there were 1917, 2060, 2078, and 1844 total residential fires in the BC Hydro service area in 2011, 2012, 2013, and 2014, respectively, and between 14.9% and 15.6% of those residential fires were related to electrical sources, which is a similar percentage to the report by Hall.

Two interesting trends reported by the University of Fraser Valley was an 11.6% decline in electricity related residential structure fires where the form of heat was electrically generated, and a 7.8% decline where electrical distribution equipment ignited an object since BC Hydro began installing smart meters. BC Hydro attributed the trends to replacement of more than 3000 preexisting unsafe meter socket conditions on customers' homes and businesses during installations (BC Hydro, 2014).

The benefits of replacing unsafe meter sockets are consistent with the report by Silver Springs Networks that "[t]he most common cause of meter temperature fires is a problem in the physical connection between the meter and the socket. This can be caused by a faulty socket housing, improper installation, or erosion caused by weather. The result is a gap that allows arcing to occur between the meter and the socket, resulting in heat and the potential for electrical fires" (2014).

New installation procedures that require inspection of meter sockets may enable safety improvements, but data from smart meters may also assist fire investigators at a fire loss and are helping power companies monitor the safety of residential equipment.

In any fire investigation, investigators are tasked to ascertain the cause and origin of the fire. This task is often highly difficult if there is extensive damage. However, if the electric power system is suspected as a cause of the fire, smart meters may be intelligent enough to provide investigators additional information that is not available from analog meters or provide an emergency alert. Furthermore, a smart meter with an internal switch may be able to turn off power to the residence and reduce the potential for a fire to develop.

As a simple example, substantial fire damage makes it difficult to locate a point of an electrical fault because of the inability to perform basic electrical characterizations (e.g., resistance measurement) caused by residual char or changes to the damaged electrical system. Furthermore, the fire origin may be at a different location from the origin of a fault. One potential source of data that may facilitate determining the source of the electrical fault may be information from smart meters with battery backups. These meters have local data storage or transmit data to a utility when a fire event affects the power delivered to a residence. For example, if the stored or transmitted data shows a sudden large current draw immediately before a smart meter loses power, it may point to an electrical short downstream from the smart meter. Such electrical conditions before or during a fire event may be analyzed after the event.

These data storage and emergency alert capabilities could be one line of protection but smart meters can monitor other characteristics that provide data before a hazardous condition exists. Temperature, vibration, and power quality monitoring are all possible and currently in various smart meter models.

Presently, some power companies use temperature data from smart meters to determine if the meters are experiencing abnormal high temperatures (2014b; Karp, 2012). If a smart meter temperature is abnormally high, an alert is sent to the power company that responds by sending technicians to analyze the smart meter that sent the alert. Power companies also use tamper detection data, which could be data on power loss to the meter, abnormal voltage levels or inconsistent meter readings compared to other meters in the system (McCullough, 2010). Although these data are traditionally used to detect power theft, there are promising hazard mitigation applications for the data, which have potential to advance prognostics for the home.

Today, various smart meter models also have the ability to measure several components of service cost including real and reactive power, harmonics, distortion, and vector apparent power. Given such advanced capabilities, detecting electrical arcs, similar to how arc-fault-circuit-interrupters (AFCI) use microcontrollers to detect electric arcs (Gregory and Manche, 2000), should be feasible for smart meters. For example, an AFCI monitors current-voltage waveforms that are particular to arcs including: voltage spikes at each half cycle of the power input as the arc ignites and extinguishes, and voltage across the arc that approaches a square wave. Similarly, an appropriately programmed and equipped smart meter can monitor voltage and current for characteristics of arcs.

Another fault condition that may be detected, by a smart meter, is a broken or high-impedance neutral (Luo et al., 2015). This fault condition at some point along the service or utility neutral wire can produce highly hazardous conditions. A broken utility neutral enables high voltages across devices that exceed device specifications and that may lead to a fire at a device such as a refrigerator compressor or television that experiences the over-voltage condition.

A residential smart meter typically monitors the current in the ungrounded conductors, but it does not monitor current through or voltage on the neutral (or grounded conductor). However, a well-developed algorithm could employ the electrical characteristics, such as equal current through the two ungrounded conductors, which may exist due to a broken neutral; hence, the smart meter may be able to determine the likely existence of a broken neutral and to take action to mitigate the damage from such a hazardous fault.

Additionally, a smart meter may have a Zigbee radio to communicate with a Home Area Network. Vibration sensors are also part of smart meters, so there is a lot of data that may utilized for prognostics or an emergency alert system. With data available to a home owner, they may be able to monitor their unoccupied residences for unusual activity such as increased power usage or loss of power that may be indicators of significant events. A fire at an unoccupied residence is likely to develop much further than a fire at an occupied residence where an occupant may be able to alert emergency response personnel or take some mitigating actions.

Prognostics and safety systems seem to be advantages but they rely on data collection of potentially personal information. Personally identifiable information includes names, contact, and biographical information, individual preferences, or any information derived from the above. The combination of personally identifiable information with energy use data obtained from the Smart Grid is also personal information (Cavoukian et al., 2010). Privacy concerns arise when there is a possibility of discovering personal information such as the personal habits, behaviors and lifestyles of individuals inside dwellings, and using the information other than for the provision of electricity (Cavoukian et al., 2010).

These privacy concerns should be addressed during the development of prognostics and alert systems enabled by the Smart Grid and one may be able to use other systems that record and/or monitor system characteristics as models. For example, vehicular data recorders store the conditions experienced by a vehicle before a crash, and AFCI systems monitor electrical characteristics of a branch circuit.

The role of smart meters will be very different than traditional analog meters. A competitive smart meter manufacturer market should incentivize manufacturers to continually add features that provide benefit to suppliers and consumers of electricity. Safety and maintenance are natural advantages to be gained from data delivered by smart meters, and one may suspect that as smart meters are upgraded through additional development cycles new features will be added to meet the goals of a smart grid.

10.6 Future Trends

The nature of the electric grid is rapidly changing at present with a growing emphasis being placed on real-time monitoring and decision making, and the concepts of failure analysis, reliability, and prognostics discussed in this chapter fall in line with this trend. Advances in sensor capabilities along with the implementation of more robust data networks make possible the acquisition of vast amounts of information, while advances in data analysis and modeling techniques make prediction of future events a tangible possibility.

As more entities adopt the approaches outlined in this chapter, a safer grid can be expected that can combat the problems associated with the nation's aging infrastructure discussed in Section 10.1. The traditional reliability approaches outlined in Section 10.2 are being replaced with more strategic prognostic techniques such as those discussed in Section 10.3. A holistic grid monitoring system discussed in Section 10.4 is being implemented, partly in response to regulatory changes in the industry, and is expected to help monitor and adapt to large scale problems that potentially span several utilities. Finally, in Section 10.5, advances in at-home metering and prognostics were discussed, where so-called smart meters are expected to help mitigate home fires and in some cases help investigate when failures do occur.

While implementation of "smart" technologies such as those discussed in this chapter are not without drawbacks, they bring benefits that outweigh potential shortcomings. The push toward predictive capabilities enabled by the smart grid will in the long run reduce cost and improve reliability. With support from both government and private utilities, this trend is expected to continue moving forward.

References

IEEE Std C57.91, 1996. IEEE Guide for Loading Mineral-Oil-Immersed Transformers. IEEE Std C57.91-1995, i.

Department of the Interior, 2003. Facilities Instructions, Standards, and Techniques Volume 3–31: Transformer Diagnostics. *In*: Department of the Interior (ed.). Springfield, VA: National Technical Information Service.

Siemens, 2010. TMDS Smart Monitor: Solutions for transformer life-cycle management. Erlangen, Germany: Siemens AG.

General Electric, 2010. Transformer Monitoring and Diagnostic Systems. G.E. Energy. General Electric Company.

ASCE, 2011. Failure to Act—The Economic Impact of Current Investment Trends in Electricity Infrastructure. American Society of Civil Engineers, Reston, VA.

IEEE, 2011. IEEE Standard for Synchrophasor Measurements for Power Systems. *IEEE Std C37.118.1-2011 (Revision of IEEE Std C37.118-2005)*, 1–61.

ASCE, 2013. 2013 Report Card for America's Infrastructure. American Society of Civil Engineers, Reston, VA.

BC Hydro, 2014. BC Hydro's new meters making electricity grid safer for customers. Vancover: BC Hydro Media Relations. Available from: <https://www.bchydro.com/news/press_centre/news_releases/2014/new-meters-making-electricity-grid-safer.html> (accessed 19.11.15.).

2014. Meter Temperature Management, Silver Springs Networks, Redwood City, CA.

NASPI, 2014a. Synchrophasor Technology Fact Sheet. In: Pacific North West National Laboratory (ed.). North American SynchroPhasor Initiative.

NASPI, 2014b. PMUs and synchrophasor data flows in North America as of March 19, 2014. In: Pacific North West National Laboratory (ed.). North American SynchroPhasor Initiative.

BPA, 2015a. TIP 332: Open Source Platform for Accelerating Synchrophasor Analysis. In: Department of Energy (ed.). Bonneville Power Administration, Portland OR.

BPA, 2015b. TIP 319: Multidimensional Learning on PMU Data for Event Detection, Characterization, and Prediction. In: Department of Energy (ed.). Bonneville Power Administration, Portland, OR.

Brown, R.E., 2008. Electric Power Distribution Reliability. CRC press.

Bruch, M., Munch, V., Aichinger, M., Kuhn, M., Weymann, M., Schmid, G., 2011. Power Blackout Risks. CRO Forum, Munich.

Cavoukian, A., Polonetsky, J., Wolf, C., 2010. Smart Privacy for the smart grid: embedding privacy into the design of electricity conservation. Identity in the Information Society. 3, pp. 275–294.

Dias, G., Ostermann, R., Layher, M., Gewher, O., Correa, S., Gazzana, D., et al., 2005. Transformer Management System. Fifth WSEAS International Conference on Power Systems and Electromagnetic Compatibility, Corfu, Greece. pp. 143–148.

Dodson, B., Dovich, R.A., Edenborough, N.B., Gee, G., Walton, K.K., Marr, R., et al., 2009. The Reliability Engineering Primer. Quality Council of Indiana, West Terre Haute, IN.

ELCON, 2004. The Economic Impacts of the August 2003 Blackout. Electricity Consumers Resource Council (ELCON).

Endrenyi, J., Aboresheid, S., Allan, R.N., Anders, G.J., Asgarpoor, S., Billinton, R., et al., 2001. The present status of maintenance strategies and the impact of maintenance on reliability. IEEE Trans. Power Syst. 16, 638–646.

Gregory, G., Manche, A., 2000. Arc detection with the AFCI. IAEI. IAEI, Richardson, TX.
Hall, J.R., 2013. Home electrical fires. National Fire Protection Association: Fire Analysis & Research. National Fire Protection Association, Quincy, MA.
Hamrick, L., 2010. Dissolved gas analysis for transformers. International Engineering Testing Association (NETA) World, pp. 1–3.
Hauser, C.H., Bakken, D.E., Bose, A., 2005. A failure to communicate: next generation communication requirements, technologies, and architecture for the electric power grid. IEEE Power and Energy Magazine. 3, pp. 47–55.
Karp, G., 2012. ComEd confirms smart meters involved in 'small fires'. Chicago Tribune, August 30.
LaCommare, K.H., Eto, J.H., 2004. Understanding the Cost of Power Interruptions to US. US Department of Energy, Berkeley, CA.
Lerner, E.J., 2003. What's wrong with the electric grid? The Industrial Physicist. 9, pp. 8–3.
Luo, G., Zhang, W., Yin, J., 2015. Automatic protection system of neutral wire-broken based on supply-side protection. International Journal of Control and Automation. 8, pp. 245–250.
Mccullough, J., 2010. Deterrent and detection of smart grid meter tampering and theft of electricity, water, or gas. Elster, Raleigh, NC.
Nowlan, F.S., Heap, H.F., 1978. Reliability-centered maintenance. DTIC Document.
Patel, M., Aivaliotis, S., Allen, E., Bakken, D., Beard, L., Bhaman, V., et al. 2010. Real-Time Application of Synchrophasors for Improving Reliability. Princeton, NJ.
Sica, F.C., Guimarães, F.G., De Oliveira Duarte, R., Reis, A.J., 2015. A cognitive system for fault prognosis in power transformers. Electr. Power Syst. Res. 127, 109–117.
Smith, A.M., Hinchcliffe, G.R., 2004. RCM—Gateway to World Class Maintenance. Butterworth-Heinemann, pp. 57–69.
Steinmetz, C.P., 1893. Complex quantities and their use in electrical engineering. Proceedings of the International Electrical Congress, Chicago, pp. 33–74.
US-Canada Power System Outage Task Force, 2004. Final Report on the August 14, 2003 Blackout in the United States and Canada: Causes and Recommendations. US-Canada Power System Outage Task Force.
Vachtsevanos, G., Lewis, F., Roemer, M., Hess, A., Wu, B., 2006. Intelligent fault diagnosis and prognosis for engineering systems. John Wiley & Sons.
Willis, H.L., Schrieber, R.R., 2013. Aging Power Delivery Infrastructures. second ed CRC Press, Boca Raton, FL, pp. 2, 5-6, 17-18, 227.

INDEX

Note: Page numbers followed by "*f*" and "*t*" refer to figures and tables, respectively.

Q

R

S

T

U

V

W

FSC
www.fsc.org
MIX
Papier | Fördert
gute Waldnutzung
FSC® C083411

Kolibri 360 GmbH
Ferdinand-Jühlke-Straße 7
99095 Erfurt, Deutschland
produktsicherheit@kolibri360.de